Optimization in Action

Optimization in Action

Proceedings of the Conference on Optimization in Action held at the University of Bristol in January 1975, organised by the Institute of Mathematics and its Applications

Edited by

L. C. W. DIXON

Hatfield Polytechnic
Hatfield, England

1976

ACADEMIC PRESS

London   New York   San Francisco

A Subsidiary of Harcourt Brace Jovanovich, Publishers

ACADEMIC PRESS INC. (LONDON) LTD.
24/28 Oval Road,
London NW1

United States Edition published by
ACADEMIC PRESS INC.
111 Fifth Avenue
New York, New York 10003

Library of Congress Catalog Card Number: 76–25697
ISBN: 0-12-218550-1

Printed by photolithography in Great Britain by
T. & A. Constable Ltd, Edinburgh

PREFACE

Since the initial conference on "Optimization" held at the University of Keele in 1968, The Institute of Mathematics and its Applications has organized a series of conferences on the theoretical aspects of nonlinear optimization, namely:

Numerical Methods for Unconstrained Optimization
N.P.L. Teddington 1971

Numerical Methods for Constrained Optimization
N.P.L. Teddington 1973.

In contrast little effort had been made to collate industrial experience using optimization techniques and my colleague, S.E. Hersom, therefore suggested to the IMA that the time was ripe for a conference concentrating on the application of optimization techniques. The IMA agreed and a committee was set up to organize the conference with the title "Optimization in Action." The committee consisted of: Dr. J.H. Wilkinson (Chairman), Mr. A.H.O. Brown, Mr. Norman Clarke, Mr. L. C.W. Dixon, Mr. J.G. Hayes, Mr. S.E. Hersom, Mr. R.F. Jackson, Mr. M.J.D. Powell, Miss C.M. Richards, Dr. H.H. Robertson, Mr. R.A. Scriven and Mr. J.K. Skwirzynski. I would like to thank them all for their efforts to ensure the success of the conference and for inviting me to act as editor of the proceedings.

The conference was held at the University of Bristol and we were all made very comfortable both in the Halls of Residence and in the Students' Union building where the lectures took place.

v

The IMA negotiated the publication of these Proceedings with Academic Press and undertook to retype all the papers in a uniform acceptable format. Again I would like to thank them for the efficient way they have performed this difficult task and to thank all the contributors for cooperating with them so closely. Special thanks are given to Ian Dawson of Rolls Royce (1971) Ltd. for lending me his personal tape recordings of the discussion session on unconstrained and constrained optimization. These form a valuable contribution to the Proceedings and I can only express my apologies to contributors that I had not the foresight to record the other four discussion sessions.

The programme of the conference consisted of six survey papers, describing the current state of the art in different areas of the subject, and eighteen case studies. In any developing subject some case studies are successful and some unsuccessful. Often unsuccessful case studies tend to be excluded from the technical literature, but experience has shown that much is often learnt by examining the reasons why a particular project ran into difficulties and for this reason it was agreed that the presentation of unsuccessful case studies would both be permitted and encouraged.

The case studies are drawn from a number of different disciplines including optical, aeronautical, chemical and electrical engineering together with building construction and town planning. This list indicates the wide range of possible applications of optimization techniques, but it is not comprehensive since they are also applicable in financial, econometric and welfare problems, though case studies in these fields were not forthcoming.

As editor I hope this book will be useful both to practising engineers who may find some guidance on whether the present optimization algo-

rithms can aid them in the solution of their problems. I also hope it will be useful to the many
final year undergraduate and postgraduate students
now studying optimization in different universities and polytechnics both in the United Kingdom
and throughout the world, who may have felt the
need of a volume containing practical case studies.

Numerical Optimization Centre,   L.C.W. Dixon
Hatfield Polytechnic
March 1976

The Institute thanks the authors of the papers, the editor, Mr. L.C.W. Dixon (Numerical Optimization Centre, Hatfield Polytechnic) and also Miss Susan Lawrence, Mrs. Maureen Downie and Mrs. Laurel Middleton of the IMA for typing the papers and preparing the diagrams for publication.

CONTENTS

Contents

Contents

A Survey of Methods for Minimizing Sums of Squares of Nonlinear Functions

Shirley A. Lill

(University of Liverpool)

SUMMARY

The survey classifies the types of sums of squares problems that arise, according to the complexity of the function to be minimized and the available level of function information. An outline of methods for each class of problem is given with an indication of readily available algorithms.

1. INTRODUCTION

The problem of minimizing a nonlinear function that has the form of a sum of squares of other functions is one of the most commonly occurring types of minimization problem. It frequently arises in the fields of engineering and applied science where the theory predicts that a certain process should satisfy some functional relationship or model and the experimenter obtains data in order to ascertain the values of the variable parameters of this model. Such problems are essentially curve fitting problems, where the form of the function is known. Other problems which can result in a sum of squares formulation are the solution of simultaneous nonlinear equations and the more general parameter estimation problems such as those described by Bard (1970).

Once a problem has been posed as finding the minimum of a sum of squares it can be tackled by either using a straightforward minimization tech-

1

nique (Murray (1972)) or the minimization process can be adapted to exploit the special nature of the function. It is this latter approach which is described in this paper.

Section 2 introduces notation and shows how a sum of squares function arises from a curve fitting problem. It also indicates points for consideration when minimizing sums of squares functions. Section 3 describes the basic approaches for solving this type of problem, whilst section 4 examines the central issue of solving the linear least squares equations at each iteration. In section 5 the question of use of derivatives of the function is considered and in section 6 some special problems are discussed. Finally, some suggestions for choosing methods are given in section 7.

Surveys of methods for minimizing sums of squares problems are also given by Powell (1972), Bard (1970), Dennis (1972) and Brown (1972). Some comparative numerical results are given by Bard (1970), Box (1966), Brown and Dennis (1972) and McKeown (1974).

2. A DISCUSSION OF THE PROBLEM

Consider determining the values of the parameters $\underset{\sim}{x} = (x_1, x_2, x_3, \ldots, x_n)^T$ which satisfy the relationship:

$$y = F(\underset{\sim}{t}, \underset{\sim}{x}), \qquad (2.1)$$

where y is the dependent variable, and $\underset{\sim}{t} = (t_1, t_2, \ldots, t_K)^T$ are the independent variables for a certain process modelled by the function F. A set of m experiments or observations is made to obtain values of y for different values of $\underset{\sim}{t}$, and these satisfy the equations:

$$y_i = F(\underset{\sim}{t}_i, \underset{\sim}{x}) + \varepsilon_i, \qquad i = 1(1)m, \qquad (2.2)$$

where the ε_i are randomly distributed independent experimental errors. The problem is to find those values of $\underset{\sim}{x}$ which give the experimental data the best fit to (2.1).

2

The best fit in the least squares sense is obtained by defining residuals:

$$f_i(\underset{\sim}{x}) = F(\underset{\sim}{t}_i, \underset{\sim}{x}) - y_i, \quad i = 1(1)m, \quad (2.3)$$

and minimizing the sum of squares of these residuals:

$$S(\underset{\sim}{x}) = \sum_{i=1}^{m} f_i(\underset{\sim}{x})^2 = \underset{\sim}{f}(\underset{\sim}{x})^T \underset{\sim}{f}(\underset{\sim}{x}), \quad (2.4)$$

where $\underset{\sim}{f}(\underset{\sim}{x}) = (f_1(\underset{\sim}{x}), f_2(\underset{\sim}{x}), \ldots f_m(\underset{\sim}{x}))^T$, with respect to $\underset{\sim}{x}$.

Note that it is possible to solve this type of problem by obtaining the best fit in some other sense such as by minimizing the maximum residual (for example Osborne (1971)) or by using a combination of least squares and minimax. It is also possible to fit smooth curves to experimental data when there is no known model function and thus interpolate intermediate values. Such techniques known as data fitting are described by Cox and Hayes (1973).

When m is greater than n, that is, there are more observations than parameters, the nonlinear least squares problem is said to be over-determined, and in many curve fitting applications for example, m will be significantly larger than n. However a very important class of problems is that of finding the solution of a set of n simultaneous equations in n unknowns (*i.e.*, $m = n$). This particular problem is covered here only as a special case of the over—determined type, and the reader is referred to Ortega and Rheinbolt (1970) for a full exposition. Similarly, the solution when the model is linear is not explicitly discussed here, except that the methods are essentially the same as those described in section 4. Finally, when m is less than n the system is said to be under-determined and only certain of the methods can be applied.

Points for consideration when choosing an algorithm for solving nonlinear least squares problems are introduced as appropriate in the text.

3

However, certain key points are listed here because of their underlying importance in the discussion of the methods:

(*a*)   Can the problem be expressed as a simple sum of squares, or are there further considerations such as constraints, errors in the t_i and so on? Can it be broken down into a simpler form?

(*b*)   Is the model a good one and are the experimental errors ε_i in equations (2.2) small so that the minimum of $S(x)$ is zero (*i.e.*, are the equations (2.3) consistent)?

(*c*)   Can analytical partial derivatives (no more than second order) of the functions $f_i(x)$ be evaluated, and at what cost, in terms of effort in obtaining the formulae and computer time in calculation?

(*d*)   Is a good estimate of the solution available? Is the sum of squares function well-behaved in the region of search, that is, are there other local minima, is the function singular and so on?

(*e*)   Are the residuals $f_i(x)$ expensive to calculate, in terms of computer time? Are any of them linear?

(*f*)   What are the sizes of *m* and *n*?

3. METHODS OF SOLUTION

3.1 *Gauss-Newton method*

The problem is to find the least value of the function:

$$S(x) = f(x)^T f(x), \qquad m \geqslant n.$$

Now it is well known that a stationary point of any function $S(x)$ occurs when the gradient $\nabla S(x) = 0$ and for that stationary point to also be a local minimum of the function the Hessian matrix of second derivatives $\nabla^2 S(x)$ must be positive definite. One way of locating such a point is to use Newton's

4

classical minimization method (Barnes (1965)) where, starting from an initial estimate of the minimum x_0, a correction d_K is applied iteratively, $K = 0,1,2,\ldots$, until convergence:

$$x_{K+1} = x_K + d_K . \qquad (3.1)$$

The correction d_K is the solution to the equations:

$$\nabla^2 S(x_K) d_K = - \nabla S(x_K) ,$$

derived from the Taylor series expansion of the function about x_K.

When $S(x)$ is a sum of squared terms and $f(x)$ is twice differentiable then the gradient can be expressed in terms of $f(x)$ and its derivatives as:

$$\nabla S(x) = 2J(x)^T f(x) , \qquad (3.2)$$

where $J(x)$ is the $m \times n$ Jacobian matrix with ijth element:

$$J_{ij}(x) = \frac{\partial f_i(x)}{\partial x_j} .$$

The Hessian of $S(x)$ is given by:

$$\nabla^2 S(x) = 2J(x)^T J(x) + 2 \sum_{i=1}^{m} \nabla^2 f_i(x) f_i(x) , \qquad (3.3)$$

where $\nabla^2 f_i(x)$ is the second derivative matrix of $f_i(x)$.

A justification for using special methods to exploit the form of $S(x)$, rather than carrying out a straightforward minimization, can at once be seen from these equations since a substantial part of the Hessian (3.3) is obtained by using only first derivatives of the residuals (*i.e.*, $J(x)^T J(x)$) thus removing the need for explicit second derivatives of the residuals which may be expensive to calculate. This observation, and the fact that near the solution, if the residuals are small (which they are for many practical problems) or nearly linear, then the second term in (3.3) is

5

negligible, has led to many algorithms using the approximation:

$$\nabla^2 S(\underset{\sim}{x}) \simeq 2J(\underset{\sim}{x})^T J(\underset{\sim}{x}) = 2A(\underset{\sim}{x}), \qquad (3.4)$$

Substituting (3.4) and (3.3) Newton's method gives a correction $\underset{\sim}{d}_K$ which is calculated from equations conventionally known as the *normal equations*:

$$J(\underset{\sim}{x}_K)^T J(\underset{\sim}{x}_K) \underset{\sim}{d}_K = - J(\underset{\sim}{x}_K)^T f(\underset{\sim}{x}_K). \qquad (3.5)$$

This algorithm may be derived directly by expanding the residuals f in a Taylor series about x_K (Kowalik and Osborne (1968)), and was first put forward by Gauss (1809). It is usually referred to as the Gauss-Newton method.

The solution of the normal equations obviously breaks down if any $J_K^T J_K$ is *singular*. However, in practice, and especially when $m >> n$ the eigenvalues of $J^T J$ are usually bounded away from zero so this problem does not arise and convergence for a region near a solution can be proved (Fox (1964), Meyer (1970) and Pereyra (1967)). Now the rate of convergence for Newton's method is second order, but in the Gauss-Newton method the error incurred by neglecting the term $\nabla^2 f$ in (3.3) reduces the rate of convergence to no more than *linear* unless $S(\underset{\sim}{x}^*)=0$ (Brown and Dennis (1972), Meyer (1970) and Osborne (1972)). Osborne (1972) shows that the reduction in $\underset{\sim}{d}_K$ at each iteration is given by:

$$\| \underset{\sim}{d}_{K+1} \| = \| (J_{K+1}^T J_{K+1})^{-1} \frac{d\bar{J}}{dt} f_K \| \, \| \underset{\sim}{d}_K \| + O(\| \underset{\sim}{d}_K \|^2) \qquad (3.6)$$

where d/dt denotes differentiation with respect to any direction t and $\bar{J}$ indicates mean values. This demonstrates that for all $K > K_0$, if the series is convergent and $0 < \gamma \leqslant 1$ where:

$$\| (J^T J)^{-1} \frac{dJ}{dt} f \| = \gamma$$

the rate of convergence is linear, whereas if $\gamma > 1$ the method is actually *divergent*. Obviously if $S(\underset{\sim}{x}^*) = 0, \gamma$ is zero so that convergence is ultimately second order. Powell (1972) and Osborne (1972) give

6

simple numerical examples demonstrating the consequences of different values of γ.

This deficiency has, in practice, led to a
wide variety of modifications and improvements to
the basic Gauss-Newton algorithm. Several suggested
alternatives are discussed below. They are introduced in the sense of development from the basic
algorithm rather than in chronological order.

3.2 Modified Gauss-Newton or Hartley method

The Gauss-Newton is often modified to prevent
divergence by using $\underset{\sim}{d}_K$ as a direction along which
to search for a lower value of S, so that (3.1)
becomes

$$\underset{\sim}{x}_{K+1} = \underset{\sim}{x}_K + \alpha_K \underset{\sim}{d}_K. \qquad (3.7)$$

α_K is a scalar which may be chosen so as to mini-
mize $S(\underset{\sim}{x}_K + \alpha_K \underset{\sim}{d}_K)$ with respect to α_K, or simply to
ensure that $S(\underset{\sim}{x}_{K+1}) \leqslant S(\underset{\sim}{x}_K)$. Kowalik and Osborne
(1968) argue that since the cost involved in cal-
culating $\underset{\sim}{d}_K$ is substantial, the further cost in
function evaluations of carrying out an accurate
linear search is justified, and for $J^T J$ bounded
above and below, Hartley (1961) proves convergence
for this version of the method. However, it is
often the case that function evaluations are at a
premium and several efficient schemes for calculating a suitable α_K to reduce the sum of squares have
been suggested. Efficient searches for nonlinear
least squares in particular are investigated by
Bard (1970) and Osborne (1972) who also supplies a
proof of convergence. However near the solution
α_K set to unity is usually successful so the convergence is ultimately that of the Gauss-Newton
method.

Although these modifications do prevent
divergence they do not overcome the problem of
solving the normal equation when $J_K^T J_K$ is singular,
nor do they ensure convergence to a solution. In
fact they may appear to converge to a point which
is not a local minimum of S and Powell (1970) gives

7

such an example. The method apparently converges to a point where $J^T J$ is singular so that $J^T f = 0$, even though it may not be singular elsewhere. The difficulty is that near such points the directions $\underset{\sim}{d}_K$ are almost orthogonal to the descent direction of $S(-\nabla S)$ so that little reduction in the value of S can be made.

3.3 Methods interpolating between Gauss-Newton and steepest descent

Powell's hybrid method (1970) was devised to solve this problem by introducing a search along the steepest descent direction whenever the Gauss-Newton correction is unsuccessful, so that:

$$\underset{\sim}{x}_{K+1} = \underset{\sim}{x}_{K+1} + \alpha_K \underset{\sim}{p}_K,$$

where
$$\underset{\sim}{p}_K = \beta_K \underset{\sim}{d}_K - \gamma_K \underset{\sim}{\nabla S}_K,$$

$\underset{\sim}{d}_K$ is the solution to (3.2), and α_K, β_K and γ_K are suitable step lengths. The algorithm is briefly as follows.

The Gauss-Newton correction, $\underset{\sim}{d}_K$, is calculated, but if this is deemed too large or does not give a reduction in the sum of squares S, then the predicted minimum of S along the steepest descent direction $\underset{\sim}{\nabla S}$ is calculated. A search for a reduced S is then made along $-\underset{\sim}{\nabla S}$ and the line joining the predicted minimum to the end point of the Gauss-Newton correction, as in Fig. 1.

predicted minimum
along steepest descent

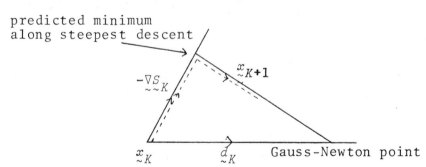

Fig. 1.   Powell's hybrid algorithm

8

Thus, the method interpolates very simply between the Gauss-Newton and steepest descent directions.

An alternative approach, due to Levenberg (1944) and Marquardt (1963), of introducing a bias towards steepest descent whilst at the same time guaranteeing the calculation of d_K is to add some positive definite matrix D_K to $J_K^T J_K$ in (3.5) and solve:

$$(J_K^T J_K + \lambda_K D_K) d_K = - J_K^T f_K, \qquad (3.8)$$

where $\lambda_K > 0$ is some variable parameter.

Providing λ_K is chosen large enough the composite matrix will be positive definite so that d_K can be calculated and in addition, as $\lambda_K D_K$ is increased, d_K is forced towards the descent direction so that a reduction in S can always be achieved. $\lambda_K D_K$ can also be considered as an approximation to the term in $\nabla^2 f$ which is ignored in (3.3). For simplicity D_K is usually chosen to be a constant diagonal matrix, either the unit matrix or a matrix which reflects the scaling of the variables (Marquardt (1963)).

The methods suggested by Levenberg and Marquardt differ slightly but the main idea behind them is that at each iteration, given a value λ_K, the equations (3.8) are successively solved for increasing values of λ_K until a d_K is obtained such that $S(x_{K+1}) < S(x_K)$ (Levenberg actually suggests finding the minimum S with respect to λ_K) when x_{K+1} is accepted as the new iterate and λ_K is decreased by some constant factor. It is easy to show that, as $\lambda_K \to 0$, d_K tends to the Gauss-Newton correction $-(J_K^T J_K)^{-1} J_K^T f_K$, and as $\lambda_K \to \infty$, d_K tends to the descent direction $-\nabla S$ and the trajectory of the end point of d_K for varying λ_K is as shown in Fig. 2.

From Fig. 2 it can be seen that increasing λ_K has the effect of reducing the size of d_K as well as altering its direction, thus α_K may in theory be set to unity in calculating x_{K+1} (3.7).

9

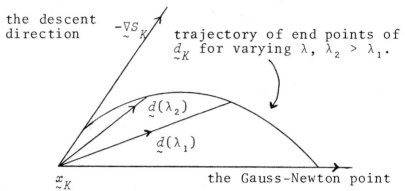

the descent direction $-\nabla S_K$

trajectory of end points of d_K for varying λ, $\lambda_2 > \lambda_1$.

$d(\lambda_2)$

$d(\lambda_1)$

x_K

the Gauss-Newton point

Fig. 2. Levenberg-Marquardt algorithm

However against this simplification must be measured the cost of re-solving (3.8) every time λ_K is altered, and so it is usually more efficient to vary α_K. A sensible strategy for choosing λ_K is to let λ_0 be some significant value, thus ensuring a reduction in S at points far removed from the solution, and to employ an over-all reduction philosophy (although λ may be temporarily increased) so that $\lambda \to 0$ as $x \to x^*$ and the second order convergence of Gauss-Newton (for $S(x^*) = 0$) can be attained (Brown and Dennis (1972), Meyer (1970) and Osborne (1972)). Computational schemes for calculating λ_K are given by Fletcher (1971), Meyer (1970) and Osborne (1972) together with limited numerical results.

To avoid the work involved in the recalculation of d_K when λ_K is changed Bard (1970) calculates the eigenvalues of $(J_K^T J_K + \lambda_K D_K)$ at each iteration and uses them to evaluate d_K. This allows the smallest possible λ_K to be chosen initially so that the condition number of the composite matrix is limited to lie within specified bounds and ensures that it can be adjusted at little extra cost if the resulting sum of squares is not acceptable. However the cost of the eigensolutions is high and a simpler method is suggested by Jones (1970) where the trajectory of Levenberg and Marquardt is replaced by a spiral which has the same end points, but is of constant slope so that points along it are simple to obtain. See Fig. 3.

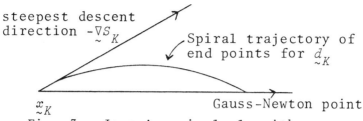

Fig. 3. Jones's spiral algorithm

Once the spiral is set up discrete points along it can be calculated simply by vector additions.

4. SOLUTION OF THE NORMAL EQUATIONS

This section outlines possible approaches to the solution of the normal equations which are central to most of the nonlinear least squares techniques discussed so far. Dropping the subscript for convenience, equation (3.5) becomes:

$$J^T J \underset{\sim}{d} = - J^T \underset{\sim}{f} . \qquad (4.1)$$

The efficiency and accuracy of the solution of these equations are crucial since they have to be solved at every iteration of the nonlinear least squares algorithm.

Traditionally (4.1) is solved by evaluating $J^T J$ and performing a Cholesky factorization (see Fox (1964)) to obtain $L L^T$ where L is a lower triangular matrix and then carrying out back substitutions to obtain $\underset{\sim}{d}$. However, forming the product $J^T J$ worsens the conditioning of the problem and leads to a loss of accuracy (Businger and Golub (1965)), and recently algorithms have been given (Gill and Murray (1972)) which avoid the explicit calculation of $J^T J$ by performing instead an orthogonal triangularization of J.

An alternative method which avoids the compu-

11

tation of $J^T J$ is due to Businger and Golub (1965). Reposing the problem as solving:

$$J\underset{\sim}{d} = -\underset{\sim}{f} \qquad (4.2)$$

in the least squares sense and factorizing J into $Q \begin{bmatrix} R \\ \vdots \\ 0 \end{bmatrix}$ where Q is an $m \times m$ orthogonal matrix and R is $n \times n$ right triangular, reduces (4.2) to:

$$\begin{bmatrix} R \\ \vdots \\ 0 \end{bmatrix} \underset{\sim}{d} = -Q^T \underset{\sim}{f}$$

since the problem is invariant to orthogonal transforms. The solution is obtained by setting $\underset{\sim}{b}$ to the first n elements of $-Q^T \underset{\sim}{f}$ and then solving the triangular system $R\underset{\sim}{d} = \underset{\sim}{b}$ to obtain $\underset{\sim}{d}$. The factorization of J into QR is very stable and an important implementation detail is that Q, which may be very large, need not be stored since $\underset{\sim}{b}$ can be built up simultaneously.

A third method, which should be used when the matrix J is of rank less than n, or cannot be guaranteed to be of full rank, is the singular value decomposition due to Golub and Reinsch (1970). Again this uses (4.2), but J is factorized into UDV^T where:

U – $m \times n$ orthogonal matrix made up of the eigenvectors of the n largest eigenvalues of JJ^T.

D – $n \times n$ diagonal matrix of singular values of $J^T J$ (non-negative square roots of the eigenvalues).

V – $n \times n$ orthogonal matrix made up of the eigenvectors of $J^T J$.

so that $\underset{\sim}{d}$ is obtained by setting:

$$\underset{\sim}{d} = -VD^+U^T\underset{\sim}{f} \quad,$$

where $D^+_{ii} = 1/D_{ii}$, $D_{ii} > 0$,
$\qquad\qquad = 0 \qquad$, otherwise.

In addition to dealing with J of rank $< n$, this method is also useful if information on the eigensolution of J is required.

When a Levenberg-Marquardt method is used the matrix $J^T J$ is replaced by the composite matrix $(J^T J + \lambda I)$ and it is possible to organize the factorizations so that λ can be adjusted without having to completely resolve the equations. For example, writing (4.2) as:

$$\begin{bmatrix} J \\ \gamma I \end{bmatrix} \underset{\sim}{d} = - \begin{bmatrix} f \\ \underset{\sim}{0} \end{bmatrix}, \quad \text{where} \quad \gamma^2 = \lambda,$$

and using the Businger-Golub approach:

$\begin{bmatrix} J \\ \gamma I \end{bmatrix}$ is factorized into $\begin{bmatrix} Q_1 & 0 \\ 0 & I \end{bmatrix} \begin{bmatrix} R_1 \\ 0 \\ \gamma I \end{bmatrix}$ where Q_1 and R_1 are independent of the choice of λ.

$\begin{bmatrix} R_1 \\ 0 \\ \gamma I \end{bmatrix}$ is then factorized into $Q_2 \begin{bmatrix} R_2 \\ 0 \\ 0 \end{bmatrix}$ and the appropriate back-

substitutions are carried out for each value of λ (the Q's are orthogonal and the R's right triangular).

When explicit derivatives of f are not available and J has to be approximated, considerable savings in computation can be made by storing and updating the factors of J used in the solution of the normal equations, rather than J itself. Details of such schemes are given in the following section.

5. TECHNIQUES FOR SOLVING SUMS OF SQUARES WITHOUT USING ANALYTICAL DERIVATIVES

All the methods discussed in section 3 rely on the matrix J of first derivatives of f being available. It is not always possible or desirable to obtain analytical derivatives of f and so alternative ways of calculating J have been derived.

The most obvious way to remove the need for analytical derivatives is to calculate estimates of J_{ij} from a finite difference formula such as:

$$\frac{\partial f_i}{\partial x_j} \simeq \bar{J}_{ij} = \frac{f_i(\underset{\sim}{x}_i + h_{ij}\underset{\sim}{e}_j) - f_i(\underset{\sim}{x}_i)}{h_{ij}} , \quad (5.1)$$

$$i = 1(1)m, \quad j = 1(1)n,$$

where $\bar{J}_{ij}$ is the ijth element of the approximate Jacobian, e_j is the jth unit vector and $h_{ij} > 0$ is some suitable interval for differencing. The main disadvantage of this approach is the cost of calculating $\bar{J}$ at each iteration, especially when m and n are large. If f has already been calculated a further $m \times n$ function evaluations are required to obtain $\bar{J}$. Further, these evaluations are necessarily grouped at points very close to each iterate and it would seem that improvements in efficiency ought to be possible by spacing the information more evenly throughout the region of search. Another problem is the difficulty in balancing the choice of h_{ij} between its being too large so that $\bar{J}$ is only a crude estimate and its being too small so that the calculation is dominated by round-off errors.

Recently, however, discrete minimization methods or methods using finite difference derivatives have been shown (Gill and Murray (1972)) to compare very favourably with other methods not requiring analytical derivatives, and for sums of squares in particular Brown and Dennis (1972) report considerable numerical success in using finite difference Gauss-Newton and Levenberg-Marquardt algorithms. They prove convergence for their versions of the algorithms (with suitable restrictions on the choice of h_{ij}) comparable to those for the analytical derivative versions.

One important advantage in estimating J in this way is that if any of the J_{ij} can be obtained analytically then it is a simple matter to incorporate them.

The alternative to estimating J by differences is to store information on the residuals and the search directions or corrections over previous

14

iterations, and use these instead of calculating
J. If the change in f between x_{K+1} and x_K is used
then J can be estimated from information which is
already calculated at each iteration so that extra
residual evaluations at intermediate points are
not required.

Many earlier methods using this approach
were based on the secant method and involved storing the information more or less directly. One
such method due to Powell (1965) approximates
derivatives along a set of n linearly independent
directions p_j rather than differencing along the
coordinate directions. The derivative of a function along any direction is given by their scalar
product so that an estimate $\bar{J}$ of the derivative of
f obtained by differencing along a set of directions p_j is in fact an estimate of:

$$\bar{J} \simeq JP$$

where P has columns p_j. Substituting this estimate
into the normal equations (3.5) gives a calculated
correction:

$$d_K = (\bar{J}_K^T \bar{J}_K)^{-1} \bar{J}_K^T f_K \simeq (P_K^T J_K^T J_K P_K)^{-1} P_K^T J_K^T f_K$$

where P_K is an $n \times n$ matrix, the set of directions
used at the Kth iteration. This gives:

$$d_K \simeq P_K^{-1} (J_K^T J_K)^{-1} J_K^T f_K$$

so to obtain the required Gauss-Newton correction
d_K is premultiplied by P_K to give:

$$q_K = P_K d_K \qquad (5.2)$$

The algorithm is essentially as follows:
using $\bar{J}_K$ and f_K solve the normal equations then
obtain q_K from (5.2). The difference in f along
q_K is then used to estimate its derivative along
q_K. A column of P_K is replaced by q_K, to obtain
P_{K+1}, the particular column being chosen to pre-
serve the rank of P_{K+1} and the corresponding column
of $\bar{J}$ is updated. Although, in practice $(\bar{J}^T \bar{J})^{-1}$ is

15

stored and updated rather than $\bar{J}$, thereby avoiding the need to solve the normal equations.

A similar approach is used by Peckham (1970), although he stores F_K, an $n \times \ell$ matrix of columns $f_i - f_{i-1}$, $i \leqslant K$, rather than $\bar{J}_K$ and always rejects the first column and adds the extra information as a new final column. Since this may lead to linear dependence, $\ell > n$ columns are stored and there is a procedure for resetting the directions randomly in the event of the method breaking down.

These methods have now been largely superseded by techniques which estimate and update J without the need to store an extra matrix of directions. An initial estimate of J_0 can be obtained by differences, then the updating formula is usually based on the property that if f is linear, it can be expressed as:

$$f = Jx,$$

where J is constant, and so from (3.7) we have:

$$f_{K+1} - f_K = \alpha_K J d_K$$

This suggests that $\bar{J}$ should be updated to satisfy:

$$f_{K+1} - f_K = \alpha_K \bar{J}_{K+1} \qquad (5.3)$$

A class of formulae satisfying (5.3) was derived by Broyden (1965) and Barnes (1965). In particular Broyden suggests using:

$$\bar{J}_{K+1} = \bar{J}_K + (f_{K+1} - f_K - \alpha_K \bar{J}_K d_K) d^T_K / \alpha_K d^T_K d_K \qquad (5.4)$$

which has the advantage that for the linear case a reduction in the error in $\bar{J}_K$ is ensured at each iteration.

By storing and updating an approximation to $J^T J$, $(J^T J)^{-1}$ or $(J^T J)^{-1} J^T$ rather than J there is a considerable saving in that the work involved in solving the normal equations is reduced or even removed. Several algorithms incorporating these ideas have been suggested. In particular Broyden

(1965), for the $m = n$ case applies Householder's formula to (5.4) to obtain a formula for updating $\bar{J}^{-1}$.

Fletcher (1968) gives a method for updating an approximation to $(\bar{J}^T\bar{J})^{-1}\bar{J}^T$. However, there are stability and accuracy problems in applying formulae of this kind and it is better numerically to approximate and update factors of J. A comprehensive survey of methods for modifying factors is given by Gill, Golub, Murray and Saunders (1972). Specific algorithms together with numerical results, have been produced by Bartels, both based on the Levenberg-Marquardt method, one using Cholesky factorization to solve the normal equations (Bartels (1973)) and the other using QR factorization (Bartels (1972)). For the Cholesky factorization the approximating matrices $\bar{J}_K$ and $\bar{L}_K$ are stored and updated so as to satisfy:

$$\bar{J}_{K+1} = \bar{J}_K + \underset{\sim}{y}_K \underset{\sim}{d}^T_K,$$

and
$$\bar{L}_{K+1}\bar{L}^T_{K+1} = \bar{J}^T_{K+1}\bar{J}_{K+1}$$

where $\underset{\sim}{y}_K$ is chosen so that (5.3) holds. In the QR factorization the matrix $\bar{J}_K$ is not necessary, only Q_K and R_K are stored and these are updated so as to satisfy:

$$Q_{K+1}R_{K+1} = Q_K R_K + \underset{\sim}{y}_K \underset{\sim}{d}^T_K.$$

Note that only n columns of Q_K need be stored so that the storage requirements of both schemes are very similar.

6. SOME SPECIAL CASES

Sums of squares problems do not all fall into categories that can be solved by the methods discussed so far. It is possible to identify different types of problems to which special techniques should be applied. In this section several different problems are listed and possible ways of dealing with them are indicated.

The major assumption of the Gauss-Newton method and all its derivatives is that the approximation for the first derivative of S, given in (3.4), is valid. This is not always the case. For instance, if there are experimental errors or the model is inaccurate then $S(x^*) \neq 0$ so that:

$$\nabla S^2(x^*) \neq 2J(x^*)^T J(x^*).$$

In the past it has often been assumed that sums of squares methods are invariably more efficient than straightforward minimization methods and the importance of the approximation (3.4) has been neglected. In addition McKeown (1974) has used the formulae (3.6) to devise a set of test problems which increasingly violate the Gauss-Newton convergence condition and for which $S(x^*) \neq 0$. He shows that, corresponding to this increase, the success and efficiency of the sums of squares methods compared to the ordinary minimization methods, diminishes. Those methods which allow for a bias towards steepest descent, such as Levenberg-Marquardt and Powell's hybrid are more successful than the basic Gauss-Newton, but not significantly better than the general methods, and usually require more house-keeping and storage.

In order to approximate ∇S more closely, Brown and Dennis (1971) suggest estimating each of the matrices $\nabla^2 f_i$ by differences and updating them at each iteration, using some suitable formula. However, the storage required for these $n\ m \times m$ matrices and the house-keeping involved in their manipulation makes this approach impractical. Alternatively Dennis (1972) outlines a method which approximates the whole of the neglected term in (3.4), namely:

$$\sum_{i=1}^{m} f_i(x) \nabla^2 f_i(x).$$

Special cases also arise when the residual functions have a particular form that can be exploited. One such type of problem is when the

18

variables separate, that is when the model equation has the form:

$$y(\underset{\sim}{t}) = \sum_{j=1}^{\ell} z_j F_j(t,x),$$

where $\underset{\sim}{z}^T = (z_1, z_2, \ldots, z_\ell)$ and $m \geqslant \ell + n$.

The ℓ parameters $\underset{\sim}{z}$ occur linearly in the formulation and there are n nonlinear parameters $\underset{\sim}{x}$. Commonly occurring examples of separable problems are exponential and rational functions, for example:

$$y(\underset{\sim}{t}) = \sum_{j=1}^{n} z_j e^{x_j t},$$

and

$$y(\underset{\sim}{t}) = \sum_{j=1}^{\ell} z_j t^{j-1} \bigg/ \sum_{j=1}^{n} x_j t^{j-1}$$

These were traditionally solved by Prony's method (see Lanczos (1957)) although this has now been shown to be numerically unstable and in some cases completely unreliable. Recently, however, approaches similar to Prony's have proved quite competitive with the basic sum of squares techniques. For example, Golub and Pereyra (1973) give a method where the linear and nonlinear parameters are separated as follows: writing $\phi(\underset{\sim}{x})$ as the matrix with elements $F_j(\underset{\sim}{t_i},\underset{\sim}{x})$ the problem becomes:

minimize $S(\underset{\sim}{x}) = (\phi(\underset{\sim}{x})\underset{\sim}{z} - \underset{\sim}{y})^T(\phi(\underset{\sim}{x})\underset{\sim}{z} - \underset{\sim}{y})$.
Writing ϕ^+ as the generalized inverse, $\phi^+ = (\phi^T\phi)^{-1}\phi^T$ when ϕ is of full rank, gives:

$$S(\underset{\sim}{x}) = (\phi\phi^+\underset{\sim}{y} - \underset{\sim}{y})^T(\phi\phi^+\underset{\sim}{y} - \underset{\sim}{y})$$

and this is minimized to obtain $\underset{\sim}{x}^*$ which is then substituted into:

$$\phi(\underset{\sim}{x}^*)\underset{\sim}{z} = \underset{\sim}{y}$$

to obtain $\underset{\sim}{z}^*$, thus solving the problem. Golub and Pereyra give results to show that the method compares very favourably with minimizing the original

19

sum of squares. The reduction in the number of iterations is marked, although the house-keeping per iteration is increased. The chief advantage is that it will solve problems which are difficult to solve accurately by other methods (see also Osborne (1974)).

Throughout the discussions it has always been assumed that $m \geqslant n$. Although the possibility of J being of rank less than n at certain points has been considered, the underdetermined case, when $m < n$ and rank J is always less than n, has not. In such cases the use of the singular value decomposition method to solve the linear least squares will ensure a solution. Alternatively a Levenberg-Marquardt approach with a suitable $\lambda > 0$ will also suffice. Osborne (1972) covers the underdetermined case in particular.

It may not always be suitable to formulate the function to be minimized as in (2.4). For example, many problems exist where it is desirable to give more weight to some observations than to others. This is simply achieved by multiplying the appropriate residual by a weighting factor w_i (see Draper and Smith (1966) and Wolberg (1967)) and setting:

$$S(\underset{\sim}{x}) = \sum_{i=1}^{m} w_i f_i(\underset{\sim}{x})^2,$$

The factor is then carried through with the calculations.

One assumption that was made in (2.4) was that the errors in the independent variables were negligible. If this is not the case a residual for each of the t_i's must be included, see Wolberg (1967) for a fuller explanation.

Sums of squares problems appear frequently as constrained problems where the values of the parameters are restricted to satisfy certain conditions, by far the most common being the non-negativi condition. Most of the techniques that are used with ordinary minimization methods can also be applied to the special sums of squares methods (see, for example Shanno (1970)) although few programmed codes exist.

In most cases, since the f_i are nonlinear any minimum that is located will be a local minimum and will not be unique. Further minima, if they exist, can usually be located by using a different initial estimate of the solution, although Brown and Gearhart (1971) recommend that a deflation technique may be useful in locating further minima once a dominant minimum is obtained.

Finally, once the solution is obtained it is often useful to be able to assess the uncertainties in the final parameters. Full details of the calculation of the variances, covariances and confidence intervals for the parameters are to be found in Wolberg (1967).

CONCLUSION

In conclusion, we turn to the questions posed in section 2 to give some indication of which combination of techniques would be best used to solve a given problem.

Firstly, if the problem is not a straightforward nonlinear least squares, as given in (2.4) or if it has some special form which could be exploited, then one of the methods mentioned in section 6 may be appropriate. Secondly, even though the function to be minimized is a sum of squared terms, in those cases where the minimum is not zero, the application of a normal minimization method is usually more suitable.

Once these possibilities are discounted the next most important point is whether or not analytical derivatives of the residuals are available. A method using analytical derivatives will generally reach the solution more quickly than a non-derivative

21

method. So, if it is possible to obtain formulae for first partial derivatives of the f_i easily, and these formulae are cheap to evaluate then the problem will be solved more quickly. Indeed in many problems the f_i are all of the same form so that in effect only one formula has to be obtained. However, it is most important that any such formulae are carefully checked, preferably against derivatives calculated numerically, since it is my experience that most user-problems in minimization are caused by incorrect derivatives. If analytical derivatives are available, or easy to calculate for some f_i, while others are not it may be worth using a finite difference method. It is also worth considering calculating second derivatives when the first derivatives are simple, and especially when some of the f_i are linear so that the corresponding second derivatives are zero. In such cases ordinary first derivative minimization methods can be used.

When a good estimate of the solution is available to start the process, or in problems where the region of search is well-behaved, then a modified Gauss-Newton method is most efficient. This should be applied with an updating technique for $\bar{J}$ if derivatives are not available, and a Cholesky or QR method for solving the normal equations when they are, although if there is a possibility of J being singular then singular value decomposition should be used. For a poor initial estimate or a difficult problem then the Levenberg-Marquardt or hybrid methods should be used. The former should be used in conjunction with the special factorization which caters for changing λ and the latter has been designed to be most efficient for non-derivative problems where an updating formula for $\bar{J}$ is employed.

Finally, the cost of each function evaluation can affect the choice of method, as can the relative sizes of m and n. A modified Gauss-Newton method will generally require more function evaluations and fewer house-keeping operations than the

Levenberg-Marquardt or hybrid approaches, so it
should be used when function evaluations are cheap.
In addition, finite difference derivatives should
not be used when function evaluations are expensive.
As m and n increase in size the time taken to solve
the problem increases. In particular as n increases
the solution usually gets more difficult, while an
increase in m affects the number of housekeeping
operations. Obviously for m less than n a method
which allows for J to be of rank less than n, such
as singular value decomposition should be used,
whereas for m very much larger than n it should
not, partly because of the number of operations
and partly because in such cases the likelihood of
J not being of full rank is very small.

REFERENCES

Bard, Y. (1970) "Comparison of Gradient Methods
for the Solution of Nonlinear Parameter Estimation
Problems", *SIAM J. Num. Anal.*, **7**, 157.

Barnes, J.G.P. (1965) "An Algorithm for Solving
Nonlinear Equations Based on the Secant Method",
C. J., **8**, 66.

Bartels, R.H. (1973) "The Cholesky Factorization
for Derivative - Free Nonlinear Least Squares",
University of Texas at Austin Report CNA-64.

Bartels, R.H. (1972) "Nonlinear Least Squares
Without Derivatives - An Application of the QR
Matrix Decomposition", University of Texas at
Austin Report CNA-44.

Box, M.J. (1966) "A Comparison of several current
optimization methods, and the use of transformations
in constrained problems", *C. J.*, **9**, 67.

Brown, K.M. (1972) "Computer Oriented Methods for
fitting tabular Data in the Linear and Nonlinear
Least Squares Sense", Proceedings of the Fall
Joint Computer Conference, p. 1309.

Brown, K.M. and Dennis, J.E. (1972) "Derivative
Free Analogues of the Levenberg-Marquardt and
Gauss Algorithms for Nonlinear Least Squares
Approximation", *Num. Math.*, **18**, 289.

Brown, K.M. and Dennis, J.R. (1971) "New Computational Algorithms for Minimizing a Sum of Squares of Nonlinear Functions", Yale University Report 71-6.

Brown, K.M. and Gearhart, W.B. (1971) "Deflation Techniques for the Calculation of Further Solutions of a Nonlinear System", *Num. Math.*, **16**, 334.

Broyden, C.G. (1965) "A Class of Methods for Solving Nonlinear Simultaneous Equations", *Math. Comp.*, **19**, 577.

Businger, P.A. and Golub, G.H. (1965) "Linear Least Squares Solutions by Householder Transformations", *Num. Math.*, **7**, 269.

Cox, M.G. and Hayes, J.G. (1973) "Curve Fitting: A Guide and Suite of Algorithms for the Non-specialist User", NPL Report NAC26.

Dennis, J.E. (1972) "Some Computational Techniques for the Nonlinear Least Squares Problem", NSF-CGMS Regional Pittsburgh Conference, July 1972.

Draper, N.R. and Smith, H. (1966) "Applied Regression Analysis", Wiley.

Fletcher, R. (1968) "Generalized inverse methods for the best least-squares solution of systems of Nonlinear equations", *C. J.*, **10**, 392.

Fletcher, R. (1971) "A Modified Marquardt Subroutine for Nonlinear Least Squares", UKAEA Research Group Report No. AERE-R6799, HMSO.

Fox. L. (1964) "An Introduction to Numerical Linear Algebra", Clarendon Press, Oxford.

Gauss, K.F. (1809) *"Theoria motus corporum coelistiam"*, *Werke*, **7**, 240.

Gill, P., Golub, G.H., Murray, W. and Saunders, M.A. (1972) "Methods for Modifying Matrix Factorizations", Stanford Report STAM-CS-72-322.

Gill, P.E. and Murray, W. (1972) "Quasi-Newton methods for unconstrained optimization", *JIMA*, **9**, 91.

Golub, G.H. and Pereyra, V. (1973) "The Differentiation of Pseudo-inverses and Nonlinear Least Squares Problems whose Variables Separate", *SIAM J. Num. Anal.*, **10**, 413.

Golub, G.H. and Reinsch, C. (1970) "Singular Value Decomposition and Least Squares Solutions", *Num. Math.*, **14**, 403.

Hartley, H.O. (1961) "The modified Gauss—Newton Method for the fitting of Nonlinear Regression Functions by Least Squares", *Technometrics*, **3**, 269.

Jones, A. (1970) "Spiral - A New Algorithm for Nonlinear Parameter Estimation Using Least Squares", *C. J.*, **13**, 301.

Kowalik, J. and Osborne, M.R. (1968) "Methods for Unconstrained Optimization Problems", Elsevier.

Lanczos, C. (1957) "Applied Analysis", Pitman.

Levenberg, K. (1944) "A Method for the Solution of Certain Nonlinear Problems in Least Squares", *Q. Appl. Math.*, **2**, 164.

Marquardt, D.W. (1963) "An Algorithm for Least Squares estimation of Nonlinear Parameters", *SIAM J. Appl. Math.*, **11**, 431.

McKeown, J.J. (1974) "Specialised versus General-Purpose Algorithms for Minimizing Functions that are Sums of Squared Terms", The Hatfield Polytechnic NOC Report 50.

Meyer, R.R. (1970) "Theoretical and Computational Aspects of Nonlinear Regression", *in* J.B. Rosen, O.L. Mangasarian and K. Ritter *eds.* "Nonlinear Programming", Academic Press.

Murray, W. *ed.* (1972) "Numerical Methods for Unconstrained Optimization", Academic Press.

Ortega, J.M. and Rheinboldt, W.C. (1970) "Iterative Solution of Nonlinear Equations", Academic Press.

Osborne, M.R. (1971) "An Algorithm for Discrete, Nonlinear, Best Approximation Problems", *in* "Proceedings of the Conference on Numerical Methods in Approximation Theory", Oberwolfach.

Osborne, M.R. (1972) "Some aspects of Nonlinear Least Squares Calculations", *in* F.A. Lootsma, *ed.* "Numerical Methods for Nonlinear Optimization", Academic Press.

Osborne, M.R. (1972) "A Class of Methods for Minimizing a Sum of Squares", *Australian Comp. J.*, **4**, 164.

Osborne, M.R. (1974) "Some Special Nonlinear Least Squares Problems", Australian National University at Canberra Report.

Peckham, G. (1970) "A New Method for Minimizing a Sum of Squares without calculating Gradients", *C. J.*, **13**, 418.

Pereyra, V. (1967) "Iterative Methods for Solving Nonlinear Least Squares Problems", *SIAM J. Num. Anal.*, **4**, 27.

Powell, M.J.D. (1965) "A Method for Minimizing a Sum of Squares of Nonlinear Functions Without Calculating Derivatives", *C. J.*, **7**, 155.

Powell, M.J.D. (1970) "A Hybrid Method for Nonlinear Equations", *in* P. Rabinowitz, *ed.* "Numerical Methods for Nonlinear Algebraic Equations", Gordon and Breach.

Powell, M.J.D. (1972) "Problems Related to Unconstrained Optimization", *in* W. Murray, *ed.* "Numerical Methods for Unconstrained Optimization", Academic Press.

Rosen, J.B., Mangasarian, O.L. and Ritter, K. *eds.* (1970) "Nonlinear Programming", Academic Press.

Shanno, D.F. (1970) "An Accelerated Gradient Projection Method for Linearly Constrained Nonlinear Estimation", *SIAM J. Num. Anal.*, **18**, 322.

Wolberg, J.R. (1967) "Prediction Analysis", Nostrand.

SMOOTHED NON-FUNCTIONAL INTERPRETATIONS OF STATISTICAL AND EXPERIMENTAL DATA

P.M. Foster

(Central Electricity Research Laboratories)

SUMMARY

In an earlier publication (Foster (1973)) a non-functional procedure was described which enabled statistical information in a coarse histogram or cumulative form to be reduced to a smooth numerical distributional form. The procedure was also directly applicable to approximation problems which required smoothed interpretations of experimental data. Certain aspects of this method have since been reformulated and further calculations and tests made. In particular, its application to non-smooth problems (*i.e.*, those in which the solution gradient or curvature varies strongly) has been considered together with the question of choosing the "best" solution. This note provides a complete description of the procedure as it exists to date.

1. INTRODUCTION

The interpretation and presentation of experimental or statistical data which are of limited quality is a problem area where the minimization of sums of squares has found considerable application. In cases where the data refer to a continuous frequency distribution they are normally obtained in either histogram or cumulative form; the problem in these cases is therefore to derive an estimate of the true distributional form from the coarse histogram or cumulative data set. For those systems which are known to conform to a

particular type of frequency distribution (e.g.,
gaussian, exponential) the problem can usually be
solved using standard probability plotting or
regression methods. However, when there exists no
preconceived analytic form for the distribution
these methods cannot be used except, perhaps, on a
trial and error basis or by using more generalized
empirical functions such as Johnson or Pearson
distributions. A similar situation also arises
with data approximation problems for which no specific analytic function describing the data is
prescribed. Thus, although polynomial or other
regression methods might be applied in these cases,
a suitable function for regression can normally
only be found by trial and error.

The disadvantages of these methods are that
they can be wasteful in time and may demand specialist skills which the casual user may not possess.
An alternative approach (Foster (1973)) has been
suggested which is both extremely flexible and
only makes small demands on the time and skills of
the user. This used the simultaneous application
of curve fitting and curve smoothing methods in a
way similar to that first used by Phillips (1962)
and Twomey (1963) for solving certain kinds of
integral equations and later by Reinsch (1967) for
smoothing spline solutions. Its advantages arise
from the use of a "non-functional" solution which
exists only as a numerical array evaluated at equispaced intervals of the independent variable. Consequently it neither makes nor requires any preconceived assumptions to be made about the form of
the solution other than that it is smooth, continuous and differentiable (at least four times)*.

---

*Integration or differentiation of the solution is
obtained using differences. This of course implies
that the "non-functional" solution is really a
piecewise polynomial function. However, since no
attempt is made to define the parameter values
associated with these polynomials the term "non-
functional" is preferred. This also conveys the
important concept that the procedure makes no pre-
judgement about the general or over-all functional
behaviour of the solution.

The shape and the degree to which the solution approximates the input data are controlled by the value of a single smoothing parameter. If required, the solution form may also be controlled by weighting terms and/or by presetting the solution values at any desired points. This enables *a priori* information to be incorporated into the type of solution derived.

Since publication (Foster (1973)) certain aspects of the method have been reformulated and further calculations and tests made. In particular, its application to non-smooth problems (*i.e.*, those in which the solution gradient or curvature varies strongly) has been considered together with the question of choosing the "best" solution. This note provides a complete description of the procedure as it exists to date.

2. PROBLEM FORMULATION

Consider a system which is sampled experimentally to provide either histogram or cumulative information on its associated frequency distribution function $D(r)$ within the range $r_1 \leqslant r \leqslant r_n$. The following integral relationships then apply.

(I) For histogram information, H_j,

$$
\left.
\begin{aligned}
I_H &= H_1 + \varepsilon_1 \qquad\qquad , \\[2em]
\int_{r_{j+1}}^{r_j} D(r')\,dr' &= H_j + \varepsilon_j, \quad j = 2,3,4,\ldots,n
\end{aligned}
\right\} \quad (1)
$$

where the ε_j's compensate for experimental errors,

$$
I_H = \int_{r_1}^{r_n} D(r')\,dr' \text{ and } H_1 = \sum_{j=2}^{n} H_j.
$$

I_H therefore denotes the total sample size. For those experiments in which this is known exactly

$$
\varepsilon_1 = \sum_{j=2}^{n} \varepsilon_j = 0.
$$

This is effectively achieved by giving the H_1 result an appropriately high weighting.

(*II*) For cumulative information, C_j,

$$\left. \begin{aligned} I_C &= C_1 + \varepsilon_1 \quad , \\ I_C + \int_{r_1}^{r_j} D(r')dr' &= C_j + \varepsilon_j, \quad j = 2,3,4,\ldots,n \end{aligned} \right\} (2)$$

where

$$I_C = \int_{r_0}^{r_1} D(r')dr'$$

and $-\infty < r_0 < r_1$. This formulation allows for that type of cumulative experiment in which the lower reference point r_0, although fixed, may not be known. For those experiments in which it is known it suffices to set $r_1 = r_0$ so that $I_C = C_1 = \varepsilon_1 = 0$.

Now consider an ordinary data approximation problem, namely the interpretation of a set of discrete measurements made upon a system described by the continuous function $F(r)$ within the range $r_1 < r < r_n$. A solution to this problem follows directly from being able to solve the previous cumulative problem. Thus:

(*III*) For experimental information, F_j, an approximation satisfying

$$F(r_j) = F_j + \varepsilon_j \qquad (3.1)$$

is obtained by solving for $D(r)$ and I_C as in *II* and then setting

$$F(r_j) = I_C + \int_{r_1}^{r_j} D(r')dr' . \qquad (3.2)$$

It now becomes clear that problems *I*, *II* and *III* all fall into the same category, the only differences between them being the particular integration ranges which are involved. Each problem can therefore be written in the quadrature form

$$\frac{\Delta r}{3} \, \underset{\sim}{A}\underset{\sim}{d} = \underset{\sim}{f} + \underset{\sim}{\varepsilon}, \tag{4}$$

where $\underset{\sim}{f}$ is an n-vector representing the values of H_j, C_j or F_j; $\underset{\sim}{\varepsilon}$ is an n-vector representing the experimental errors ε_j plus errors introduced by the quadrature; $\underset{\sim}{d}$ is an $(m + 1)$-vector in which the first m values denote $D(r)$ evaluated at $(m - 1)$ equispaced intervals Δr in the range r_1 to r_n, and the $(m + 1)$th value represents either I_H or $I_C{}^*$; $\underset{\sim}{A}$ is an n by $(m + 1)$ quadrature matrix operator which is chosen to numerically integrate $\underset{\sim}{d}$ over the particular ranges specified by the problem. The term $\Delta r/3$ is simply a scalar multiplier associated with $\underset{\sim}{A}$.

3. THE SMOOTHING SOLUTION

For quadrature errors to be negligibly small and to ensure a detailed and smooth solution, the dimension of $\underset{\sim}{d}$ must be large (typically between 20 and 100). Hence, in general, $(m + 1) > n$, thus making equation (4) under-determined and insoluble by the traditional least squares approach. However, if in addition to satisfying equation (4), the solution is assumed to be smooth in the sense that the sums of the squares of the gradients or curvatures of $\underset{\sim}{d}$ are also minimized, then smoothed estimates of $\underset{\sim}{d}$ can be derived. The problem now becomes,

$$\text{minimize } \frac{1}{m} \sum_{i=1}^{m+1} (\underset{\sim}{W}_d \underset{\sim}{D}\underset{\sim}{d})^2_i \tag{5}$$

with respect to the elements of $\underset{\sim}{d}$ such that

$$\frac{1}{n} \sum_{j=1}^{n} (\underset{\sim}{W}_f \underset{\sim}{\varepsilon})^2_j = S, \tag{6}$$

*N.B. For the histogram problem I_H is a redundant variable. Thus, although it has been included in the problem formulation for consistency, it is not used as a variable in the problem solution.

where $\underset{\sim}{D}$ is a first or second order difference operator depending upon whether the sum of the squares of the gradients or curvatures is to be minimized, $\underset{\sim}{W}_d$ and $\underset{\sim}{W}_f$ are diagonal weighting matrices associated with the difference and error terms, respectively, and S is a positive quantity to be defined below.

Assuming that the data represent a sample test drawn from a population with standard deviations σ_j, then the data weights should be set as follows,

$$\left.\begin{array}{ll} (\underset{\sim}{W}_f)_{jj} = \sigma_j^{-1}, & j = 1,2,3,\ldots,n, \\ (\underset{\sim}{W}_f)_{ij} = 0, & i \neq j \end{array}\right\} \qquad (7)$$

The term

$$\sum_{j=1}^{n} (\underset{\sim}{W}_f \underset{\sim}{\varepsilon})_j^2$$

then becomes the sum of n standard normal variates and therefore has the properties of a χ^2-distribution of degree n^*. According to Wilson and Hilferty (1931), $(\chi^2/n)^{\frac{1}{3}}$ is approximately normally distributed with mean $1 - 2/9n$ and variance $2/9n$, for $n > 30$. This therefore enables the quantity S, equation (6), to be defined within given confidence limits. Thus, for a 95% confidence limit S is given by

$$S = \left(1 - \frac{2}{9n} \pm 2\sqrt{\frac{2}{9n}}\right)^3, \quad n > 30 \qquad (8)$$

For smaller n, however, tables of the χ^2-distribu-

---

*Footnote. It may be argued that one degree of freedom is lost in setting the value of the smoothing parameter α which is defined later. It should also be noted that the number of degrees of freedom will also be reduced if the values of σ_j are estimated rather than known. For these reasons equation (8) is only approximate although its accuracy can be improved if required by replacing n with a better estimate for the number of degrees of freedom.

tion must be used.

Unlike previous applications of the smoothing technique, a diagonal weighting matrix W_d has been incorporated into the smoothing process. If d_T represents the true solution then ideally these smoothing weights should be set as follows

$$
\left.
\begin{aligned}
(W_d)_{jj} &= |(Dd_T)_j|^{-\frac{1}{2}}, \quad j = 1,2,3,\ldots,m. \\
(W_d)_{ij} &= 0 \qquad\qquad, \quad i \neq j
\end{aligned}
\right\} \tag{9}
$$

The precise reason for choosing these values will be given later. For the moment though, it suffices to say that it ensures that smoothing is uniform relative to the true solution and not accentuated at local regions of the solution where the gradient or curvature is large. In practice only a rough estimate of d_T will be available so that W_d can never be set with any great accuracy. However, an example is given later which suggests that this is not a serious limitation of the method. Indeed it often suffices to simply set the W_d weights all equal to unity.

The solution to equations (5) and (6) is obtained by minimizing the functional

$$
\frac{1}{m} \sum_{i=1}^{m+1} (W_d Dd)^2_i + \beta \left[\frac{1}{n} \sum_{j=1}^{n} (W_f \varepsilon)^2_j - S \right] \tag{10}
$$

with respect to the elements of d, β being an undetermined Lagrangian multiplier. The following smoothing solution is then obtained

$$
d(\alpha) = \frac{3}{\Delta r} \left[A^T W_f^2 A + \alpha D^T W_d^2 D \right]^{-1} A^T W_f^2 f \tag{11}
$$

where $\alpha = 9n/\beta m \, \Delta r^2$, $\beta \neq 0$, α being referred to as the smoothing parameter since it controls the extent to which the solution is smoothed. It may also be considered as a scaling factor which operates upon the smoothing weights, W_d^2, and thereby determines their absolute value for any given problem.

For each solution $d(\alpha)$ there clearly corresponds a value $S(\alpha)$ satisfying equation (6). To within known confidence limits therefore the best solution $d(\alpha_c)$ can be bounded by solutions which have S values set by equation (8) *et seq*. Alternatively $d(\alpha_c)$ could be chosen subjectively and its corresponding S value used for a significance test. Irrespective of which procedure is used it is assumed that the sole effect of the smoothing process is to smooth out distortions in the solution produced by random data errors. However, this may not always be the case. By rewriting equation (11) as

$$d(\alpha) = \left[A^T W_f^2 A \right]^{-1} \left[\frac{3}{\Delta r} A^T W_f^2 f - \alpha D^T W_d^2 D d(\alpha) \right], \quad (12)$$

it can be seen that the smoothing process itself introduces terms which can distort the solution. An effective method of suppressing this might be to choose W_d so that for $d(\alpha) \simeq d_T$, the smoothing terms $D^T W_d^2 D d(\alpha)$ are small. Noting that D is very nearly symmetric so that $D^T \simeq D$, it can then be seen that with W_d set in accordance with equation (9) this condition is met. The smoothing process can then be said to be compatible with d_T. However, if, as in previous work involving smoothing solutions, W_d is omitted (equivalent to setting the smoothing weights equal to unity) the smoothing process will be incompatible and will be more likely to distort or over-smooth the solution. This will be particularly noticeable for problems in which α_c is large and/or d_T contains large variations in gradient or curvature. Examples are given in the next section which illustrate these points.

When *a priori* information about the behaviour of d_T is available it may sometimes be desirable to be able to preset the values of some of the elements of $d(\alpha)$. For example, if the elements d_k are preset for given values of k, then minimizing equation (10) with respect to the remaining elements produces the following modified solution,

$$\underset{\sim}{d}{}'(\alpha) = \frac{3}{\Delta r}\left[\underset{\sim}{A}{}'^{T}W_{\underset{\sim}{f}}^{2}\underset{\sim}{A}{}' + \alpha \underset{\sim}{D}{}'^{T}W_{\underset{\sim}{d}}^{2}\underset{\sim}{D}{}'\right]^{-1}\left\{\underset{\sim}{A}{}'^{T}W_{\underset{\sim}{f}}^{2}\underset{\sim}{f}\right.$$

$$\left. - \frac{\Delta r}{3}\sum_{k}\left[\underset{\sim}{A}{}'^{T}W_{\underset{\sim}{f}}^{2}\underset{\sim}{A}{}_{k} + \alpha \underset{\sim}{D}{}'^{T}W_{\underset{\sim}{d}}^{2}\underset{\sim}{D}{}_{k}\right]d_{k}\right\} \tag{13}$$

where $\underset{\sim}{d}{}'(\alpha) \equiv \underset{\sim}{d}(\alpha)$ modified to remove its kth elements d_k, $\underset{\sim}{A}{}' \equiv \underset{\sim}{A}$ modified to remove its kth columns $\underset{\sim}{A}_k$ and $\underset{\sim}{D}{}' \equiv \underset{\sim}{D}$ modified to remove its kth columns $\underset{\sim}{D}_k$.

Presetting the elements of $\underset{\sim}{d}$ has different practical consequences depending upon the particular application in mind. Thus, for problem types *I* and *II* it forces the distributional solutions to have given *values* at the selected points. For problem type *III* it forces the solution to have given *gradients* at the selected points. However, if required, the solution values may also be controlled in this type of problem by appropriate weighting of the data at the selected points.

4. EXAMPLES

Examples have already been given (Foster (1973)) which demonstrated qualitatively the usefulness of smoothing solutions for these three types of problem. The examples in this section have therefore been chosen mainly to illustrate the effects of incompatible smoothing and choosing a "best" solution, these being the two most important features to have arisen from the present work. These examples were all computed using an updated version of the Fortran IV computer program SMUVIT.

The first three examples use histogram data relating to the standard normal distribution for r = -3 (1) 3. Fig. 1 shows smoothing solutions evaluated at r = -3 (.2) 3* using exact histogram data values. For direct comparison with later examples these data were normalized so as to correspond to a total sample size of 50. No attempt was

---

*N.B. To improve clarity not all solution points have been plotted. This also applies to other illustrations used in this section.

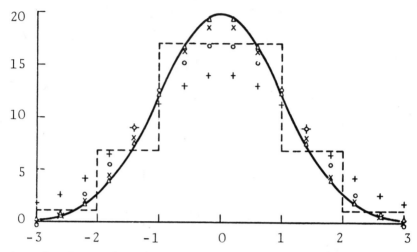

Fig. 1. Reproduction of a normal distribution from
exact histogram data. Sample size 50,
smoothing parameter values 1, 10, 100 and
1000 (Δ, $\times$, o and + respectively)

made to set the smoothing weights correctly, *i.e.*,
they were all set equal to unity. The results
demonstrate how for increasing α the solutions
become distorted or over-smoothed. They also
suggest that, given accurate data, a good approximation to the true solution can be derived for
small α irrespective of whether smoothing is compatible or not. However, this is only valid for
problems which are locally smooth, *i.e.*, the gradient or curvature of the true solution does not
vary greatly between data points. Indeed it would
be unreasonable to expect any empirical method to
reproduce the solution to a problem which was not
locally smooth unless some additional information
were available or assumed for the non-smooth
regions. With the present method such information,
if available, could probably be used to preset the
smoothing weights or the solution values at suitable
points, thereby ensuring that the correct type of
behaviour is built into the solution.

Fig. 2 illustrates the effects data errors
have upon this same problem. These data were

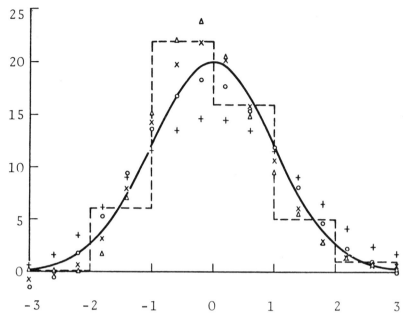

Fig. 2. Reproduction of a normal distribution from random histogram data. Sample size 50, smoothing parameter values 1, 10, 100 and 1000 (Δ, ×, o and +, respectively)

generated by taking a random sample of size 50 from a relatively large standard normal population. In this particular case they have a mean weighted square residual of about 0.5 relative to the exact data. Since the distribution of these data about their mean values should be the same as for a binomial distribution, their standard deviations (required for setting W_f, equation (7)) should be approximately given by the square root of the data values themselves, *i.e.*,

$$\sigma_j \simeq H_j^{\frac{1}{2}}, \quad j = 2,3,4,\ldots,n.$$

Where the data value was zero its standard deviation was arbitrarily set equal to 1. Likewise the data value H_1 (*i.e.*, the total sample size) was also allocated a standard deviation equal to 1. H_1 then received adequate weighting to ensure that

37

all the solutions derived had identical total sample sizes to within about 0.01% accuracy.

The results in Fig. 2 demonstrate how the effects of data errors, which are significant for small α, are progressively smoothed out with increasing α. Thus, for $\alpha \gtrsim 100$ the differences between these solutions and those derived using exact data are relatively small. Not unexpectedly the solution $\underset{\sim}{d}(\alpha_1)$ for which $S(\alpha_1) = 1$ is found to occur near this point at $\alpha_1 = 400$. On statistical grounds this might therefore be expected to provide a reasonable estimate of the best solution. However, it has already been noted from Fig. 1 that for α this large, the solutions become significantly distorted by the smoothing process itself. Attempts at controlling this distortion using the smoothing weights were made but the results were still unsatisfactory, It appeared that even when smoothing was compatible α still had to be chosen too large to smooth out the effects arising from data errors. A two stage approach was therefore adopted in which the data were first smoothed to remove random errors. This was accomplished by computing the smoothed data vector $\Delta r/3 \underset{\sim}{A}\underset{\sim}{d}(\alpha_1)$, where $\underset{\sim}{d}(\alpha_1)$ is defined above (it should be noted that this involves no additional work by the user since these values are automatically provided by SMUVIT for any smoothing solution calculated). These smoothed data were then used as input for the second stage in which a set of solutions $\underset{\sim}{d}(\alpha_2)$ were computed for decreasing values of α_2, $\alpha_2 < \alpha_1$. These were then found to converge to a particular solution which was a good approximation of the true solution.

The philosophy behind the approach simply assumes that the random errors removed by the first smoothing operation outweigh any additional errors produced by the smoothing operation itself. The smoothed data therefore provide an improved data set, free of random errors, from which solutions can be derived for α sufficiently small for smoothing distortion effects to be negligible. The results of the approach are illustrated in Fig. 3.

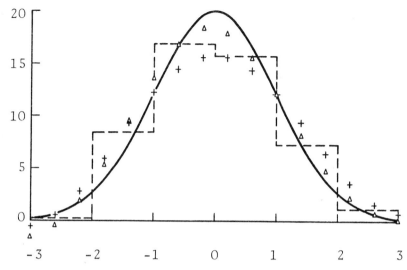

Fig. 3. Reproduction of a normal distribution
         from smoothed random histogram data.
         Sample size 50, smoothing parameter values
         10 and 400 (Δ and +, respectively)

The smoothed solution and its associated smoothed
data derived in the first stage are given by the
crosses and the histogram, respectively. The
final solution derived in the second stage is
given by the triangles. Agreement with the true
curve (solid line) is seen to be good although
there was a tendency for the solution to become
negative in one of the tail regions. However, if
required this could be prevented using the presetting facility, e.g., by presetting the last one or
two values of the solution in this region equal to
zero.

It was mentioned above that use of the smoothing weights was unable to control solution distortion with this problem. The implication was
that the effects of the smoothing weights were
insignificant in comparison with the over-all
smoothing required to counter the effects arising
from data errors. This is understandable because

the smoothing parameter α had to be increased by approximately two orders of magnitude beyond those values at which distortion effects were known to be negligible. However, the variation in curvature, and therefore smoothing weights, over the entire solution was only about one order of magnitude. Under these conditions therefore the effects due to α would predominate.

An example in which this situation is reversed will now be considered. Fig. 4 illustrates the results of the same two stage approach when applied to a data set generated by taking a random sample of size 50 from a standard log-normal population.

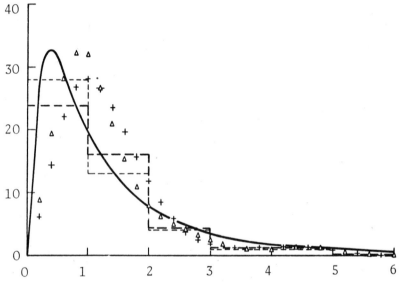

Fig. 4. Reproduction of a log-normal distribution from smoothed random histogram data. Sample size 50, smoothing parameter values 1 and 4 (Δ and +, respectively)

The histogram corresponding to the original data is given by the pecked line and the remaining information is defined similarly to that in the previous example. It should be noted that the

40

solutions have been preset to zero at the origin. This was essential since they would have otherwise assumed a form similar to a decreasing exponentially shaped function. It can be seen that although the final solution (given by triangles) resembles the true solution, the procedure is not able to predict the position of the peak at all well. The reason for this is associated with the fact that the problem is not globally smooth. The curvature of the true solution varies by approximately three orders of magnitude over the range of interest, being greatest close to the origin. Since the nature of the smoothing procedure is to minimize curvature, the computed solution will tend to be over-smoothed close to the origin, thereby displacing its peak to the right.

In order to control this type of behaviour it is necessary to supply an approximate indication of the relative variation in curvature which is present by setting the smoothing weights more realistically. In the absence of any knowledge about the form of the true solution these weights must be estimated in a purely subjective way, such as from a free-hand interpretation of the data or from an empirical function (or piecewise function) crudely fitted to the data. Of course, if *a priori* information were available then this too should be used in setting these weights. The following example illustrates how a free-hand interpretation supplied by an independent observer might be used to improve the smoothing solution of the previous problem.

Using the free-hand interpretation given by the broken curve in Fig. 5 the following estimates of mean curvature and mean smoothing weights squared were calculated.

41

r_i	$\dfrac{d^2}{dr^2}\, \underset{\sim}{d}_i$	$(\underset{\sim}{W}_d)^2_i$
0-0.4	950	$\left.\begin{array}{c} \\ \end{array}\right\}\, 1$
0.4-0.8	870	
0.8-2.0	32	30
2.0-5.0	1	10^3
5.0-6.0	0.1	10^4

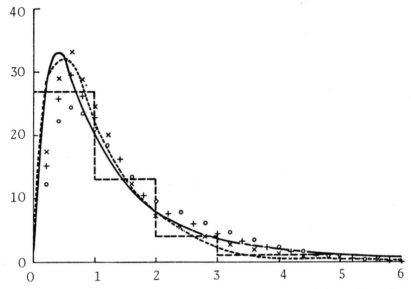

Fig. 5. Reproduction of a log-normal distribution from random histogram data using smoothing weights estimated from a free-hand interpretation (given by broken line). Sample size 50, smoothing parameter values 1.3, 3.2 and 6.4 (×, + and o respectively)

The mean curvature values for each range were calculated using the second difference expression $4(y_{-1} - 2y_0 + y_1)/(x_1 - x_{-1})^2$, the range being chosen to give an accuracy to within about an order of magnitude. It would be unreasonable to expect an accuracy any better than this with the present problem. The mean smoothing weights squared were calculated using equation (9), but for computational convenience were then normalized to the values shown. Because of their large variation in value the effect of smoothing weights upon the solution dominated over that arising from α. Not surprisingly therefore the solution derived in the second stage of the process was not significantly different from that derived in the first stage. It seemed pointless to attempt to illustrate this and so instead Fig. 5 was chosen to illustrate the range in solutions obtained in the first stage which spanned the full 95% confidence limit of $S(\alpha)$. In this case this corresponded to $S(\alpha)$ values of 0.3, 0.9 and 2.0, respectively, for α equal to 1.3, 3.2 and 6.4. Clearly, in the absence of any reason for choosing otherwise, the central solution would normally be taken as the "best", this corresponding to the expected mean value of $S(\alpha)$. Comparison with Fig. 4 shows that use of the smoothing weights has produced a significant improvement in the solution, both in the peak position and in the shape of the exponentially decreasing tail. However, the solution still fails to reproduce the sharpness of the true peak. This is understandable though, because the problem is not locally smooth in this region and there are inadequate data to expect peak shape to be defined by an empirical method.

By way of a final example Fig. 6 illustrates the smoothing solution to a data approximation problem in which the data errors are known to have a standard deviation of about 10. No attempt has been made to set the smoothing weights correctly, *i.e.*, they have all been set equal to unity. The solution (circles) is seen to provide a good approximation of the true solution (solid line) in spite of the relatively large scatter present in the data (squares). Deviations which do occur are mainly confined to the tails of the solution where its shape is more sensitive to data variations. The data were taken from an example used by Termonia and Deltour (1974) to illustrate a data smoothing technique. Although it is thought that

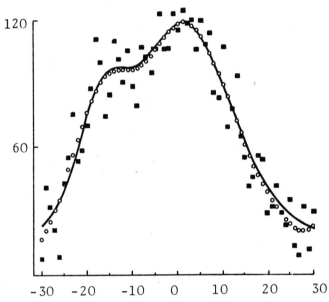

Fig. 6. Reproduction of a double peaked curve from scattered data. ■ data points, —— true curve, taken from Termonia and Deltour (1974), o smoothing solution for α = 10

the example may be slightly flattering to both this and the present technique it does enable some comparison to be made between the two methods. Perhaps the most significant advantage of the present method is that it always produces smoothly varying solutions whereas data smoothing solutions usually retain a degree of raggedness due to the effects of data scatter.

5. DISCUSSION

The present work has been concerned with improvements in the method and application of the smoothing procedure previously described by the author (Foster (1973)). One improvement has been the reformulation of the histogram problem (problem type I) so that total sample size is now also taken into consideration by the procedure. This is essential for that type of histogram problem in which total sample size is known accurately. Another notable improvement has been the inclusion of smoothing weights into the procedure. This enables the smoothing operation to be made compatible with the type of solution to be derived and is essential for the globally non-smooth type of problem in which large variations in curvature occur within the solution.

Some consideration has also been given to the question of choosing the "best" solution to a problem. This solution can be defined from statistical considerations of the mean square residual value. However, in certain cases where excessive smoothing is necessary to overcome the effects due to data errors, the "best" solution may be distorted by the smoothing operation itself. It is suggested that this may be overcome by first using the procedure to smooth the data and thereby remove random errors (at the expense of adding smaller smoothing errors). This then enables relatively little smoothing to be used in the second stage in which the smoothed data are used to calculate the final solution. It should be noted that if the standard deviations of the data (used to set the data weights) are related to the data values in some way, then this approach also provides improved estimates of the data weights for use in the second stage.

The fact that good results were obtained with such different distribution shapes as the standard normal and the log-normal examples suggest that smoothing solutions should be generally useful for most problems. It is of interest to note that the two stage approach significantly improved the solution for the standard normal distribution whereas the use of smoothing weights had little effect. However, with the log-normal distribution the situation was completely reversed. Although these differences can be accounted for by the different nature of the two problems it does suggest that in general both the two stage approach and the smoothing weights should always be used together. In this way a good solution should be ensured irrespective of the problem type.

It will be clear that with an empirical method such as this the quality of the final solution is strongly dependent upon the quality of the data. However, the presetting facility and the smoothing weights do enable the solution to be influenced by *a priori* or intuitive information if required. Ideally the problem should be locally smooth so that the solution curvature varies relatively slowly between adjacent data points. If certain regions are not locally smooth then purely empirical methods cannot be expected to give good results in these regions except with the help of local *a priori* or intuitive information. In principle the smoothing approach attempts to fit the data in much the same way that one might fit it by eye, *i.e.*, it attempts to draw the smoothest possible curve through the data. However, it is clearly more objective in the way in which it does this and is therefore able to provide confidence limits on the solution. In comparison with other empirical methods it has the advantage that it requires no skills other than those which might be required to provide an initial free-hand interpretation of the data. For this reason it should prove useful to users who have relatively little mathematical expertise.

6. ACKNOWLEDGEMENT

This work was carried out at the Central Electricity Research Laboratories, Leatherhead, England, and the paper is published by permission of the Central Electricity Generating Board.

7. REFERENCES

Foster, P.M. (1973) "A Non-functional Method for Reducing Cumulative or Histogram Data to a Smooth Distributional Form or for Constructing a Smooth Approximation to Experimental Data", *J. Comp. Phys.*, **12**, 143-149.

Phillips, D.L. (1962) "A Technique for the Numerical Solution of Certain Integral Equations of the First Kind", *J. Assoc. Comput. Mach.*, **9**, 84-97.

Reinsch, C.H. (1967) "Smoothing Spline Solutions", *Numerische Mathematik*, **10**, 177-183.

Termonia, U. and Deltour, J. (1974) "The Optimization of the Least Square Smoothing of Experimental Data", *J. Phys. D: Appl. Phys.*, **7**, 2157-2165.

Twomey, S. (1963) "On the Numerical Solution of Fredholm Integral Equations of the First Kind by the Inversion of the Linear System Produced by Quadrature", *J. Assoc. Comput. Mach.*, **10**, 97-103.

Wilson, E.B. and Hilferty, M.M. (1931) "The distribution of chi-squared", *Proc. Nat. Acad. Sci.*, **17**, 684.

PARAMETERIZATION OF NONLINEAR LEAST SQUARE FITTING PROBLEMS

A.R. Curtis

(Atomic Energy Research Establishment, Harwell)

SUMMARY

Experience with the following two problems is described.

(i) Automatic choice of chemical reaction rate constants (and other parameters) to fit the results of kinetics simulation to observe time courses of reagent concentrations.

(ii) Curve-fitting in routine radio-immuno-assays to calibration data, which for reasons of economy offer very little redundancy.

In these two very different problems, it was found desirable to impose a nonlinear transformation

$$a = f (b) \qquad\qquad (A)$$

between the parameter vector a naturally arising in the problem and an artificial parameter vector b (whose dimension may be less than or equal to that of a). The general purpose nonlinear least squares subroutine works in terms of b, and the special purpose function subroutine for computing residuals uses (A) to recover a. This device has been used in the past to ensure that simple constraints are satisfied. We have exploited it to obtain additional advantages including:

(i) improved conditioning of the problem

relative to metric assumptions built in to the least square subroutine;

(ii) avoidance of "wild" values of q for which the computation of fitting functions fails or presents difficulties;

(iii) simple standardization of initial values for repetitive fitting problems.

* * *

It is part of the folk-lore of data fitting that if a non-negativity constraint is to be applied to a particular parameter, a_r say, then this can be done by setting $a_r = b_r^2$ and taking b_r as parameter in an unconstrained least-square fitting process. (Note that this appears to be slightly more powerful than the alternative $a_r = \exp(b_r)$ since $a_r = 0$ can be achieved for a finite value of b_r; in practice, the difference may be illusory since derivatives with respect to b_r will tend to zero with b_r, causing numerical difficulties.) The principle can obviously be carried further, setting, e.g., $a_r = \sin^2 b_r$ to enforce $0 \le a_r \le 1$.

The first of the two applications to be described depends only on a systematic application of this elementary idea. Having developed an efficient method (Curtis and Chance (1974)) of solving the initial value problem for chemical mass action kinetics, it was desired to use this as a sub-calculation in choosing values of the rate coefficients for specified chemical reactions so as to give best least-square fit to observed time courses of the concentrations of some of the reacting chemical species. Although for the problem as described it is possible to generate derivatives of observed concentrations with respect to fitted

49

rate coefficients by solving a larger initial value problem, instead a nonlinear least square routine not requiring derivatives was chosen. This gives flexibility by permitting the user of the program to code patch subroutines to obtain special effects. The minimization routine chosen was VA05A (Hopper (1973)) from the Harwell Subroutine Library; this was found much more reliable than the alternative VA02A (Hopper (1973)), which requires only about half as much workspace.

The numerical values of chemical rate coefficients vary widely because the flux of an nth order reaction is the product of a rate constant with n chemical concentrations (and must have a significant value at some time in the course of the run to permit the rate constant to be determined), while the concentrations themselves vary widely; moreover the units of concentration and time may be chosen so as to give very different numbers. Thus we must cater for different rate coefficients spanning several orders of magnitude, even in one problem, with a greater range from problem to problem; and the user of the program often cannot supply an initial guess closer than a factor of 100 or so. Fortunately, he will usually be well satisfied with a relative accuracy of a few per cent. on the fitted value.

All these considerations as well as the essential positivity, suggested using the logarithms of the rate coefficients as the parameters to be fitted. Thus, if we use the notation b_r, $r=1$ to m, for the parameters actually varied by VA05A to find the best fit, we first form the numbers $c_r = \exp(b_r)$, $r=1$ to m, and store them as "rate coefficients". Then we call a user-supplied patch subroutine (which may be null), by means of which constraints can be imposed (e.g., if an equilibrium ratio is known, this determines one rate coefficient in terms of others). Then we carry out the simulation run, returning discrepancies between computed and observed concentrations at specified times to VAO5A as residuals. The advantages of this approach are worth listing.

(i) Positivity is automatically maintained.

(ii) If the user cannot supply an initial guess for c_r, we start from $b_r = 0$, giving $c_r = 1$; this has often been successful.

(iii) Routines such as VAO5A expect equal changes in the various components to be about equally important; our device comes near to achieving this in practical problems, since a given proportional change in a rate coefficient is usually what matters.

(iv) For similar reasons, different rate coefficients which can be determined from the data are found to comparable relative accuracies, corresponding to comparable absolute accuracies for the b_r. Computed standard deviations on the b_r, if small, can be interpreted as relative accuracies on the c_r; computed correlation coefficients are also easy to interpret.

This device has also proved successful when, because of a user patch, the parameter is not in fact being interpreted as a rate coefficient. Two straightforward practical examples are worth mentioning.

(i) A particular c_r is interpreted as the unknown initial concentration of one of the reacting species.

(ii) Transparency at a particular wavelength, observed as a function of time, is known to depend linearly on the concentrations (x and y, say) of two reacting species, but their relative contributions to the absorption coefficient are unknown. One of the c_r, c_m say, is used to construct a third "concentration"

$$z = (x + c_m y)/(1 + c_m)$$

51

which is compared with the observations, the value of b_m being fitted as well as the logarithms of genuine rate coefficients.

The considerations outlined above may seem a little elementary, although they played an important part in the development of a successful method (Curtis and Chance (1974)). In the next problem, which presented itself shortly afterward, a more complicated parameterization was necessary. The problem is to fit, routinely, a curve of the general character shown in Fig. 1, to four pairs of data points such as those marked with crosses (at x=1, 3,6 and 10); the data points marked with dots should be ignored for the time being. The context is the routine assay, by radio-immunological methods, of human placental lactogen (HPL) (Harding, Thomas and Curtis (1973)). Because the serum used in the assay is unstable, a few standard HPL solutions are included with each batch of samples for assay, and two assays done on each, giving the points marked with crosses. Those marked with dots would normally represent assays (also in pairs) on the unknown samples of the batch, only the y values being given, and the x values to be read from the fitted curve. The requirement was to automate the process of fitting a suitable curve, and printing the x values corresponding to each unknown sample. In the case shown in Fig. 1, the points marked with dots were also obtained from known standard solutions; about 10 such sets of data were made available to enable the fitting program to be developed, but for reasons of acceptability in routine use the number of standards has to be limited to four.

Other information was also given: the standard solutions would always have x values of 1,3,6, 10, and greatest accuracy was required in the region $2 \leq x \leq 8$ approximately; the curve could shift sideways with aging test serum, and its vertical scale could change; for large x, the curve would approximate to a rectangular hyperbola, whose horizontal asymptote might be above or below the

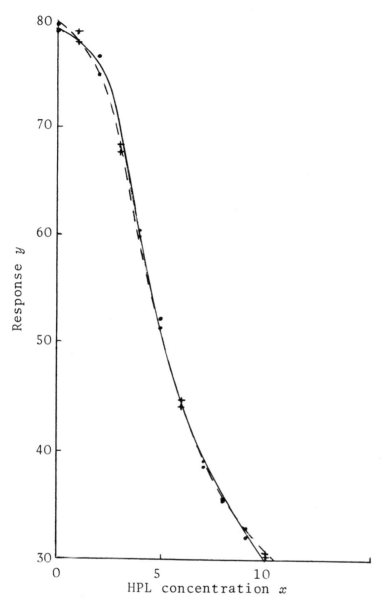

Fig. 1.   Typical fit: + standard data points;
· additional data points; —— best 5-
parameter fit to all points; - - - best
3-parameter fit to standard points

x-axis; for small x, with the curve shifted well to the right, a fairly sharp turnover could be expected from another horizontal asymptote for $x \to -\infty$; close control of the x scale was attempted during manufacture of the test serum, but some variation might occur. Because of the small number of data points in routine use, it was necessary to devise a functional form for the fitted curve which had the correct qualitative form for all combinations of parameter values. The function

$$y = a_1 + a_2 / \left[1 + a_3 \ln \left\{ 1 + \exp(\frac{x-a_4}{a_5}) \right\} \right] \quad (1)$$

has this property if the five parameters a_1 to a_5 are positive; but it has too many parameters. Roughly speaking:

a_1 is the lower asymptotic value of y;

a_2 is the difference between upper and lower asymptotes;

a_3 is a shape parameter, controlling sharpness of the upper bend;

a_4 controls the horizontal position;

a_5 controls the horizontal scale.

Note that for large x we have approximately

$$y \simeq a_1 + \frac{a_2}{1 + \frac{a_3}{a_5} (x-a_4)} \quad (2)$$

which is the required rectangular hyperbola approximation; while for small x

$$y \simeq a_1 + \frac{a_2}{1 + a_3 \exp (\frac{x-a_4}{a_5})} \quad (3)$$

Because there are too many parameters, we have to apply at least one, and preferably more, equality constraints among them to reduce the effective number. We used the following procedure: first we found five-parameter fits to all the sets of data supplied, using VA05A for nonlinear least squares and treating all the data points (crosses and dots) as known. This gave the full-line curve in Fig. 1 for the data points marked there. Then we examined the fitted parameter values, seeking constraints to apply. We found that the value of a_5 was never very different from 1.0 (*i.e.*, 0.1 times the range of x), and since the manufacturing process attempted to control the scale of x we tried adopting this fixed value for a_5. Discarding the points marked with dots, we carried out four-parameter fits (a_1 to a_4) on four data points for each set; for the case of Fig. 1 this gave a curve passing through the mean of each pair of points, which is not shown because it is scarcely distinguishable from the dashed curve. These fits were quite acceptable for all the data sets supplied, extremely so for most; however, comparing the two fits on a few of the data sets suggested that the second curves were unduly distorted by unlucky data errors, so some smoothing was judged desirable. This was applied by observing that the ratio a_1/a_2 did not vary greatly, so we tried fixing this ratio at 0.14, leaving only the three parameters a_2, a_3, a_4 to be chosen for best fit to the four pairs of data points. This gave the dashed curve in the case of Fig. 1, and over-all gave results for the dot points as good as the second curve, although naturally the fit to the standard (cross) points was worse.

The 3-parameter method was finally adopted in preference to the 4-parameter one on the basis of statistical analysis of their performance on sets of data known to the originator of the problem but not supplied to the analyst. However, it was still necessary in routine use to make the whole process more rugged, for example by guarding against intermediate sets of parameter values causing exponential overflow for $x = 10$. In the end we chose b_s, $s = 1$ to n, as the actual parameters used by VA05A, where $n = 3, 4$ or 5 is the number of parameters being fitted, and defined the a_r in

55

terms of the b_s as follows, where Y is the largest data y value:

$$a_1 = 0.3Y \frac{\exp(b_4) - 1}{\exp(b_4) + 1} \tag{4}$$

$$a_2 = Y \exp(b_3) \tag{5}$$

$$a_3 = \exp(b_2) \tag{6}$$

$$a_4 = 10 - 8 \exp(b_1) \tag{7}$$

$$a_5 = (1 - 0.1 \, a_4)[0.2 + 0.8 \exp(b_5)] \tag{8}$$

For $n = 3$ or 4 we added the equation

$$b_5 = \ln[(1 + 0.025 \, a_4)/(1 - 0.1 \, a_4)] \tag{9}$$

which has the effect

$$a_5 = 1.0. \tag{10}$$

For $n = 3$ we replaced (4) by

$$a_1 = 0.14 \, a_2. \tag{11}$$

The reason for our somewhat complicated choice (8) is that the exponential in (1) is largest $x = 10$, when its argument becomes

$$u = \frac{10 - a_4}{a_5} = \frac{10}{0.2 + 0.8 \exp(b_5)} \tag{12}$$

which cannot exceed 50. This completely cured overflow problems which had previously been encountered on some sets of data (covering, in fact, a difference x range, with (7) and (8) scaled in proportion). The initial approximation is given by $b_s = 0$ in all cases. Other constraints implied by our parameterization are:

$$|a_1| < 0.3Y, \quad a_2 > 0, \quad a_3 > 0, \quad a_4 < 10, \quad a_5 > 0.$$

A uniform increment of 0.05 in each of the b_s was specified to VA05A for numerical different-iation to estimate the Jacobian matrix, and a bound of 3.0 on the value of each $|b_s|$. The accuracy requested in minimizing the sum

$$\Sigma \left\{ \frac{y(x_i) - y_i}{Y} \right\}^2 \tag{13}$$

was 10^{-6} max $(0.2, n_D-n)$, where n_D is the number of data points. This caused a very close approxi-mation to the best fit to be obtained, without demanding too much time.

REFERENCES

Curtis, A.R. and Chance, E.M. (1974) "CHEK and CHEKMAT, two chemical reaction kinetics programs", AERE Report R-7345, HMSO.

Harding, B.R., Thomson, R. and Curtis, A.R. (1973) "A new mathematical model for fitting an HPL radioimmunoassay curve", *J. Clin. Path.*, **26**, 973-976.

Hopper, M.J. (1973) "Harwell Subroutine Library: A Catalogue of Subroutines (1973)", AERE Report R-7477, HMSO.

Use of Optimization Techniques in Optical Filter Design

Heather M. Liddell

(Queen Mary College, University of London)

SUMMARY

Optimization techniques have been used for many years in optical design work, particularly by lens designers and multilayer filter designers. This paper will describe two areas of interest to the latter.

The reflectance and transmittance of a multilayer filter is a very complicated nonlinear function of the refractive indices and thicknesses of the materials forming the multilayer. Apart from a few specialized designs, the general synthesis of such a device has only become possible over the past ten years by application of optimization techniques; in particular, those devised for functions which are sums of squared terms and simple constrained techniques. Often a hybrid approach to the problem has been found successful where a fairly crude exhaustive search, followed by a more sophisticated technique will often produce a novel design for a particular problem which may be close to the "global" optimum. However, physical intuition on the part of the designer can result in a much quicker solution which meets the design specification to within the desired tolerance level. The problem becomes very much more complicated in a system such as a colour television camera where one has several filters and other components all of which have design variables to be optimized.

The second problem area concerns the determination of optical constants (refractive index n, absorption k and film thickness d) of the films within the multilayer and the dispersion of these quantities with respect to wavelength. Even if one considers a single absorbing film on a substrate, the equations for reflectance and transmittance in terms of n, k and d are still very cumbersome, and if measurements are made at an angle of incidence, there are two components of polarization to consider. Thus, if one is to obtain values of n, k and d from measurements of reflectance and transmittance, one must again use optimization techniques. The problem is further complicated by the fact that there are ranges of wavelengths for which the equations become very ill conditioned. Univariate search, Newton-Raphson, Nelder and Mead's Simplex and Powell's and Peckham's methods for functions which are sums of squares have been applied to this problem by various workers.

ACKNOWLEDGEMENTS

I should like to express my gratitude to Dr. Brian Ford for presenting this paper, as I was unable to attend the conference. I also wish to thank the editor of *Optica Acta* for permission to reproduce Figs. 4 to 8 inclusive, and the editor of *Applied Optics* for permission to reproduce Figs. 2 and 3.

INTRODUCTION

In the field of optical design, optimization techniques have been used by lens designers and by multi-layer filter designers for many years; it is the latter application which will be described in this paper.

A multi-layer filter consists of a number of thin films deposited on a substrate, and has the property of being able to reflect some wavelengths and transmit others; the particular wavelengths or range of wavelengths for which the filter is

highly reflecting or transmitting may be altered
by changing the characteristics of the component
films. From the mathematical point of view, we
consider a "thin" film as one whose thickness is
of the order of the wavelength of light, and whose
extent is infinite compared to its thickness; it
is characterized by its refractive index, its
absorption coefficient and its thickness. We
assume that the film is homogeneous, isotropic and,
as far as the optical filter designer is concerned,
usually either non-absorbing or weakly absorbing.
Thin film multi-layers may be used in the ultraviolet, visible and infrared regions of the spectrum, covering a total range from about 100nm to
100μm, although in some parts of this range the
materials used are strongly absorbing. Optical
filters are usually designed to operate over all
or part of the visible region, *i.e.*, 400nm - 700nm.

The reflectance and transmittance at a wavelength λ of an *N*-layer filter deposited on a substrate are given by the following equations:

$$R = \left| \frac{\mu_0 E_0 - H_0}{\mu_0 E_0 + H_0} \right|^2 \tag{1}$$

$$T = \frac{4 n_0 n_{sub}}{\left| \mu_0 E_0 + H_0 \right|^2} \tag{2}$$

where

$$\begin{pmatrix} E_0 \\ H_0 \end{pmatrix} = \left[\prod_{m=1}^{N} \begin{pmatrix} \cos\delta_m & \frac{i}{\mu_m}\sin\delta_m \\ i\mu_m\sin\delta_m & \cos\delta_m \end{pmatrix} \right] \begin{pmatrix} 1 \\ \mu_{sub} \end{pmatrix} \tag{3}$$

E_0 and H_0 are the magnitudes of the electric and
magnetic vectors in the medium of incidence;
$\mu_m = n_m$, the refractive index of the *m*th layer at
normal incidence (the subscript *sub* refers to the
substrate); and at an angle of incidence, θ_0, for
$m = 1,2,\ldots,N$, *sub* $\mu_m = n_m/\cos\theta_m$ for the *p*-component
of polarization, $\mu_m = n_m\cos\theta_m$ for the *s*-component
of polarization, θ_m is the angle of incidence in
the *m*th layer, related to the angle of incidence
in the medium of incidence (θ_0) by Snell's law;

δ_m, the phase change of the beam of electromagnetic radiation on traversing the mth film is given by

$$\delta_m = \frac{2\pi n_m d_m \cos\theta_m}{\lambda} \qquad (4)$$

where d_m is the physical thickness of the film. If the film is absorbing, n_m is a complex quantity whose real and imaginary parts represent the refractive index and absorption coefficient, respectively. Some of these quantities are illustrated in Fig. 1.

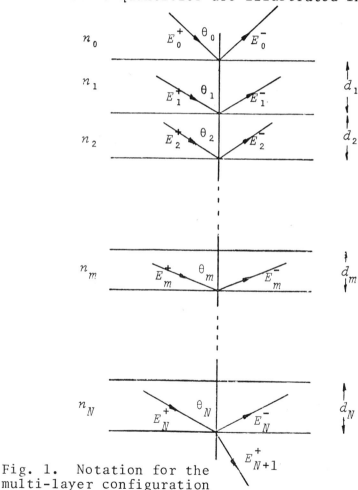

Fig. 1. Notation for the multi-layer configuration

From the above equations it can be seen that the reflectance and transmittance of a multi-layer filter are complicated nonlinear functions of wavelength and of the refractive indices and thicknesses of the materials used in the multi-layer. In order to design a filter to have a specified reflectance or transmittance response, we characterize the fit between the specified response and that corresponding to a particular multi-layer configuration by a suitably chosen "merit function" (objective function) which may be minimized with respect to the design variables. To obtain a perfect fit, the minimum value of the merit function should be zero, normally this does not happen, but as long as a design falls within the allowed tolerance requirements in the specification it is considered satisfactory. In any case, practical problems of manufacture introduce errors into the design, so it is unnecessary for the designer to waste large amounts of computer time producing a very low global minimum; it is more important to consider ease of manufacture and susceptibility to error of a particular design. It should perhaps be emphasized that in general, much more time and effort has been spent by designers on physical considerations than on the mathematical techniques used.

Multi-layer filters have two characteristics which can be considered as design variables, namely refractive index and thickness of the component films. Unfortunately, only a small number of materials are suitable for use in multi-layers so, from the practical point of view, it is easier to use fixed values of refractive index and to treat thicknesses only as the design variables. Also, from the manufacturing point of view, one cannot provide a large range of materials in the evaporating plant in order to produce a single filter; film thickness is a much easier quantity to control.

62

AUTOMATIC DESIGN METHODS WHICH USE UNCONSTRAINED OPTIMIZATION TECHNIQUES

For some design problems, the designer may have a fairly good approximation which may be used as a starting design which may then be refined; in others, no starting design is available, so the designer must be able to reach a satisfactory minimum of the merit function from any starting point in the parameter space of the design variables. In some of the early methods an "exhaustive search" technique was used, whereby the parameter space was divided into a mesh of points; each of the N design variables was evaluated at p different values, so $(p)^N$ function evaluations were required to cover every node of the mesh, and the lowest value of the merit function selected as the optimum. Clearly, unless p and N are both small, the computer time involved was formidable! Elsner (1964) applied a variation of this method to the design of an 11-layer high reflection coating; the values of the refractive indices and some layer thicknesses were fixed, while the remaining thicknesses were allowed to take values $k\lambda_0/4$, $k = 1,2,\ldots,6$; every time the calculated reflectance R obeyed the relation

$$P_1(\lambda) \leqslant R(\lambda,\underset{\sim}{x}) \leqslant P_2(\lambda) \qquad \lambda_0 \leqslant \lambda \leqslant \lambda_1 \qquad (5)$$

the functions $P_1(\lambda)$, $P_2(\lambda)$ were adjusted to provide stiffer criteria. ($\underset{\sim}{x}$ is the vector of design variables). Shatilov and Tyutikova (1963) developed an evolutionary design method; at each stage a layer was added to the design and its thickness adjusted until the best approximation to the specification was obtained - the method is therefore a type of univariate search technique with one linear search (using an equal interval mesh) for each design variable. Their merit function took the form

$$F(\underset{\sim}{x}) = \int_{\lambda_1}^{\lambda_2} (R_0(\lambda) - R(\lambda,\underset{\sim}{x}))^2 d\lambda \qquad (6)$$

where $R_0(\lambda)$ is the given reflectivity at wavelength λ, $R(\lambda, \underset{\sim}{x})$ the calculated reflectivity and $\underset{\sim}{x}$ a vector representing the film thicknesses $x_1, x_2, \ldots, x_N$. The integral was evaluated using Simpson's rule. They reported that systems of up to eight layers were designed without the aid of a digital computer! Ermolaev, Minkov and Vlasov (1962) treated both refractive index and thickness as design variables, using the merit function

$$F(\underset{\sim}{x}) = \int_{\lambda_1}^{\lambda_2} \rho(\lambda) [R(\underset{\sim}{x}, \lambda) - R_0(\lambda)]^K d\lambda \qquad (7)$$

$\underset{\sim}{x}$ is the vector of design variables, ρ a weighting function and $R_0(\lambda)$ the specified curve. K is a positive integer constant normally having the value 2. The optimization was performed using a steepest descent method, which included a parabolic interpolation for the minimum along each direction of search. They also attempted to find the global optimum by calculating a series of merit functions which differed from the minimum function by just one parameter; if any of these calculations produced a value of F which was smaller than the previous minimum found, the minimization procedure was repeated from this new point.

Dobrowolski (1965) produced a rather more elaborate and efficient automatic design method which incorporated features of both the exhaustive search and evolutionary methods; the advantage of the latter is that the multi-layer configuration need not be specified in advance. He allowed two or three layers to be added at a time and performed an exhaustive search to find the optimum parameters for these layers only; after each addition of layers, the whole configuration at that stage was refined using a univariate search technique, with a quadratic interpolation for the minimum along each direction of search. The merit function used was

$$F(\underset{\sim}{x}) = \left[\frac{1}{m} \sum_{j=1}^{m} \frac{|P_0(\lambda_j) - P(\lambda_j, \underset{\sim}{x})|^K}{T(\lambda_j)} \right]^{1/K} \qquad (8)$$

64

where $P_0(\lambda_j)$, $P(\lambda_j, x)$ and $T(\lambda_j)$ represent the desired value, the calculated value for a configuration with design variables represented by the vector x, and the tolerance of the response (reflectance or transmittance) at wavelength λ_j, and K is an integer taking one of the values 1, 2, 4 or 16; the higher values of K are useful in a particular problem where the response at some wavelengths is much further from the desired value than it is at others, and the normalization with respect to tolerance is convenient when the merit function is nonhomogeneous. When all values are just within tolerance, $F = 1$, so by setting $F = 0.5$ as the termination test, allowance is made for a higher state of correction at some wavelengths than at others. Two other features of the program are of interest: (*i*) If both thicknesses and refractive indices are treated as design variables, a refractive index list of available materials is read in at the outset, and the index fixing routine by-passed until the desired performance is almost obtained, then the index values are replaced by the nearest value in the list; (*ii*) the consolidation process is activated if the indices of two adjacent layers approach the same value - the two layers are then replaced by a single layer of intermediate index. Fig. 2 shows a block diagram of the program. This method has been used for many years to produce successful designs; one recent application of interest to those of us concerned with the design of filters for use in colour TV systems is the production of a set of filters for tristimulus colorimetry, designed to simulate the colour mixture functions of the C.I.E. standard observer. These designs are shown in Fig. 3; note that some standard glass absorption filters are also incorporated in order to "block" certain wavelengths.

Pelletier, Klapisch and Giacomo (1971) have developed an automatic design method which appears to have some advantage over the previous methods when applied to the design of achromatic reflectance coatings (high reflectors, beam splitters,

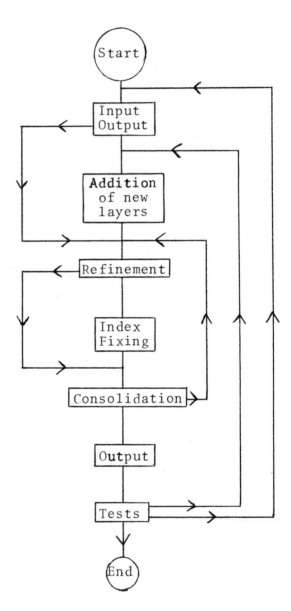

Fig. 2. Block diagram for Dobrowolski's automatic design method (Dobrowolski (1965))

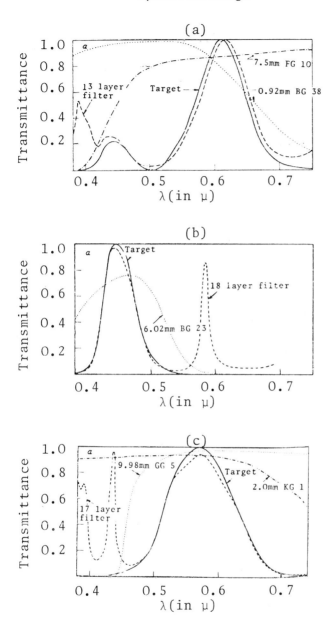

Fig. 3. Filters produced for use in tristimulus colorimetry (Dobrowolski (1970))

anti-reflection coatings) for an extended spectral range; their merit function takes the form

$$F(\underset{\sim}{x}) = \Omega_1 \left\{ \frac{1}{L} \sum_{j=1}^{L} |R(\lambda_j, \underset{\sim}{x}) - \bar{R}|^K \right\}^{1/K} + \Omega_2 |\bar{R} - R_0| \quad (9)$$

where $R(\lambda_j, \underset{\sim}{x})$ is the calculated reflectivity at wavelength λ_j for a filter with film thicknesses x_i, $i = 1, 2, \ldots, N$, $\bar{R}$ is the mean of these computed values for the L design wavelengths and R_0 is the desired reflectivity for all λ_j. K, Ω_1, and Ω_2 may be adjusted to suit the needs of a particular specification. The first term in (9) ensures that the amplitude of fluctuation of the design from its mean value is minimized. The spectral range of the design may be split into several sections, each having different values of Ω_1 and Ω_2; other characteristics may be used in place of reflectance. Nelder and Mead's simplex method (1964) was used for the minimization; the authors chose this method because it requires no evaluation of derivatives of the merit function, nor does it assume the function is differentiable and continuous over the range of interest, and has the additional advantage that the user can ensure that the design variables take only positive values.

A METHOD WHICH MAKES USE OF A TECHNIQUE FOR MINIMIZING SUMS OF SQUARES

One of the disadvantages of most of the methods described in the previous section is that the optimization techniques employed are inefficient. It is relatively simple to formulate most problems of filter design with merit functions which are sums of squared terms, so that one of the specialized techniques for this type of function may be used. This was done by Heavens and Liddell (1968) who used Powell's (1965) method to minimize

$$F(\underset{\sim}{x}) = \sum_{j=1}^{m} (R_0(\lambda_j) - R(\underset{\sim}{x}, \lambda_j))^2 \quad (10)$$

where x is the vector of design variables, R_0 the specified response and R the computed response at wavelength λ_j. The method may be applied either to refine an initial approximation to the design or to produce a completely automatic design; in the latter case a hybrid approach was adopted in order to avoid converging to an unsatisfactory minimum, whereby a preliminary scan of the parameter space is made before refinement. Two forms of scan have been used; an exhaustive search method, which becomes impracticable for filters with more than six layers, or a type of univariate search technique, called the N^2-scan which may be described as follows : consider a system of N layers; the thickness of each layer is varied in turn using a fixed interval search method, then for the next iteration the second layer is varied first up to the Nth and finally the first, and so on; in the final iteration, the Nth layer is varied first, then the first, second, ... and finally the N-1th layer. If q values of thickness are investigated for each layer, a total of $q * N^2$ function evaluations will be made compared to q^N for the complete scan process. A block diagram for the program is shown in Fig. 4.

This method has been applied successfully to a large number of design problems during the past 8 years. One example which compares well with results produced by other methods is a broad band high reflection coating. The "classical stack" which is a multi-layer consisting of a number of alternating high and low index layers with equal optical thicknesses (refractive index * physical thickness) has a very high reflectance over a limited region, but this region cannot be extended merely by increasing the number of layers in the stack. Fig. 5 shows two 13-layer broad band coatings produced by the "Least Squares" method of Heavens and Liddell; one is an "automatic" design, the other a refinement of a combination of two stacks. Another more recent application which was reported by Clapham (1971) was the design of achromatic beam splitters. The specification

69

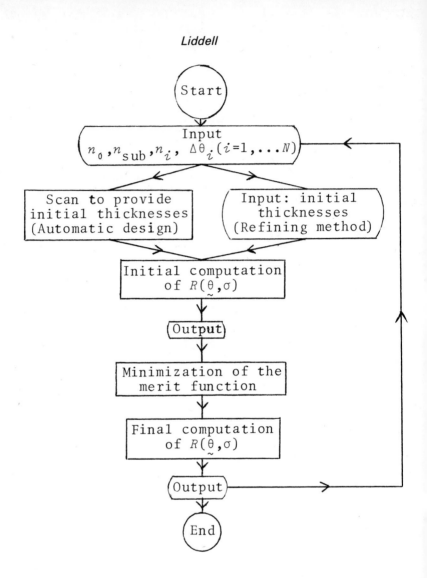

Fig. 4.   Block diagram for the Least Squares method (Heavens and Liddell (1968))

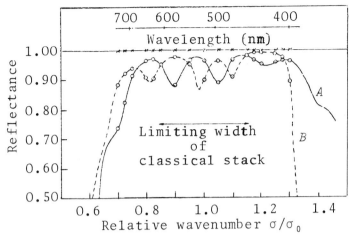

Fig. 5. High reflection coating designs: curve *A*,
automatic design; curve *B*, refined design
(Heavens and Liddell (1968))

required approximately equal reflectance and transmittance throughout the visible spectrum for a surface enclosed in a cemented cube for unpolarized
light at 45° angle of incidence. An intuitive
design was developed by Clapham and the Least
Squares program used to refine this design and
also to produce an automatic design. The results
are shown in Fig. 6. An error analysis revealed
that the automatic design was more sensitive to
errors in thickness so, in practice, both designs
exhibited similar performance.

A restart facility has been incorporated in
later versions of the program, which has been
found useful for certain problems; allowance for
dispersion of refractive index and for absorption
has also been made. Powell's method was used initially because it was readily available from the
Harwell Subroutine library; however, Marquardt's
method (1963) could be used instead, since it is
not too difficult to provide derivatives of the
functions involved.

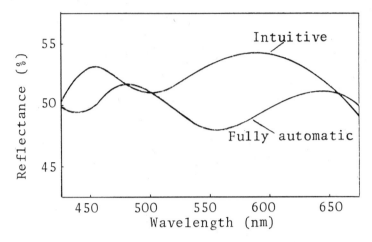

Fig. 6. Theoretical performances of intuitive and automatic designs for an achromatic beam splitter (Clapham (1971))

USE OF CONSTRAINED OPTIMIZATION TECHNIQUES

The type of constraints which appear in design problems are usually simple limits on layer thicknesses. The Least Squares program has the ability to accomodate these by applying a simple transformation of variables, *i.e.*, for $A \leqslant d_i \leqslant B$,

$$\sin^2 y_i = \left[\frac{d_i - A}{B - A}\right] \tag{11}$$

A rather different approach has been adopted by Chen (1973) who was concerned with the design of low pass filters (high transmittance at low frequencies (long wavelengths), followed by a sharp cut-off to high reflectance at high frequencies (short wavelengths)) which have a Tschebysheff "equiripple" characteristic in the passband. This response is often used as a prototype for electrical filter design work, and there was reason to believe that it was a response which could be realized by a suitable multi-layer configuration. An exact design can be produced if refractive indices are not specified in advance, but this is not a

72

practical procedure, so Chen considered a fixed index configuration with symmetric thicknesses about the central layer. Most automatic design methods are applied by fitting the multi-layer transmittance to the specified response at fixed frequencies; however, the positions of the transmittance minima in the pass band shift as the layer thicknesses are altered, and these minimum values are necessary design points if one is fitting to a Tschebysheff curve. Chen's Turning Value method allows these frequencies to shift by including them in the set of design variables, although they are in fact dependent on the other design variables, the layer thicknesses. The following equations must be satisfied

$$T(w, d_1, d_2, .. d_{n+1}) = \frac{1}{1+h^2}, \quad \text{at} \quad w = w_i$$

where h is the "ripple" width in the pass band,

$$\left[\frac{\partial T}{\partial w}\right] = 0 \quad \text{at} \quad w = w_i, \quad (i = 1, 2, .. \frac{n-1}{2}). \quad (12)$$

To specify the rate of cut-off,

$$T(w_p) = 1$$

$$T(w_r) = \varepsilon.$$

Here w_p the transmittance maxima nearest the band edge, w_r a frequency in the high reflectance (stop) band, and ε are chosen on the basis of the Tschebysheff response. Fig. 7 illustrates this specification. The equations may be simplified by ignoring the transmittance minimum at the low frequency end of the spectrum, since this value is not sensitive to changes in layer thickness. The merit function

$$M = \sum_{i=2}^{(n-1)/2} (a_i \delta T_i)^2 + \sum_{i=2}^{(n-1)/2} (b_i (\frac{\partial T}{\partial w})_i)^2$$

$$+ (c \delta T_p)^2 + (d \delta T_r)^2 \quad (13)$$

is then formed (δT_i, δT_p, δT_r are differences between desired and calculated values of transmit-

tance and the a_i, b_i, c and d are weighting functions) and minimized using Rosenbrock's method
(1960); the latter can easily accommodate limits
on layer thicknesses. The method has proved
successful when applied to the design of both low
pass and high pass filters in the visible and
infrared regions of the spectrum. Fig. 8 shows
the result obtained for a 17-layer filter.

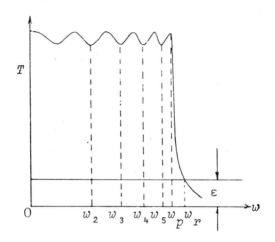

Fig. 7. Design specification for Chen's Turning
Value method (Seeley, Liddell and Chen
(1973))

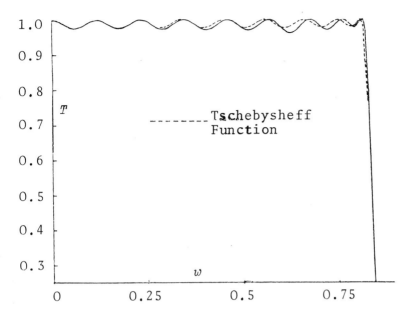

Fig. 8. Computed performance of a 17-layer filter
designed by Chen, compared with the
Tschebysheff response (Seeley, Liddell
and Chen (1973))

DETERMINATION OF OPTICAL CONSTANTS OF THIN FILMS

Another problem of interest to the optical
filter designer is that of determining accurate
data for the optical constants (refractive indices
and absorption coefficients) of the materials in
thin film form, which can be very different from
the bulk values, and their dispersion behaviour
with wavelength. Even if one considers a single
absorbing film deposited on a non-absorbing substrate, the equations for reflectance and transmittance in terms of n, k and d are fairly cumbersome;
they may be expressed as follows (Born and Wolf
(1965)):

$$R = \frac{l^2 e^{2v\eta} + m^2 e^{-2v\eta} + 2lm\cos(\phi_{12} - \phi_{01} + 2u\eta)}{e^{2v\eta} + l^2 m^2 e^{-2v\eta} + 2lm\cos(\phi_{12} + \phi_{01} + 2u\eta)} \quad (14)$$

for both polarizations, and for the p-polarization component

$$T = \frac{n_2\cos\theta_2}{n_0\cos\theta_0}\left[\frac{f^2 g^2 e^{-2v\eta}}{e^{2v\eta}+l^2 m^2 e^{-2v\eta}+2lm\cos(\phi_{12}+\phi_{01}+2u\eta)}\right] \quad (15)$$

θ_0, θ_2 are the angles of incidence in the medium of incidence and the substrate, and

$$l^2 = \frac{[(n^2-k^2)\cos\theta_0-n_0 u]^2 + [2nk\cos\theta_0-n_0 v]^2}{[(n^2-k^2)\cos\theta_0+n_0 u]^2 + [2nk\cos\theta_0+n_0 v]^2}$$

$$m^2 = \frac{[(n^2-k^2)\cos\theta_2-n_2 u]^2 + [2nk\cos\theta_2-n_2 v]^2}{[(n^2-k^2)\cos\theta_2+n_2 u]^2 + [2nk\cos\theta_2+n_2 v]^2}$$

$$\tan\phi_{01} = \frac{2n_0\cos\theta_0[2nku-(n^2-k^2)v]}{[(n^2+k^2)\cos^2\theta_0+n_0^2(u^2+v^2)]}$$

$$\tan\phi_{12} = \frac{2n_2\cos\theta_2[2nku-(n^2-k^2)v]}{[(n^2+k^2)\cos^2\theta_2+n_2^2(u^2+v^2)]}$$

$$f^2 = \frac{4(n^2+k^2)^2\cos^2\theta_0}{[(n^2-k^2)\cos\theta_0+n_0 u]^2 + [2nk\cos\theta_0+n_0 v]^2}$$

$$g^2 = \frac{4n_2^2(u^2+v^2)}{[(n^2-k^2)\cos\theta_2+n_2 u]^2 + [2nk\cos\theta_2+n_2 v]^2}$$

where

$$u^2-v^2 = n^2-k^2-n_0^2\sin^2\theta_0$$

$$uv = nk$$

$$\eta = \frac{2\pi d}{\lambda}$$

To obtain the corresponding expressions for the s-component of polarization, $n_i/\cos\theta_i$, $i = 0,2$, is replaced everywhere by $n_i\cos\theta_i$; the equations are somewhat simplified if normal incidence measurements are used for reflectance and transmittance, but this is not always possible in practice. Various optimization techniques have been used for this problem of determining n and k from measured values of R and T, or from R, T and d; Bennett and Booty (1966) used a univariate search technique, Abelès and Thèye (1966) used a Newton type method

for normal incidence measurements over a range of
wavelengths; Ward, Nag and Dixon (1969) applied
the Nelder and Mead simplex method (1964) to the
problem of determining n, k and d, using measurements of R and T at several angles of incidence.
All these techniques were developed for a single
film deposited on a substrate; Hansen (1973) produced a method for calculating the optical constants of any film within a multi-layer, which
included an optimization process based on the generalized secant method, which is somewhat similar
to the methods described by Powell (1964) and
Peckham (1970). However any method which calculates the optical constants by considering single
wavelength values of R and T has a serious defect;
there are a number of ranges of values of n, k, d
and λ for which the equations are non-unique, *i.e.*,
several values of n and k will give rise to the
same values of R and T; in the case of weakly
absorbing films, these values may be very close.
The consequences of this may be illustrated by
results produced by Liddell and Staerck (1972) for
an aluminium oxide film (Table I), where Powell's
method had been used to minimize

$$F(\underset{\sim}{n},\underset{\sim}{k}) = \sum_{\lambda_i} [(R_{exp} - R_{calc}(n_i, k_i, d))^2$$
$$+ ((R+T)_{exp} - (R+T)_{calc}(n_i, k_i, d))^2] \tag{16}$$

the summation being performed over various wavelengths in the visible and ultraviolet range of
the spectrum. The results were satisfactory numerically, but the physical behaviour of n is unacceptable. One way of attempting to overcome this
problem of non-uniqueness of solution was to
reformulate the equations in order to build in a
dispersion formula relationship for the refractive
index values. A modification of Sellmeier's formula

$$n^2 - 1 = \frac{\lambda^2}{A + B\lambda^2}$$

is known to give a good fit to the known data for
dielectric (non-absorbing) materials, so Liddell

Table I

Results of optical constant determination for an aluminium oxide film (d = 153.7nm) using a single wavelength technique

λ(nm)	n	k	$\Delta R(\%)$	$\Delta T(\%)$
215	1.926	0.003	0.00	-0.01
225	1.789	0.001	0.00	-0.01
250	1.790	0.001	0.00	0.00
300	1.800	0.001	0.08	0.00
350	1.680	0.000	-0.05	-0.21
400	1.685	0.001	-0.08	-0.29
450	1.702	0.000	0.00	0.00
502	1.735	0.000	0.01	0.09
546	1.659	0.000	0.00	0.00
633	1.589	0.000	0.02	-0.02

(1974) decided to apply it to the particular case of weakly absorbing films; the objective function used to determine these optical constants took the form

$$F(A,B,\underset{\sim}{k}) = \sum_{\lambda i} [\ (R_{exp}-R_{calc})^2$$
$$+ (R_{exp}+T_{exp}-R_{calc}-T_{calc})^2\ ] . \tag{18}$$

Both Peckham's method (1970) and Powell's method have been used to minimize (18), giving similar results. Those corresponding to the aluminium oxide film quoted earlier are shown in Table II(a) - they proved to be about average for the set of seven films studied. The discrepancy between the experimental and calculated values of R and T was slightly larger than one would expect from consideration of experimental error of measurement, but this was reduced to a very much smaller level by allowing the film thickness value to vary to a limited extent about the experimental value. This meant that a function of 13 variables was minimized, using an objective function based on 20 independent measurements of R and T. The improved results are shown in column (b) of Table II. In some examples,

Table II

Results of optical constant determination of an aluminium oxide film using the dispersion formula technique.

λ(nm)	(a) Homogeneous film with thickness d_a = 153.7nm				(b) Homogeneous film with optimized thickness d_b = 158.6nm			
	n	k	ΔR(%)	ΔT(%)	n	k	ΔR(%)	ΔT(%)
215	1.808	0.003	-1.2	1.2	1.805	0.003	0.1	-0.1
225	1.792	0.001	-0.1	0.1	1.785	0.001	-0.1	-0.1
250	1.760	0.001	1.2	-1.2	1.749	0.001	0.2	-0.2
300	1.723	0.001	-0.8	0.8	1.707	0.001	0.2	-0.2
350	1.703	0.000	-0.7	0.7	1.684	0.000	-0.1	-0.1
400	1.690	0.000	-0.1	0.1	1.669	0.000	-0.1	-0.1
450	1.682	0.000	0.4	-0.4	1.660	0.000	0.2	-0.2
502	1.675	0.000	0.2	-0.2	1.653	0.000	0.2	-0.2
546	1.672	0.000	0.0	0.0	1.649	0.000	0.2	-0.2
633	1.666	0.000	-0.4	0.4	1.643	0.000	-0.2	0.3
Final merit function	$0.486 * 10^{-3}$				$0.247 * 10^{-4}$			

notably the thinnest films, evidence was found of slight inhomogeneity in the film structure, so a simple graded index model was also incorporated into the physical formulation of the problem. Preliminary results by Pelletier (private communication (1974)) indicate that formula (17) is oversimplified a more accurate model for the dispersion behaviour is given by

$$n = A + \frac{B}{\lambda^2} + \frac{C}{\lambda^4}$$

or

$$n = A + \frac{B}{\lambda^2} + \frac{C}{\lambda^4} + \frac{D}{\lambda^6}$$

although neither is perfect for all weakly absorbing films.

CONCLUSION

Two problems in optical filter design have been discussed in this paper; in both cases, various optimization techniques have been used to obtain suitable solutions. Generally, more difficulties arise in finding a suitable physical model for the problem than in applying the numerical techniques, and the accuracy of the solution is usually limited by experimental rather than mathematical considerations. In practice, one may be faced with a problem which involves the simultaneous design of several filters and other components within an over-all system - an example of this is the colour television camera, consisting of a number of lenses, filters and electronic components, all having design variables which can be used to optimize the performance. Thus, the field of optical design produces a number of challenging practical problems for those of us who are interested in the application of optimization techniques.

REFERENCES

Abelès, F. and Thèye, M.L. (1966) *Surface Science*, **5**, 325.

Bennett, J.M. and Booty, M.J. (1966) *Appl. Opt.*, 5, 41.

Born, M. and Wolf, E. (1965) "Principles of Optics", Oxford, Pergamon.

Clapham, P.B. (1971) *Optica Acta*, 18, 563.

Dobrowolski, J.A. (1965) *Appl. Opt.*, 4, 937.

Dobrowolski, J.A. (1970) *Appl. Opt.*, 9, 1396.

Elsner, Z.N. (1964) *Opt. & Spectr.*, 17, 446.

Ermolaev, A.M., Minkov, M. and Vlasov, A.G. (1962) *Opt. & Spectr.*, 13, 259.

Hansen, W.N. (1973) *J. Opt. Soc. Amer.*, 63, 793.

Heavens, O.S. and Liddell, H.M. (1968) *Optica Acta*, 15, 129.

Liddell, H.M. and Staerck, J.A. (1972) *in* "Proc. Electro Optics 1972 International, Brighton", Chicago: Ind. Scient. Conf. Management.

Liddell, H.M. (1974) *J. Phys. D: Appl. Phys.*, 7, 1588.

Marquardt, D.W. (1963) *J. SIAM*, 11, 431.

Nelder, J.A. and Mead, R. (1964) *Comput. J.*, 7, 308.

Peckham, G. (1970) *Comput. J.*, 13, 418.

Pelletier, E., Klapisch, M. and Giacomo, P. (1971) *Nouv. Rev. d'Optique Appliquee*, 2, 247.

Powell, M.J.D. (1965) *Comput. J.*, 7, 303.

Rosenbrock, H.H. (1960) *Comput. J.*, 3, 175.

Seeley, J.S., Liddell, H.M. and Chen, T.C. (1973) *Optica Acta*, 20, 641.

Shatilov, A.V. and Tyutikova, L.P. (1963) *Opt. & Spectr.*, 14, 426.

Ward, L., Nag, A. and Dixon, L.C.W. (1969) *J. Phys. D: Appl. Phys.*, 2, 301.

An Application of Optimization Techniques to the Design of an Optical Filter

J.J. McKeown and A. Nag*

*(The Numerical Optimization Centre,
The Hatfield Polytechnic and
Grubb Parsons Ltd.)*

*now with Durham University

SUMMARY

A thin-film optical filter is composed of a substrate material upon which are deposited layers of material whose thickness is small compared with that of the substrate. The object of the system is to maximize the transmission of light over a given range of wavelengths; its efficiency depends on the refractive indices and thicknesses of the layers.

A study of this design problem, seen in mathematical programming terms, is described. The proportion of reflected light was minimized directly over the chosen range by minimizing the sums of squares of percentage reflected energy at selected wavelengths. The design variables were the layer thicknesses, and they were constrained to be non-negative. Designs were generated which compared favourably with those produced by standard methods; and these advantages were realised in practice.

1. INTRODUCTION

The rapid development in detector technology over the last few years has led to considerable interest in its application to infrared technology, especially in the military or defence area.

82

In the latter part of 1972, the Ministry of Defence approached Grubb Parsons regarding the design and manufacture of broad-band anti-reflection coatings in the 8 to 14 μ region of infrared radiation. It was stated that the aim should be to increase the transmission of germanium optical systems without increasing scattered radiation. It was further stipulated that the coatings must satisfy the military environmental specifications. In other words, they must survive arctic as well as tropical conditions for several months. Such a requirement imposes considerable restrictions on the number of materials that can be used for the manufacture of such coatings.

There are two main reasons for using anti-reflection coatings on the surfaces of the elements of an optical system. These are: (*i*) most of the incident radiation reflected at each surface is lost to the transmitted beam and, if there are many elements in an optical system with high refractive indices such as germanium or silicon lenses, this loss will reduce the transmitted beams' intensity; (*ii*) the multiple reflections from the surfaces of the elements cause unwanted radiation to fall on the image plane, thus reducing the contrast and definition of the image. For example, the reflectance of a glass surface with a refractive index of 1.5 is 4%, with transmittance of about 92%, whereas a germanium surface of refractive index 4.0 has a single surface reflectance of 36% and a parallel-plate transmission of 47% in the infrared. Thus an optical system with glass elements requires anti-reflection coatings for image improvement whereas the same system with germanium elements requires such coatings to produce increased transmission. There are several methods which can be used to produce anti-reflection coatings. These are:

(*i*) evaporation in high vacuum
(*ii*) reactive sputtering
(*iii*) deposition from organic solutions
(*iv*) chemical vapour deposition
(*v*) centrifugal coating.

The vacuum evaporation method is the most widely used as this is a method by which uniform thin films with a desired thickness can be prepared with considerable accuracy. However, method (ii), reactive sputtering, is an attractive alternative, since several authors have pointed out that in general sputtering films show stronger adhesion than evaporated films. Because no facilities were currently available for the latter process, the materials chosen had to be suitable for deposition by vacuum evaporation; but it was intended that designs generated in the course of the present study should be the basis of a possible investigation into the advantages of reactive sputtering from the point of view of the military specification.

This paper deals with an application of numerical optimization techniques to the solution of the problem of designing coatings to meet the above requirement; but it should be mentioned that analytical solutions are available for some special cases (see Liddellp.58-80). These designs produce zero reflectance at a number of wavelengths; for example, two-layer coatings can be designed for which this is true at two chosen wavelengths. However, for the purposes of the present specification, it was felt that an acceptably good transmission over the wide band required was not likely to be obtained by designs such as this. Because of the difficulty of designing a multi-layer system to approximate a chosen response over a given wavelength, it was decided that numerical optimization should be used; the next sections describe this approach and the results achieved with it.

In view of the environmental conditions discussed earlier, and because of production difficulties and even health hazards, it was necessary to restrict the choice of coating material to a small set. Unfortunately, the names and refractive indices of these materials are classified information at this time, and the authors apologize for the restrictions that this limitation has placed on the description which follows.

2. OPTIMIZATION

2.1 *The specification:*

Given (*i*) a set of film materials of known
refractive index;

(*ii*) a continuous range of wavelengths;

the problem was to select the configuration
(defined by the order in which the materials were
deposited on the substrate) and the film thicknesses, such that the transmission of radiation
through the system was as large as possible
throughout that range of wavelengths. The minimum
thickness of film that could be deposited was stipulated. The elements of the problem are shown in
Fig. 1.

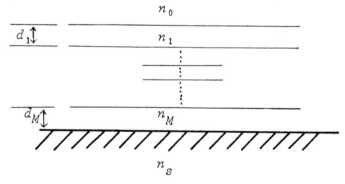

Fig. 1.

2.2 *Formulation*

The problem as expressed in 2.1 required
further definition before optimization techniques
could be applied. Fig. 2 shows the response of an
ideal design.

T is the percentage transmission of light.
Although the figure shows a sharp cut-off outside
the range (λ_1, λ_2) in fact the problem does not
specify a criterion for behaviour outside this
interval. The following decisions were made.

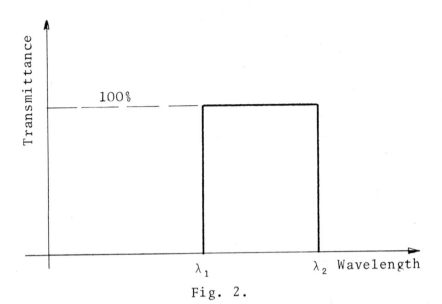

Fig. 2.

(i) The objective function to be minimized should be the sum of the squares of the deviation of the response of a trial design from the ideal response curve, measured at a pre-set number of points in the interval (λ_1, λ_2).

(ii) Any material should not appear more than once in a given configuration.

For the purpose of (i), the ideal response curve was a transmission of 100% at every wavelength in the given range, including the endpoints. The point of (ii) was that it restricted the choice of configurations to a finite set of permutations of the given materials, and so allowed all such configurations to be investigated.

The problem could now be formulated (for M materials and the range (λ_1, λ_2) divided into k intervals):

$$\min_{C \varepsilon C}\left\{\min_{\underset{\sim}{d}} \sum_{i=1}^{k+1}[\,100(1 - T_i(\underset{\sim}{d}, C)]^2\right\} \qquad (1a)$$

s.t. $\qquad d_i = 0.0$
or $\qquad\quad d_i \geqslant d_e$ $\left.\begin{array}{c}\\\\\end{array}\right\}$ $i = 1,2,\ldots m.$ $\qquad$ (1b)

Here, C is a configuration and $\mathcal{C}$ is the allowable set of such configurations, and T is related to λ, $\underline{d}$ and C as follows

$$T_i = \left| \frac{2n_a}{n_a(M_{11}^i + n_s M_{12}^i) + (M_{21}^i + n_s M_{22}^i)} \right|^2 \qquad (1c)$$

$$M^i = \prod_{p=1}^{M} \begin{bmatrix} \cos\phi_p^i & j\dfrac{\sin\phi_p^i}{n_p} \\ jn_p\sin\phi_p^i & \cos\phi_p^i \end{bmatrix} \qquad (1d)$$

$$\phi^i = 2\pi n_p d_p \lambda_0 / \lambda_i . \qquad (1e)$$

The effect of configuration C is to decide the order of multiplication in expression (1d).

The quantity λ_0 is referred to as a monitoring wavelength, so that d_p is a dimensionless factor of λ_0.

2.3 *The optimization algorithm*

The problem as formulated above is a constrained, least squares problem. The constraints are simple, being no more than lower bounds on the values of the variables (although the fact that a material may not appear at all in a configuration introduces a complication that will be discussed in a later section). A simple transformation would therefore have sufficed to render it effectively unconstrained. There are, however, objections to this procedure. For example, if the obvious transformation $y_i^2 = (d_i - d_e)$ is used, then singularity will occur if the solution is constrained; on the other hand, trigonometric transformations carry the risk of introducing spurious local minima. The decision to use a constrained algorithm was therefore made. The algorithm used was the Recursive Quadratic Programming Algorithm of Biggs (1973). The algorithm can be summarized as follows. Consider the problem:

$$\min_{\underset{\sim}{x}} F(\underset{\sim}{x})$$

s.t. $\quad l_i(\underset{\sim}{x}) = 0 \quad i = 1,\ldots m$

$$b_j(\underset{\sim}{x}) \geqslant 0 \quad j = 1,\ldots q$$

where $\underset{\sim}{x} = (x_1,\ldots x_n)^t$ and F, l_i and b_j are all twice continuously differentiable.

If g is the vector of values of all violated constraints at a point $\underset{\sim}{x}$, then a penalty function can be defined as follows:

$$P(\underset{\sim}{x},r) = F(\underset{\sim}{x}) + \frac{1}{r}g^t(\underset{\sim}{x})\,g(\underset{\sim}{x})$$

where r is a penalty parameter.

If $P(\underset{\sim}{x},r_l)$ is minimized for a sequence of r_l tending to zero, then the sequence of solutions x_i^* so generated can be shown to tend to x^*, a solution of the constrained problem, if $P(\underset{\sim}{x},r_l)$ is convex for each l. However, P can be a difficult function to minimize, particularly for small values of r. The Biggs algorithm in fact minimizes a quadratic approximation to F, subject to a stationarity condition on P. The derivative of P is:

$$\nabla P(\underset{\sim}{x},r) = \underset{\sim}{f} + \frac{2}{r}A^t\underset{\sim}{g}$$

where f is $\partial F/\partial \underset{\sim}{x}$, and A is the Jacobian matrix of the violated constraints: $A_{ij} = \partial g_i/\partial x_j$.

At any point $\underset{\sim}{x}^+ = \underset{\sim}{x} + \underset{\sim}{p}$, we have the expansion:

$$\nabla P(\underset{\sim}{x} + \underset{\sim}{p},r) = \underset{\sim}{f} + G\underset{\sim}{p} + \frac{2}{r}A^t\underset{\sim}{g} + \frac{2}{r}\left\{A^tA + \sum g_i \, H^i\right\}\underset{\sim}{p} \cdots$$

where G is the second derivative matrix of F, and H^i is the second derivative matrix of the ith violated constraint. If the assumption is made that

F and the g_i are of sufficiently small curvature, then second derivatives of both can be ignored, and the stationarity condition is:

$$\underset{\sim}{f} + \frac{2}{r}(A^t\underset{\sim}{g} + A^t A\underset{\sim}{p}) = 0.$$

Introducing B as an approximation to $\underset{\sim}{G}$, and using AB^{-1} to premultiply the last equation, we have:

$$A\underset{\sim}{p} = -\frac{r}{2}(AB^{-1}A^t)^{-1}AB^{-1}\underset{\sim}{f} - \underset{\sim}{g}$$

and the auxiliary problem becomes:

$$\min_{\underset{\sim}{p}} {\scriptstyle\frac{1}{2}}\underset{\sim}{p}^t B\underset{\sim}{p} + \underset{\sim}{f}^t\underset{\sim}{p}$$

s.t. $\qquad A\underset{\sim}{p} = -\frac{r}{2}\hat{\underset{\sim}{\lambda}} - \underset{\sim}{g}$

(where $\hat{\underset{\sim}{\lambda}} = (AB^{-1}A^t)^{-1}AB^{-1}\underset{\sim}{f}$) .

The solution to this Quadratic Program is:

$$\underset{\sim}{p} = B^{-1}(A^t(AB^{-1}A^t)^{-1}(AB^{-1}\underset{\sim}{f} - \frac{r}{2}\hat{\underset{\sim}{\lambda}} - \underset{\sim}{g}) - \underset{\sim}{f}) .$$

This defines the lth iteration, and is the basis of the Biggs algorithm, as implemented as Fortran Program *OPRQP*. (NOC Optima Package).

2.4 *Analysis of results*

The first set of designs to be produced had $M = 3$, $k = 29$. There were therefore 6 possible configurations. However, one of the materials had the same index as the substrate, and this eliminated two configurations. The remaining four were run using *OPRQP*, with first derivatives estimated by central differences. In fact, the four configurations led to 6 distinct designs with the following main characteristics (Table I). (Unfortunately the actual designs cannot be given).

Table I

Response of Design set 1

Design number	RMS reflected energy
	%
1	5.5
2	7.8
3	7.6
4	1.2
5	1.27
6	1.24

The last three designs had a common configuration, and the differences between them were brought about by the nature of the constraints (1b). The feasible regions are shown shaded in Fig. 3.

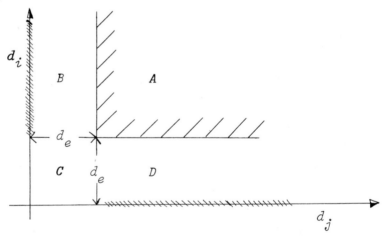

Fig. 3.   Feasible region (hatched)

This is an unusual case, and had to be handled as follows: first, the problem was run with the whole positive orthant as the feasible region, except for the outside layer which was always

90

constrained $>l_e$ for environmental reasons. If the design fell into one of the feasible regions, it was accepted. If, however, it fell into one of the infeasible regions B, C or D, it was re-run with the neighbouring constraints enforced separately, and the better design accepted. For example, if it fell into region B, the problem would be run once with $d_j = 0$, $d_i \geqslant d_e$; and then with d_i, $d_j > d_e$. (Note that this is not the same as taking the nearest constrained points). If a first solution had ever fallen into region C (which never actually happened),three further runs would have been needed.

Returning to the designs 4, 5 and 6, these could be described in terms of Fig. 1 as:

4   region B, $d_3 < d_e$, $d_i \geqslant d_e$, $i = 1,2$

5      $d_i \geqslant d_e$ $i = 1,2,3$

6      $d_3 = 0$, $d_i \geqslant d_e$, $i = 1,2$

It can be seen that the best design of the two feasible ones had only two layers, and that this was the best design achieved. Fig. 4 shows the response curve of this design, which has very much better transmittance than the plain substrate in the specified range.

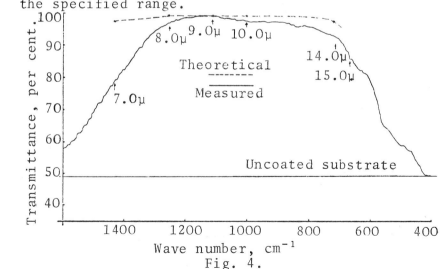

Fig. 4.

Table I does not show the other local optima that certainly existed in the design space. For example, the configuration corresponding to design 1 gave rise to a local optimum quite close to that referred to in the table, with an RMS value of 6.5%. The existence of such local optima is made likely by the nature of equation (1 d) and (1 e) where it can be seen that the effect of a layer of index n_p and thickness d_p on the transmission of light of wavelength λ is the same for all integral values of m in the equation:

$$\phi_p = 2\pi(n_p d_p \lambda_0 / \lambda + m)$$

i.e., the layer has a periodic length $\lambda/n_p\lambda_0$ at wavelength λ. Clearly, the function F is not therefore simply periodic, since it encompasses a range of wavelengths; but it would not be unexpected to find multiple local minima. The best available way to deal with this was to use a number of different starting points on each configuration, and also to put an upper limit (which in fact never proved binding) on the allowed values of the d_i.

Following the results summarized in Table I, a further set of designs was produced, involving different materials and specifications. The best design proved to have four layers, with $n_1 = n_3$ and $n_2 = n_4$; this design had a RMS reflectance of only 0.56%, and was unconstrained. The sensitivity analysis of this design was carried out as described in the next section.

2.5 Sensitivity analysis

In the region of a local stationary point, the function can be represented to $O(\underset{\sim}{d} - \underset{\sim}{d}^*)^3$ accuracy as:

$$F(\underset{\sim}{d}) = F(\underset{\sim}{d}^*) + \tfrac{1}{2}(\underset{\sim}{d} - \underset{\sim}{d}^*)G(\underset{\sim}{d} - \underset{\sim}{d}^*) \qquad (2)$$

where * denotes the stationary point. In the case of a "stationary" point located by a numerical algorithm, there will in general be some inaccuracy, so that d^* will be approximate and $\dfrac{\partial F(\underset{\sim}{d}^*)}{\partial \underset{\sim}{d}}$ will not

92

be quite zero. Assuming, however, that it is small, the sensitivity of the solution to small changes in d will be governed by the curvature of the function, *i.e.*, by the matrix

$$G \equiv \left(\frac{\partial^2 F}{\partial d_i \partial d_j} \right).$$

The contours of the function in the neighbourhood of d^* will be as sketched in Fig. 5 if d^* is a minimum, *i.e.*, G pos. def.

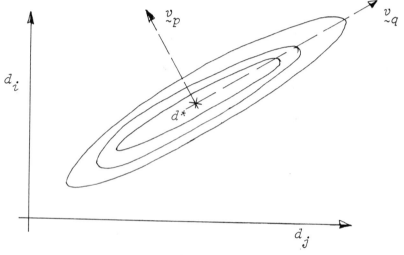

Fig. 5.

If (2) is an adequate representation of these contours in a sufficiently small neighbourhood of d^*, then the directions of the principal axes of the elliptical contours will be given by the eigenvectors of G, while the lengths of the axes will be inversely proportional to the square roots of the corresponding eigenvalues; in other words, if v_i, the eigenvectors, were chosen as the axes of coordinates, and d^* as the origin, a typical contour for some fixed value of F would be:

$$F = \sum \alpha_i^2 / \mu_i$$

93

where $\underset{\sim}{\alpha}$ is the new coordinate system, $\underset{\sim}{\mu}$ is the vector of eigenvalues. Thus, if $\underset{\sim}{G}$ is known at a solution (or possible solution), an eigenvalue analysis will enable the following investigation to be made.

(i) The positive definiteness of $\underset{\sim}{G}$ can be checked, i.e., $\mu_i > 0$, $i = 1,2,\ldots N$.

(ii) The sensitivity of the design to small changes in the design variables can be found: thus a variation along the eigenvector corresponding to maximum eigenvalue will cause the largest change in F, while that corresponding to minimum eigenvalue gives the least sensitive direction.

(iii) By taking small steps along the eigenvectors (from d^*) one can compare the actual changes in F with the change predicted if (2) were exact. This is a very powerful check on the accuracy of both the solution and the sensitivities.

In the least squares case, an approximation to the Hessian matrix can be found from the expression:

$$\underset{\sim}{G} = 2\left\{ \underset{\sim}{J}^t \underset{\sim}{J} + \sum_1^k f_i \underset{\sim}{G}_i \right\}$$

where f_i are the subfunctions, $\underset{\sim}{J}$ is the Jacobian matrix $\partial f_i/\partial x_j$, and $\underset{\sim}{G}_i$ is the Hessian matrix of the ith subfunction. If the second term on the RHS is sufficiently small (Gauss-Newton approximation) we can use the matrix $2J^tJ$ to approximate the Hessian, and this makes second derivatives of F unnecessary.

In the case of the 4-layer design already mentioned, the approximate Hessian was as follows:

$$2\underset{\sim}{J}^t\underset{\sim}{J} = \begin{bmatrix} 1.767749 & -0.0161477 & -1.579881 & 0.5466093 \\ & 1.232962 & 1.323857 & 0.5350043 \\ & \text{SYM} & 3.151352 & 0.5394221 \\ & & & 1.222820 \end{bmatrix}$$

This matrix was analyzed to produce the following eigenvalues:

4.615187,   2.169595,   0.5596029,   0.03049764

These give a condition number (μ_{max}/μ_{min}) of 151.3, and a ratio of maximum principal axis to minimum principal axis of 12.3 for each of the contours in the neighbourhood of d^*. The problem is therefore quite well conditioned, and the solution should be well determined. To check this, a step of fixed length was made along each eigenvector. The true function values were computed at the points thus generated, and these were compared with the quadratic approximation (with a correction for non-zero gradient). The results are summarized in Table II.

Table II

Step: 0.005

	True function	Quadratic Prediction
difference	+0.005	+0.005
	-0.005	-0.005
	+0.1913612-03	+0.1156885-03
	+0.1731148-04	+0.1150708-03
	+0.9640148-04	+0.5431801-04
	+0.8847739-04	+0.5416175-04
	+0.8013258-04	+0.1396904-04
	+0.7019557-04	+0.1401110-04
	+0.5056797-05	+0.7326962-06
	+0.5309805-05	+0.7921856-06

The first point about Table II is that all the differences ($F - F^*$) are positive: so a local minimum is established to at least an accuracy of 0.005 unit (in fact, to a thickness of 0.040). However, the correspondence between measured and predicted values is unusually poor, a fact which suggests that the local minimum lies in a sharply curving valley in the four-dimensional space of the problem.

95

2.6 Non-sums-of-squares objectives

Although this paper is being presented in a section devoted to sums-of-squares problems, a description of the thin-film project would not be complete without a mention of two other forms of objective function used. Because the "errors" between the ideal design and the one actually achieved always had the same sign, it was legitimate to use simply the sum-of-errors as an objective function, *i.e.*,

$$\sum_{i=1}^{k} 100(1 - T_i).$$

One of the configurations was therefore optimized with respect to this. An equally plausible criterion of a design was the value of the greatest error measured over the selected points in the range of wavelengths. Accordingly, one run was made using this as the objective function (Min-Max run). The results are shown in Fig. 6. The three curves are quite similar; but the least-squares approach provides the flattest response curve, although all three curves coincide at two points in the range.

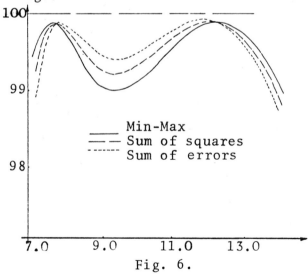

Fig. 6.

3, IMPLEMENTATION

A number of the designs described in section
2 were made and tested. The thin films were deposited in a Badzer 510 vacuum plant in a vacuum of
2×10^{-6} torr, the materials being evaporated with
the aid of a 270° electron beam source. The thickness of the deposited film was monitored by measur-
ing the change in the reflectance curves obtained;
Fig. 7 shows a typical curve.

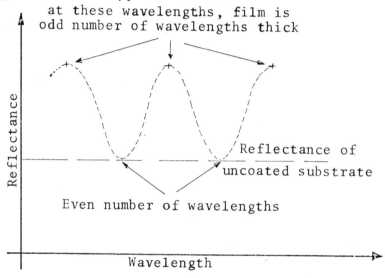

Fig. 7.

The error in monitoring, which was determined
after several trial runs with stabilized electronic
system and quartz crystal rate monitor, was found
to be 0.05λ, where λ was the monitoring wavelength.
This justified the choice of lower limit on thick-
ness used in the optimization.

Fig. 4 shows how the performance of one of
the manufactured designs compared with that predicted by the theory on which the optimization was
based. It is clear that the agreement is better
than 5% over the whole range 8 to 14µ, and considerably better in the lower part of this range.

Fig. 8.

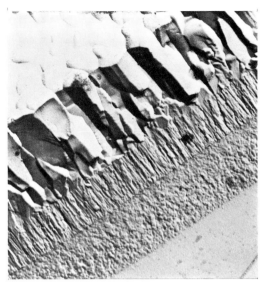

Fig. 9

The curves coincide at about 9.0μ, and show an increasing divergence above this value. These discrepancies are acceptable, and can be attributed
to the following causes.

(*i*) Dispersion and absorption not allowed
for in the mathematical model.

(*ii*) Flaws in the film material.

The state of the films was examined using a
scanning electron microscope; Figs. 8 and 9 show
some aspects of the thin-film systems. These elec-
tromicrographs show some thickness variation which
may arise from drying out of the freshly packed
source material; they show clearly defined boundary
layers. In fact, X-ray fluorescence studies showed
no evidence of migration of material across boundaries. However, the structure varies with the
material. Fig. 10 shows evidence of cracks (probably caused by different thermal coefficients of
expansion) and pin holes. The latter were probably
caused by spattered particles from the electron
beam source, possibly resulting from the high evaporation temperature of one of the layer materials.

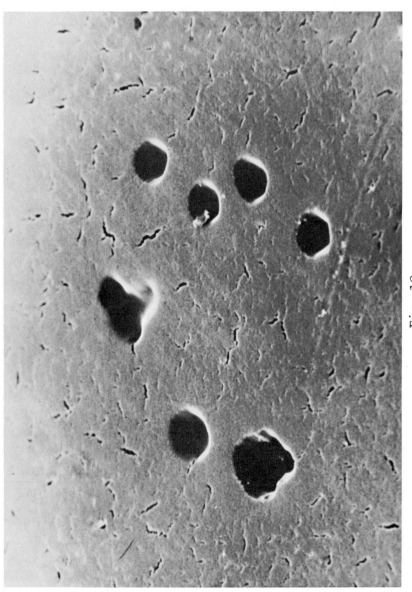

Fig. 10.

ACKNOWLEDGEMENT

This project was funded by the Ministry of Defence under CVD; and the authors are also grateful to Grubb Parsons Ltd. for permission to publish the above results.

REFERENCE

Biggs, M.C. (1973) "Constrained minimisation using recursive equality quadratic programming", NOC TR 24, Hatfield Polytechnic.

LEAST SQUARES FITTING OF MCCONOLOGUE ARCS

D.R. Divall

(Portsmouth Polytechnic)

SUMMARY

This paper considers the representation of discrete data in the plane by a smooth curve built up from sections. Over each section x and y are given as cubics in a parameter which is chosen to approximate the length of arc. Fitting is performed by the method of least squares and to ensure a satisfactory minimum is obtained constraints of a linear nature are imposed.

The resulting optimization problem is sparse and an algorithm of the hypercube type has been coded to exploit this. Each step of the algorithm solves a quadratic program in which an approximation to the sum of squares, obtained using first derivative information, is minimized subject to the problem constraints and a localizing constraint.

Results on test cases are shown to be satisfactory but the computation is rather slow.

1. INTRODUCTION

This paper is concerned with the approximation of discrete data in the plane by a smooth curve. The original motivation sprang from the problem of recovering a curve which had been reduced to a sequence of data by a digitizing device. Such problems occur in map making although

---

*This material forms part of a thesis presented to the University of Sussex.

the methods are of wider application.

Throughout, the data will be denoted by
(x_i, y_i), $i = 1, \ldots, m$. Certain assumptions are
made which are suggested by the original problem.

A1: the order in which the data lie along the
     fitted curve is given.

This assumption provides a means of distin-
guishing the two curves in Fig. 1 which might
otherwise be equally acceptable. The order of the
points will be taken as the natural order of the
subscripts.

A2: the data may contain noise but this may be
     regarded as being orthogonal to the fitted
     curve.

As a consequence noise cannot cast doubt on
the application of the ordering assumption A1.

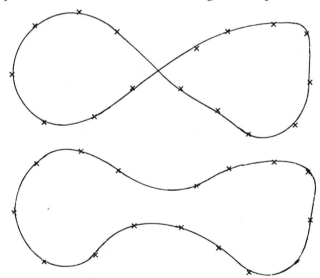

Fig. 1. Alternative curves for a data set which
         can be distinguished by the ordering
         assumption A1

The fitted curve will be represented by a pair of parametric Cartesian equations, giving x and y in terms of t. This permits the formation of loops, closed curves and other departures from the behaviour of a single-valued function. The particular choice outlined in the next section ensures continuity of position and tangent, and invariance under rotation and translation of axes. The method is easily adapted to other forms of curve.

2. THE MCCONOLOGUE ARC

The selected curve was introduced by McConologue (1970) and applied to the interpolation of data. The curve, like a spline, is piecewise in nature and is composed of sections which join smoothly at a set of points called *knots*. The position of these knots and the direction of the curve as it passes through them will determine the curve. Let (ξ_k, η_k), $k = 1, \ldots, n+1$ denote an ordered set of knots and let the associated slopes be denoted by $\tan\theta_k$, $k = 1, \ldots, n+1$. For each pair of knots (ξ_k, η_k) and (ξ_{k+1}, η_{k+1}) an arc is calculated from equations (1) to (8) below. Here Δ denotes the forward difference operator, c_k denotes $\cos\theta_k$ and s_k denotes $\sin\theta_k$. For clarity the subscript is omitted from g, f and e.

$$g = (\Delta\xi_k)^2 + (\Delta\eta_k)^2 , \tag{1}$$

$$f = (c_k + c_{k+1})\Delta\xi_k + (s_k + s_{k+1})\Delta\eta_k , \tag{2}$$

$$e = 7 - c_k c_{k+1} - s_k s_{k+1} , \tag{3}$$

$$T_k = 3(\sqrt{(f^2 + 2eg)} - f)/e , \tag{4}$$

$$B_k = 6\Delta\xi_k/T_k^3 - 3(c_k + c_{k+1})/T_k^2 , \tag{5}$$

$$C_k = 6\Delta\eta_k/T_k^3 - 3(s_k + s_{k+1})/T_k^2 , \tag{6}$$

$$x_k(t) = \xi_k + tc_k + t^2\Delta c_k/(2T_k) + B_k(T_k t^2/2 - t^3/3) , \tag{7}$$

$$y_k(t) = \eta_k + ts_k + t^2\Delta s_k/(2T_k) + C_k(T_k t^2/2 - t^3/3) . \tag{8}$$

The parametric equations

$$
\left.
\begin{array}{l}
x = x_k(t) \\
y = y_k(t)
\end{array}
\right\}
\tag{9}
$$

define a curve Γ_k of which the portion correspon-ding to the restriction

$$
0 \leqslant t \leqslant T_k
\tag{10}
$$

forms the building block for McConologue's arc.

It is easily verified that at $t = 0$, Γ_k passes through (ξ_k, η_k) with slope $\tan\theta_k$: while at $t = T_k$, Γ_k passes through (ξ_{k+1}, η_{k+1}) with slope $\tan\theta_{k+1}$. Thus the union of the n arcs defined by (9) and (10) is an arc with continuity of position and tangent. Moreover, the particular choice of T_k ensures that over the range (10), t is an approximation to the length of the curve Γ_k mea-sured from (ξ_k, η_k).

The arc is defined by $3n + 3$ parameters, but for the application it is convenient to reduce the freedom in the parameter system to give an effect analogous to fixing the knots of a natural spline. This requires the prior division of the data into n sections: thus a set of integers

$$
0 = m_1 < m_2 < \ldots < m_{n+1} = m
\tag{11}
$$

is given and the data points with subscripts $m_k+1, \ldots, m_k+1$ are associated with Γ_k. For $2 \leqslant k \leqslant n$ (the interior knots), a *local trend line* is obtained by fitting a straight line by least squares to the four adjacent data points, *i.e.*, those with subscripts m_k-1 to m_k+2. The kth knot (ξ_k, η_k) is constrained to lie on a line passing through the mean of the points with subscripts m_k and m_k+1 and orthogonal to the local trend line. This is illustrated in Fig. 2. For a closed curve it is necessary that $\xi_1 = \xi_{n+1}$, $\eta_1 = \eta_{n+1}$ and $\theta_1 = \theta_{n+1}$ and for this "coalesced knot" a similar procedure using the four adjacent points,

105

$1,2,m_{n+1}-1$ and m_{n+1}, defines the local trend line and the knot constraint.At the end of an open curve, the local trend line is defined from the three end points at each extremity and the knot constrained to lie on a line orthogonal to this and offset slightly to lie outside the data set. The net result is a system of linear equations which have the form

$$D_1 \underset{\sim}{\xi} + D_2 \underset{\sim}{\eta} = \underset{\sim}{c}, \qquad (12)$$

where D_1 and D_2 are diagonal matrices, constraining the knots.

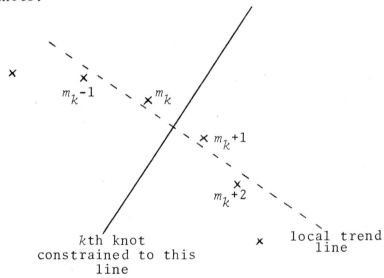

Fig. 2. The local trend line and knot constraint

3. THE FITTING CRITERION

Assumption A2 suggests that the curve be chosen to minimize the sum of squares of the perpendicular distances from each data point to the curve. If (x_i, y_i) has been assigned to Γ_k then the perpendicular distance minimizes, with respect to t_i, the quantity $r_{2i-1}^2 + r_{2i}^2$, where

$$r_{2i-1} = x_k(t_i) - x_i \qquad (13)$$

and $\qquad r_{2i} = y_k(t_i) - y_i \; . \qquad (14)$

Moreover, to ensure that this relates to the required portion of Γ_k, t_i must satisfy

$$0 \leqslant t_i \leqslant T_k \; . \qquad (15)$$

Additionally, assumptions A1 and A2 require that the feet of the perpendiculars lie in prescribed order along the curve. Thus if the ith and $(i+1)$th points have both been assigned to Γ_k then t_i and t_{i+1} must satisfy

$$t_i \leqslant t_{i+1} \; . \qquad (16)$$

The curve fitting problem then reduces to the optimization:

$$\text{minimize} \quad z = \sum_{i=1}^{2m} r_i^2 \qquad (17)$$

with respect to $\{\xi_k\}$, $\{\eta_k\}$, $\{\theta_k\}$ and $\{t_i\}$

subject to $\quad (a)$ the linear equations (12),

$\qquad\qquad\quad (b)$ the linear inequalities (16)

and $\qquad\qquad (c)$ the nonlinear bounds (15).

The inequality constraints (b) and (c) are not expected to be active at the solution: they act as a guide to ensure that a suitable local minimum of (17) is obtained.

4. THE METHOD OF HYPERCUBES

The method adopted for the solution of this optimization is a variation on an algorithm proposed by Fletcher (1972) and akin to the Marquardt (1963) algorithm. It is designed to accommodate linear constraints but with a small modification has been able to deal with (15) also. For the description of the method the single symbol $\underset{\sim}{p}$ will be used to represent the set of parameters $\{\xi_k\}$, $\{\eta_k\}$, $\{\theta_k\}$ and $\{t_i\}$.

The process is iterative and each step must commence with a vector of parameters p_0 which is feasible for the constraint system and a hypercube parameter $d(>0)$, the purpose of which will be outlined later. The corresponding value of z will be denoted by z_0 and those of T_k by T_k^0. Implicit differentiation of (1) to (8) yields the corresponding gradient vector and Jacobian matrix,

$$g = [\partial z / \partial p_i] \left.\begin{array}{c} \end{array}\right\} \text{ all derivatives} \tag{18}$$

and $\quad J = [\partial r_i / \partial p_j]$ evaluated at p_0 $\qquad$ (19)

and thus an approximation to the hessian of z

$$H = 2J^T J. \tag{20}$$

The main burden of the step is the solution of the *quadratic subproblem*:

minimize $q = z_0 + g^T(p - p_0) + (p - p_0)^T H (p - p_0)/2$ (21)

subject to $\quad (a)$ the problem constraints (12), (16) and (15) with T_k replaced by T_k^0

and $\qquad (b)$ the *hypercube constraint*

$$\| p - p_0 \|_\infty \leq d. \tag{22}$$

The special form of the hypercube constraint means that some problem constraints may become superfluous for the current step. Such constraints are easily located and need not be included in (a).

The hypercube parameter is introduced to ensure that over the region considered in the quadratic subproblem q is an acceptable approximation to z. In order that this may be effective it is necessary to adjust d as the iteration proceeds. Occasionally a step must be repeated with a reduced value for d, but in general the vector p_1 which solves the quadratic subproblem is used to initiate a new step. It is necessary, however, to consider the nonlinear part of (15), since the values of T_k calculated at p_1 may no longer satisfy (15). If (15) ceases to be satisfied for a

particular Γ_k then a linear transformation is applied to all the associated $\{t_i\}$ so that (15) is
just satisfied. In practice the intrinsic character of the $\{t_i\}$ and $\{T_k\}$ ensures that such adjustment is necessary only on a few early steps. The
modified ρ_1 is then used to replace ρ_0.

An initial vector is easily obtained by
requiring that the knots be located at the mean of
their adjacent data points with directions given
by the local trend line. The parameters $\{t_i\}$ are
then selected to satisfy (15) and so that their
differences are proportional to distance between
data points. Termination is controlled by monitoring the steepest feasible gradient, for which
suitable bounds are yielded by the solution of the
quadratic subproblem. A detailed discussion of
termination, the adjustment of d, the elimination
of superfluous constraints and a general convergence theory is given by Divall (1974).

5. SOLUTION OF THE QUADRATIC SUBPROBLEM

The main burden of the calculation consists
in solving the quadratic subproblem. It will be
apparent that the constraint system is very specialized: indeed it is a combination of simple
bounds and constraints containing only two non-
zeros per row. Moreover, J contains only seven
non-zeros per row so that considerable sparsity is
evident. Recently methods for sparse matrices
have been applied to linear programming (Beale
(1971), Mitra (1974) for example) and a natural
method of exploiting the sparsity of the quadratic
subproblem is to use Wolfe's (1959) method. In
this method a solution is obtained by solving a
linear program with certain forbidden basis combinations. The code adopted for the linear program is that of Reid (1974) which employs, with
due regard to sparsity, Bartel's (1971) method of
maintaining and updating in a stable manner a triangular decomposition of the basic matrix.

6. SEQUENTIAL PROCESSING

The basic algorithm outlined so far may be applied as it stands, but in practice it has proved helpful not to do so. In the belief that the curve should be determined by its local properties, the basic algorithm is applied in stages to a sequence of small subsets of the entire data set. Thus initially the sections Γ_1, Γ_2 and Γ_3 are considered in isolation and a fit obtained. For an open curve, Γ_1 and Γ_2 are then regarded as fixed but not Γ_3: at the second stage Γ_3, Γ_4 and Γ_5 are considered - but the knot $(\xi_3, \eta_3, \theta_3)$ is held constant to ensure a smooth join to Γ_2. The process of fitting a group of three sections overlapping the previous group by one section continues until Γ_n has been fitted. A similar sequence of overlapping groups is used for a closed curve, save that Γ_1 is not fixed until the final stage. At the final stage Γ_1 is considered with Γ_n and $(\xi_2, \eta_2, \theta_2)$ is held constant. Clearly the initial parameter set for an overlapping section is advantageously chosen to be the final parameter set for that section at the previous stage.

The choice of groups of three sections overlapping by one at each stage is empirical: trials have been made with various sizes of group and overlap and compared with the global fit obtained by using the basic algorithm on the entire data set. Pleasingly the results were equally satisfactory to the eye. However, the shortest overall time was obtained with the "three and one" combination described.

7. SOME TEST RESULTS AND DISCUSSION

Figs. 3 and 4 show the kind of results obtained on moderate sized data sets. The behaviour of the algorithm on each minimization for these figures is summarized in Tables I and II.

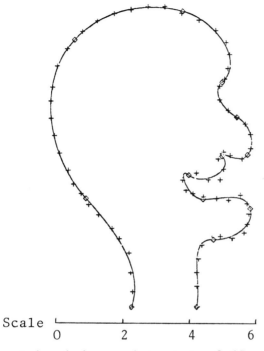

Scale
0   2   4   6

Fig. 3. A hand drawn data set of 65 points.
◇ denotes a knot; + denotes a data point

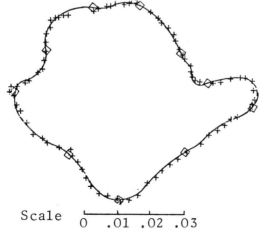

Scale   0 .01 .02 .03

Fig. 4. A digitized map of Anglesey comprising 84
points. ◇ denotes a knot; + denotes a
data point

111

Table I

(a) Number of Iterations*	(b) Number of Parameters	(c) Number of Equality Constraints (12)	(d) Number of Inequality Constraints (16)	(e) Initial Sum of Squares	(f) Final Sum of Squares
11 (8)	37	4	22	78.8	0.032
13 (10)	29	6	14	8.46	0.018
22 (18)	23	6	8	0.183	0.0045
22 (22)	24	6	9	0.415	0.088
14 (6)	26	6	11	2.01	0.046
22 (10)	16	5	9	7.50	0.030

Behaviour for the open curve of Fig. 3. The set comprises 65 points in 12 sections. 88 seconds were required on an IBM 360/165.

Each row corresponds to fitting a group of 3 sections.

*In brackets is the number after which the sum of squares was constant to 4 significant figures.

Table II

(a) Number of Iterations*	(b) Number of Parameters	(c) Number of Equality Constraints (12)	(d) Number of Inequality Constraints (16)	(e) Initial Sum of Squares	(f) Final Sum of Squares
8 (4)	37	4	22	0.41×10^{-4}	0.15×10^{-5}
15 (8)	37	6	22	0.52×10^{-4}	0.81×10^{-5}
8 (4)	37	6	22	0.84×10^{-4}	0.34×10^{-5}
22 (6)	38	6	23	0.11×10^{-3}	0.14×10^{-4}
20 (8)	37	8	22	0.13×10^{-3}	0.92×10^{-5}

Behaviour for the Anglesey map of Fig. 4. The set comprises 84 points in 12 sections. 220 seconds were required on an IBM 360/165.

Each row corresponds to fitting a group of 3 sections.

*In brackets is the number after which the sum of squares was constant to 4 significant figures.

Testing on larger data sets was curtailed because of the long run times required. Consideration of column (a) in the Tables reveals one cause of the run times: close examination has shown that the terminating criterion, which was constructed to give satisfactory conclusions even on poorly conditioned problems has required unnecessary iterations. The bracketed figures indicate a more appropriate number of iterations. Thus with a laxer termination requirement the time for the Anglesey map might have come down to about 90 seconds with little loss to the fit obtained, but this is still long. It is worth recording that some of the H matrices encountered were quite ill-conditioned (one was found with a spectral condition number of 10^8), this effect can be exacerbated by poor scaling of the data.

Without doubt the method can be made more efficient only by improving the efficiency of the optimizing code. The number of iterations is reassuringly small (on the available evidence it is never worse than the number of parameters) and therefore an appropriate place to seek efficiency is in the solution of the quadratic subproblem. Bartels (1971) has indicated that the use of stabilized linear programming methods may involve an overhead of up to 60%, but it is not desirable to seek speed at the expense of stability. More pertinently, although second and subsequent iterations begin with the basis which solved the previous subproblem, there was little evidence of a progressive attenuation of time per iteration as the solution was neared. Perhaps this indicates that in the later stages of the subproblem the overheads in rejecting forbidden basis combinations become a dominant part of the calculation. If this is so, then Wolfe's method is an inappropriate method of tackling the subproblem. Further work is required to determine this.

It has also transpired that in order to obtain satisfactory fitting the number of data points per section is quite small. As a result

the sparsity, which led to the choice of Wolfe's
method, is less marked than at first envisaged:
typically the matrices contain about 20% non-zeros.
At this density the gains from sparse methods are
not marked and the way is open to consider other
methods for the subproblem.

If a more efficient method were available
then the algorithm would be quite powerful. The
problem of subdividing the data into sections is
as yet untouched. Fig. 5 shows that this can be
important: the data are identical to Fig. 3, the
sole difference being that one knot (indicated by
the arrow) has been moved.

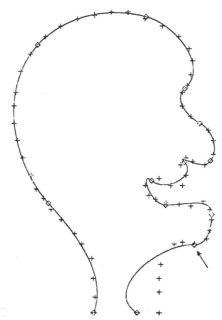

Fig. 5.   The open curve of Fig. 3 with one knot
           moved

ACKNOWLEDGEMENTS

The possibilities of the McConologue arc
were brought to my notice by A.R. Curtis. A great

115

deal of practical help and encouragement was given by J.K. Reid. The Anglesey data were supplied by the Mapping and Charting Establishment.

REFERENCES

Bartels, R.H. (1971) "A stabilization of the simplex method", *Num. Math.*, **16**, 414-434.

Beale, E.M.L. (1971) "Sparseness in Linear Programming", *in* "Large Sparse Sets of Linear Equations", *ed.* J.K. Reid, Academic Press.

Divall, D.R. (1974) "Topics in the approximation of functions and data of several independent variables", Thesis, University of Sussex.

Fletcher, R. (1972) "An efficient, globally convergent, algorithm for unconstrained and linearly constrained optimisation problems", *Math. Prog.*, **2**, 133-165.

McConologue, D.J. (1970) "A quasi-intrinsic scheme for passing a smooth curve through a discrete set of points", *Comp. J.*, **13**, 392-396.

Marquardt, D.W. (1963) "An algorithm for least squares estimation of nonlinear parameters", *Siam. J. Num. Anal.*, **11**, 431-441.

Mitra, G. (1974) "Sparse inverse in the factored form and maintaining sparsity during simplex iteration", *in* "Software for Numerical Analysis", *ed.* D.J. Evans, Academic Press.

Reid, J.K. (1974) "Sparse Linear Programming Using the Bartels-Golub Decomposition", Harwell Technical Report (*to appear*).

Wolfe, P. (1959) "The simplex method for quadratic programming", *Econometrica*, **27**, 382-398.

A View of Unconstrained Optimization

M.J.D. Powell

(Atomic Energy Research Establishment, Harwell)

SUMMARY

The development of computer programs for unconstrained optimization is surveyed. It seems that most of the useful work in this area during the last five years has been given to improving, extending and understanding available numerical methods instead of devising new algorithms. However we do consider a recent algorithm (Wolfe (1974a)) that offers the exciting possibility of being able to treat first derivative discontinuities routinely. Also we consider algorithms that are suitable when the number of variables is very large. Here it is usually important to take advantage of any sparseness in the matrix of second derivatives. Finally the relations between the theoretical and practical aspects of the subject are discussed.

1. INTRODUCTION

We consider algorithms that calculate the least value of a function of several variables

$$F(\underset{\sim}{x}) = F(x_1, x_2, \ldots, x_n), \qquad (1.1)$$

given a starting point $\underset{\sim}{x}^{(1)}$ in the space of the variables. Usually the function is specified by a computer subroutine that evaluates $F(\underset{\sim}{x})$ for any $\underset{\sim}{x}$. In addition some algorithms require the first derivative vector

$$g(\underset{\sim}{x}) = \nabla F(\underset{\sim}{x}) \qquad (1.2)$$

117

to be calculated, and a few even require the
second derivative matrix

$$G_{ij}(\underset{\sim}{x}) = d^2 F(\underset{\sim}{x})/dx_i dx_j; \qquad i,j=1,2,\ldots,n, \quad (1.3)$$

also.

The oldest common methods for solving the
unconstrained minimization problem are the method
of steepest descents, the method of adjusting the
variables one at a time, and the Newton-Raphson
iteration that requires exact second derivatives,
see Kowalik and Osborne (1968) for instance. However, in 1959, Davidon published an algorithm that
shows that, if second derivatives are not calculated, then enormous gains in efficiency can be
obtained by allowing for them. Therefore most of
the techniques that are leading now have been
developed during the last fifteen years. My own
view of this development is given in Sections 2
and 6.

Most of the algorithms that are in current
use were proposed before 1971, so during the last
four years much of the research in unconstrained
optimization has been given to understanding, comparing, improving and extending these algorithms.
Two improvements that are particularly valuable
are the gains in accuracy that can be obtained by
representing approximations to second derivative
matrices in factored form, and the development of
good finite techniques for the line search calculation that has to be done on every iteration of
many algorithms. These two improvements are considered in Section 3.

Also there has been much useful work recently on the solution of constrained optimization
problems by techniques that solve a sequence of
unconstrained problems, see Bertsekas (1974) for
instance. In these augmented Lagrangian methods
the objective functions of the unconstrained problems have discontinuities in their second derivatives. Some earlier work in this area (Fletcher
(1973)) converts a constrained problem to a single

unconstrained calculation, whose objective function
has a discontinuous first derivative. The first
derivative discontinuities occur away from the
solution, but if one can manage discontinuities at
the solution the technique of Pietrzykowski (1969)
may be applied. These are a few of many examples
where we would like to have good algorithms for
minimizing functions with derivative discontinuities. A technique that seems to be very suitable
was discovered recently by Wolfe (1974a). It is
described in Section 4.

Another important area in which few algorithms have been developed is the minimization of
functions when n, the number of variables, is
large. Then some structure is needed to solve the
problem in a moderate amount of computer time.
Some available techniques are surveyed in Section
5.

The last section of this paper considers our
theoretical knowledge of algorithms for unconstrained optimization, and it discusses the contribution that the theory has made to the development of useful algorithms.

2. THE DEVELOPMENT OF MODERN ALGORITHMS FOR UNCONSTRAINED OPTIMIZATION

In 1962 I attended a seminar, chaired by
R.W.H. Sargent at Imperial College, where it was
claimed that, due to the amount of computer time
required, it was often not possible to calculate
the least value of a function of ten or more variables. This observation gave me great pleasure
for, by applying a version of Davidon's (1959) algorithm, I had already obtained the minimum of a
function of one hundred variables in a few minutes
on a Ferranti Mercury computer. The same algorithm
completed another calculation in about 30 seconds
for a colleague, who consulted me because a traditional algorithm made little progress on his minimization problem in 30 minutes on a faster machine.
These experiences are typical of the advances made

119

in unconstrained optimization, that were begun by Davidon's work.

The property that is characteristic of the modern algorithms is that they all work well when the objective function (1.1) is quadratic. Apart from the obvious remark that, if a minimization algorithm cannot treat quadratic functions well, then one cannot expect it to be successful at minimizing more general functions, there is a fundamental reason why the characteristic property is helpful. To explain it we suppose that we have an algorithm that searches for the least value of $F(x)$ by a method that includes safeguards to ensure that the objective function decreases, and that subject to the safeguards the algorithm tries to take quite large steps in the space of the variables. Further we suppose that $F(x)$ has continuous second derivatives. Then, if the algorithm is able to take large steps, it is probably making good progress, and, if it has to take small steps, then the continuity of second derivatives makes the objective function appear to the algorithm as almost quadratic, so good progress should be made in this case also due to the characteristic property. This argument cannot be applied to an algorithm that is designed to treat linear functions well, because the automatic control of step-lengths depends on the curvature of the objective function. Further, because the gradient vector (1.2) is zero where $F(x)$ is least, the objective function does not behave like a linear function near the required solution.

Davidon's (1959) algorithm (see also Fletcher and Powell (1963)) treats quadratic functions well by building up an estimate of the inverse of the second derivative matrix (1.3). The algorithm is iterative, and calculates a sequence of points $x^{(k)}$ ($k = 1,2,3,\ldots$) such that the condition

$$F(x^{(k+1)}) < F(x^{(k)}) \qquad (2.1)$$

is satisfied. The gradients $g^{(k)} = g(x^{(k)})$ ($k = 1,2,3,\ldots$) are required. We let $H^{(k)}$ be the estimate of $[G(x^{(k)})]^{-1}$. If there is no error in this estimate and if $F(x)$ is quadratic, then the least value of $F(x)$ occurs when x is on the line through $x^{(k)}$ with direction

$$d^{(k)} = -H^{(k)}g^{(k)}. \qquad (2.2)$$

Therefore we let $x^{(k+1)}$ be the vector

$$x^{(k+1)} = x^{(k)} + \lambda^{(k)}d^{(k)}, \qquad (2.3)$$

where $\lambda^{(k)}$ is a step-length parameter that gives condition (2.1), and which, in the original version of the algorithm, is intended to be the value of λ that minimizes the function of one variable

$$\phi^{(k)}(\lambda) = F(x^{(k)} + \lambda d^{(k)}). \qquad (2.4)$$

The matrix $H^{(k+1)}$ is calculated from $H^{(k)}$ and the differences

$$\left.\begin{array}{c} \delta^{(k)} = x^{(k+1)} - x^{(k)} \\ \gamma^{(k)} = g^{(k+1)} - g^{(k)} \end{array}\right\}. \qquad (2.5)$$

Provided that each value of $\lambda^{(k)}$ minimizes the function (2.4), this algorithm obtains the least value of a quadratic function in at most n iterations. This statement may be proved (Fletcher and Powell (1963)) by showing that the search directions satisfy the conjugacy condition

$$(d^{(k)}, Gd^{(i)}) = 0, \qquad i \neq k, \qquad (2.6)$$

where G is the constant second derivative matrix of $F(x)$, and where the notation (a,b) is the scalar product of the vectors a and b.

The conjugacy condition (2.6) is relevant because, if it holds for a sequence of n directions $\{d^{(1)}, d^{(2)}, \ldots, d^{(n)}\}$, then searching along each direction in turn from any starting point leads to the minimum of a positive definite quadratic function.

Zoutendijk (1960) was the first person to apply conjugate directions to the minimization of general functions. He notes that, if $x^{(i+1)}$ is defined by equation (2.3) for $k = i$, and if $\lambda^{(i)}$ is non-zero, then condition (2.6) is equivalent to the equation

$$(d^{(k)}, g^{(i+1)} - g^{(i)}) = 0, \quad i \neq k, \quad (2.7)$$

when $F(x)$ is a quadratic function. This equation is very useful because it does not include any second derivatives. Therefore Zoutendijk proposes a gradient algorithm that calculates search directions $d^{(k)} (k = 1, 2, \ldots, n)$, which satisfy equation (2.7) for $1 \leqslant i \leqslant k-1$, and which satisfy the downhill condition

$$(d^{(k)}, g^{(k)}) < 0. \quad (2.8)$$

Usually it is necessary to use more than n iterations, and for $k > n$ the values of i in condition (2.7) refer to the most recent steps taken by the algorithm, because we may not impose more than $(n - 1)$ conjugacy conditions on $d^{(k)}$.

Another important method that satisfies equation (2.6) in the quadratic case is due to Fletcher and Reeves (1964). Some analysis is needed to prove that equation (2.6) is obtained, and, besides the requirement that $F(x)$ be quadratic, it depends on each value of $\lambda^{(k)}$ being calculated to minimize the function (2.4). The initial search direction is the vector $d^{(1)} = -g^{(1)}$, and the subsequent search directions are obtained from the formula

$$d^{(k)} = -g^{(k)} + \frac{\| g^{(k)} \|^2}{\| g^{(k-1)} \|^2} d^{(k-1)}, \quad (2.9)$$

where the vector norms are Euclidean. This algorithm is particularly useful when the number of variables is large, because no matrices have to be stored. Therefore it is considered further in Section 5.

An important generalization of Davidon's algorithm was suggested by Broyden (1967). He

122

noticed that one can introduce a free parameter into the formula that is used to calculate $H^{(k+1)}$ from $H^{(k)}$ and the data (2.5). Specifically we may use the formula

$$H^{(k+1)} = H^{(k)} - \frac{H^{(k)}\underset{\sim}{\gamma}^{(k)}\underset{\sim}{\gamma}^{(k)T}H^{(k)}}{(\underset{\sim}{\gamma}^{(k)},H^{(k)}\underset{\sim}{\gamma}^{(k)})}$$

$$+ \frac{\underset{\sim}{\delta}^{(k)}\underset{\sim}{\delta}^{(k)T}}{(\underset{\sim}{\delta}^{(k)},\underset{\sim}{\gamma}^{(k)})} + \beta^{(k)}\underset{\sim}{w}^{(k)}\underset{\sim}{w}^{(k)T},$$

(2.10)

where $\beta^{(k)}$ is the free parameter, where $\underset{\sim}{w}^{(k)}$ is the vector

$$\underset{\sim}{w}^{(k)} = \frac{H^{(k)}\underset{\sim}{\gamma}^{(k)}}{(\underset{\sim}{\gamma}^{(k)},H^{(k)}\underset{\sim}{\gamma}^{(k)})} - \frac{\underset{\sim}{\delta}^{(k)}}{(\underset{\sim}{\delta}^{(k)},\underset{\sim}{\gamma}^{(k)})}, \quad (2.11)$$

and where the value $\beta^{(k)} = 0$ gives Davidon's original formula. Huang (1970) showed that, if $F(\underset{\sim}{x})$ is quadratic and if each iteration calculates $\lambda^{(k)}$ to minimize the function (2.4), then, given $\underset{\sim}{x}^{(1)}$, the sequence of points $\underset{\sim}{x}^{(k)}$ ($k = 1,2,3,\ldots$) is independent of the parameters $\beta^{(k)}$ ($k = 1,2,3,\ldots$). Later Dixon (1972) proved the remarkable theorem that this result is true even when $F(\underset{\sim}{x})$ is a general function. Therefore the optimal choice of $\beta^{(k)}$ depends on factors like the difference between the calculated value of $\lambda^{(k)}$ and the value that minimizes the function (2.4). Many people now agree that it is best to use the value of $\beta^{(k)}$ that gives the formula

$$H^{(k+1)} = \left[I - \frac{\underset{\sim}{\delta}^{(k)}\underset{\sim}{\gamma}^{(k)T}}{(\underset{\sim}{\delta}^{(k)},\underset{\sim}{\gamma}^{(k)})}\right]H^{(k)}$$

$$\left[I - \frac{\underset{\sim}{\gamma}^{(k)}\underset{\sim}{\delta}^{(k)T}}{(\underset{\sim}{\delta}^{(k)},\underset{\sim}{\gamma}^{(k)})}\right] + \frac{\underset{\sim}{\delta}^{(k)}\underset{\sim}{\delta}^{(k)T}}{(\underset{\sim}{\delta}^{(k)},\underset{\sim}{\gamma}^{(k)})}.$$

(2.12)

We return to this subject, including consideration of the "rank-one updating formula", in the next section. Some other recent work on gradient methods

for unconstrained optimization is mentioned in
Sections 4, 5 and 6.

The methods considered so far require the
calculation of $F(x)$ and $g(x)$ at $x^{(k)}$. When the
second derivatives (1.3) are available also, then
the classical Newton-Raphson iteration

$$x^{(k+1)} = x^{(k)} - [G^{(k)}]^{-1} g^{(k)} \qquad (2.13)$$

is sometimes very effective, especially when $x^{(k)}$
is close to the point that minimizes $F(x)$. However it may happen that this iteration fails to
reduce the value of $F(x)$. Therefore it is usual
to modify equation (2.13) by including some parameters whose values are calculated to satisfy
inequality (2.1). For example we may replace
equation (2.13) by the formula

$$x^{(k+1)} = x^{(k)} - \lambda^{(k)} [G^{(k)} + \mu^{(k)} I]^{-1} g^{(k)}, (2.14)$$

where the values of $\lambda^{(k)}$ and $\mu^{(k)}$ are at our dis-
posal. Methods of this type are reviewed by Hebden
(1973), and he suggests a particular algorithm
which works well in practice, even when the starting point $x^{(1)}$ is far from the solution. We note
in Section 5 that, if the elements (1.3) are estimated by numerical differences, then this approach
may be preferable to a gradient method when the
second derivative matrix is sparse.

The estimation of derivatives by numerical
differences gives some good algorithms for unconstrained optimization that only require values of
$F(x)$ to be calculated. One of the first papers on
this subject (Stewart (1967)) is based on Davidon's
(1959) algorithm, and it describes an automatic
method for choosing the step-lengths h_i in the formula

$$dF(x)/dx_i \simeq \{F(x+h_i e_i) - F(x)\}/h_i, \quad i=1,2,\dots,n, (2.15)$$

where e_i is the ith coordinate vector. However,
because a similar algorithm proposed by Gill,
Murray and Pitfield (1972) applies equation (2.12)
instead of the formula used by Davidon to calculate

$H^{(k+1)}$, this algorithm is often faster than Stewart's method. This algorithm, instead of choosing the step-lengths h_i automatically, requires the user to provide suitable step-lengths.

Another approach to minimization without calculating any derivatives is to use a more direct method to construct search directions that satisfy the conjugacy condition (2.6) in the quadratic case. Some useful algorithms of this type have been proposed, most of them being developed from the one described by Powell (1964). The construction that is used to satisfy equation (2.6) without calculating any derivatives is as follows.

Given $d^{(i)}$ and a starting point, $z^{(1)}$ say, in the space of the variables, we let $z^{(2)}$ be the point on the line through $z^{(1)}$ with direction $d^{(i)}$ that minimizes $F(x)$. We make a displacement away from this line of search to a point, $z^{(3)}$ say. We let $z^{(4)}$ be the point on the line through $z^{(3)}$ with direction $d^{(i)}$ that minimizes $F(x)$. Then, if we let $d^{(k)} = z^{(4)} - z^{(2)}$, equation (2.6) holds when the objective function is quadratic.

Powell's (1964) method is still used frequently to solve real problems, and some claim that it is the best one for their applications. However, a feature that keeps the search directions $d^{(1)}, d^{(2)}, \ldots, d^{(n)}$ independent often impairs efficiency when the number of variables is greater than about ten. Therefore Brent (1973) proposed a different method for maintaining linear independence, which makes an eigenvector calculation every n iterations. This technique often saves many function evaluations, but the routine calculation increases significantly, so it is recommended only when the total computer time is dominated by the evaluations of $F(x)$.

Most comparisons of algorithms based on numerical examples show that, for minimization without derivatives, the methods that estimate derivatives numerically tend to be faster than conjugate direction methods. However both techniques have disadvantages. Therefore in the next five years some faster and more reliable algorithms could be developed, possibly along the lines described in a recent report (Powell (1974)).

3. MATRIX FACTORIZATIONS AND APPROXIMATE LINE SEARCHES

The gradient methods of Section 2 that employ the sequence of matrices $H^{(k)}$ ($k = 1,2,3,\ldots$) require an initial matrix $H^{(1)}$. Usually $H^{(1)}$ is set to the unit matrix, because if it is positive definite, if Davidon's algorithm or formula (2.12) is used, and if each line search ensures that the scalar product $(\delta^{(k)}, \gamma^{(k)})$ is positive, then all the matrices $H^{(k)}$ are positive definite. Thus each search direction (2.2) is downhill with respect to $F(\underset{\sim}{x})$, and each matrix $H^{(k)}$ defines a metric.

However the retention of positive definiteness is a theoretical result, and in practice the accumulation of numerical rounding errors may destroy it. This does not happen often, but it is liable to occur in the extreme cases reported by Bard (1968), especially if Davidon's algorithm is used. This possibility can be avoided completely by using matrix factorizations, instead of working with the actual elements of the matrices $H^{(k)}$ ($k = 1,2,3,\ldots$) (Gill and Murray (1972)). The use of factorizations is one of the main improvements that have been made to optimization algorithms in the last four years.

Various forms of factorization are available, and the one that is most usual is to work with a positive diagonal matrix $D^{(k)}$ and a lower triangular matrix $L^{(k)}$, instead of with $H^{(k)}$ ($k=1,2,3,\ldots$). The elements of $D^{(k)}$ and $L^{(k)}$ are related to $H^{(k)}$ by the equation

$$[H^{(k)}]^{-1} = L^{(k)} D^{(k)} L^{(k)T}. \tag{3.1}$$

This equation leaves some freedom in $D^{(k)}$ and $L^{(k)}$, which is fixed by letting the diagonal elements of

$L^{(k)}$ be equal to one. Instead of applying formula
(2.2), it is appropriate to obtain $\underset{\sim}{d}^{(k)}$ by solving
the equation

$$L^{(k)}D^{(k)}L^{(k)T}\underset{\sim}{d}^{(k)} = -\underset{\sim}{g}^{(k)}, \qquad (3.2)$$

which also requires only of order n^2 operations,
because we have triangular matrices.

We could define $L^{(k+1)}$ and $D^{(k+1)}$ by the
equation

$$[H^{(k+1)}]^{-1} = L^{(k+1)}D^{(k+1)}L^{(k+1)T}, \qquad (3.3)$$

but we wish to calculate them from $L^{(k)}$ and $D^{(k)}$.
Therefore we note that, if $H^{(k+1)}$ is the matrix
(2.10), and if $H^{(k)}$ and $H^{(k+1)}$ are both positive
definite, then there exist vectors $\underset{\sim}{u}^{(k)}$ and $\underset{\sim}{v}^{(k)}$
such that the equation

$$L^{(k+1)}D^{(k+1)}L^{(k+1)T} = (I + \underset{\sim}{u}^{(k)}\underset{\sim}{v}^{(k)T})L^{(k)}D^{(k)}L^{(k)T}$$
$$(I + \underset{\sim}{v}^{(k)}\underset{\sim}{u}^{(k)T}) \qquad (3.4)$$

is satisfied (Brodlie, Gourlay and Greenstadt
(1973)). Thus $L^{(k+1)}$ and $D^{(k+1)}$ can be computed
from $L^{(k)}$ and $D^{(k)}$ quite economically (Goldfarb
(1973)).

Two advantages of matrix factorizations are
the following ones. First, rounding errors cannot
make the matrices $L^{(k)}D^{(k)}L^{(k)T}(k = 1,2,3,...)$
lose positive definiteness, provided only that the
diagonal elements of $L^{(k)}$ are non-zero, and the
diagonal elements of $D^{(k)}$ are positive. The other
advantage may occur if some second derivative mat-
rices $G(\underset{\sim}{x})$ are singular or have both positive and
negative eigenvalues. In this case there are lines
in the space of the variables along which the curvature of $F(\underset{\sim}{x})$ is practically zero. If the search
from $\underset{\sim}{x}^{(k)}$ to $\underset{\sim}{x}^{(k+1)}$ is along one of these lines,
then the positive definite matrix (3.3) will have
a very small eigenvalue, in which case the matrix
$H^{(k+1)}$ has some large elements, and therefore includes some rounding errors that in absolute terms
are much larger than usual. Because the sequence

127

of matrices $H^{(i)}$ $(i = 1,2,3,...)$ is obtained by modification, some of the large rounding errors are inherited by $H^{(i)}$ for $i \geqslant k + 2$. However, because $L^{(k)}$ and $D^{(k)}$ are not excessively large, we do not find a similar loss of accuracy in $L^{(i)}$ and $D^{(i)}$ for $i \geqslant k + 2$. Thus the use of matrix factorizations can improve the numerical stability.

Personally I give this last remark little weight, because the variable metric algorithms work well from poor initial estimates of the second derivative matrix. Therefore they have excellent self-correcting properties, which can cope with occasional rounding errors that are larger than usual, provided that positive definiteness is maintained.

Positive definiteness is important, because if $H^{(i)}$ is singular, and if no rounding errors occur, then for $k \geqslant i$ every search direction $d^{(k)}$ is orthogonal to the null space of $H^{(i)}$. Thus the search for the least value of $F(x)$ fails to cover the full space of the variables. In other words recovery from a singular situation is dependent on fortuitous rounding errors.

Singularity or loss of positive definiteness in the matrices $H^{(k)}$ $(k = 1,2,3,...)$ is liable to occur in Davidon's algorithm, because the following property is obtained. It is that the scalar products $(g^{(k)}, H^{(k)} g^{(k)})$ $(k = 1,2,3,...)$ decrease strictly monotonically, if every iteration sets $\lambda^{(k)}$ to the value that minimizes the function (2.4). Therefore, if $H^{(1)}$ is set to the unit matrix, but the scale of the variables is such that it would be more appropriate to let $H^{(1)}$ be 10^6 times the unit matrix, then all the values of $(g^{(k)}, H^{(k)} g^{(k)})$ are abnormally small until $g^{(k)}$ becomes small. A technique that is very useful when the scaling of $H^{(1)}$ is poor is described by Oren (1974). Other good reasons for preferring equation (2.12) to Davidon's formula are given by numerical results, see Gill, Murray and Pitfield (1972) and Sargent and Sebastian (1972), for example.

The remainder of this section is given to the problem of fixing $\lambda^{(k)}$ by making only a few calculations of the function (2.4). One of the most successful methods in practice for variable metric algorithms was proposed by Fletcher (1970). He lets $\lambda^{(k)}$ be the first number in the sequence $(0.1)^j (j = 0,1,2,\ldots)$ that satisfies the condition

$$F(\underset{\sim}{x}^{(k)} + \lambda^{(k)} \underset{\sim}{d}^{(k)}) \leqslant F(\underset{\sim}{x}^{(k)}) + 0.0001\lambda^{(k)}(\underset{\sim}{g}^{(k)}, \underset{\sim}{d}^{(k)}).$$
$$(3.5)$$

The value $\lambda^{(k)} = 1$ is tried first, because eventually we expect it to provide superlinear convergence to the solution (Broyden, Dennis and Moré (1973)). Because we ensure that the scalar product $(\underset{\sim}{g}^{(k)}, \underset{\sim}{d}^{(k)})$ is negative by keeping positive definite second derivative approximations, condition (3.5) is a little stronger than inequality (2.1). This helps to avoid inefficient oscillatory behaviour. Usually inequality (3.5) holds when $\lambda^{(k)}$ is 1 or 0.1, because continuity of the first derivative provides the equation

$$F(\underset{\sim}{x}^{(k)} + \lambda^{(k)} \underset{\sim}{d}^{(k)}) = F(\underset{\sim}{x}^{(k)}) + \lambda^{(k)}(\underset{\sim}{g}^{(k)}, \underset{\sim}{d}^{(k)}) + o(\lambda^{(k)}).$$
$$(3.6)$$

Another property of inequality (3.5) is that very occasionally it can prevent $\lambda^{(k)}$ from being the value of λ that minimizes the function (2.4). I used to think that this was a disadvantage, but now I believe that it is a helpful feature, because it avoids the kind of difficulty shown by Box (1966), where exact line searches can lead to infinity even when the required $\underset{\sim}{x}$ is finite.

I prefer to extend Fletcher's method by adding another condition on $\lambda^{(k)}$ to ensure that $\lambda^{(k)}$ is not too small. A convenient technique (Wolfe 1969)) is to check that the line search increases the directional derivative by a definite amount. This condition is the inequality

$$(g\{\underset{\sim}{x}^{(k)} + \lambda^{(k)} \underset{\sim}{d}^{(k)}\}, \underset{\sim}{d}^{(k)}) \geqslant c_1(\underset{\sim}{\dot{g}}^{(k)}, \underset{\sim}{d}^{(k)}),$$
$$(3.7)$$

where c_1 is a constant from the open interval $(0,1)$. If we write inequality (3.5) in the form

129

$$F(\underset{\sim}{x}^{(k)} + \lambda^{(k)}\underset{\sim}{d}^{(k)}) \leqslant F(\underset{\sim}{x}^{(k)}) + c_2\lambda^{(k)}(\underset{\sim}{g}^{(k)},\underset{\sim}{d}^{(k)}), \quad (3.8)$$

and if the gradient of $F(x)$ is continuous, then conditions (3.7) and (3.8) do not conflict provided that $c_2 < c_1$. Further, if a trial value of $\lambda^{(k)}$ does not satisfy condition (3.8), then this value of $\lambda^{(k)}$ can be set to an upper bound in the search for a suitable value of $\lambda^{(k)}$, and, if $\lambda^{(k)}$ does satisfy condition (3.8) but does not satisfy condition (3.7), then it can be set to a lower bound in the search. Thus inequalities (3.7) and (3.8) provide a nice bracketing method for calculating an acceptable value of $\lambda^{(k)}$, where $0 < c_2 < c_1 < 1$. In order to be able to treat quadratic functions well, and in order that we may let $\lambda^{(k)} = 1$ in the final stages of the calculation, it is necessary that c_2 be less than 0.5. The values $c_1 = 0.5$ and $c_2 = 0.0001$ are usually suitable.

An important observation in Fletcher's (1970) paper is that, if the value $\lambda^{(k)} = 1$ is tried first, good convergence properties are obtained in the whole calculation without using a linear search technique that is exact when $F(x)$ is a quadratic function. When his algorithm is applied to a quadratic function, superlinear convergence is obtained instead of termination. Therefore, when using quadratic functions to help the development of optimization algorithms for general functions, there may be no advantage in seeking termination instead of superlinear convergence. Indeed Fletcher's results suggest that making the line searches exact for quadratic functions reduces efficiency when the algorithm is applied to general functions, where efficiency is measured by the total number of function and gradient evaluations that are required to minimize $F(x)$ to a prescribed accuracy.

Because the remarks of the last paragraph were not accepted generally at the end of the 1960s, the "rank-one formula" for calculating $H^{(k+1)}$ from $H^{(k)}$ (see Davidon (1968) for instance)

caused great excitement. This formula may be
obtained by letting $\beta^{(k)}$ have the value

$$\beta^{(k)} = \frac{(\underset{\sim}{\gamma}^{(k)}, H^{(k)}\underset{\sim}{\gamma}^{(k)})(\underset{\sim}{\delta}^{(k)}, \underset{\sim}{\gamma}^{(k)})}{(\underset{\sim}{\gamma}^{(k)}, H^{(k)}\underset{\sim}{\gamma}^{(k)}) - (\underset{\sim}{\delta}^{(k)}, \underset{\sim}{\gamma}^{(k)})} \qquad (3.9)$$

in equation (2.10). The exciting property is
that, if for all values of k the denominator of
expression (3.9) is non-zero, if $F(\underset{\sim}{x})$ is a quadratic function, and if the step-lengths $\lambda^{(k)}$ are set
to any non-zero values whatsoever, then either
$\underset{\sim}{g}^{(k)}$ is zero for some $k \leqslant n$ or $H^{(n+1)}$ is equal to
the inverse of the true second derivative matrix.
Further, the truth of the theorem does not depend
on the directions $\underset{\sim}{d}^{(k)}$ being defined by equation
(2.2) (see Wolfe (1967) for instance). However,
now that we know that we can obtain good variable
metric algorithms that do not possess quadratic
termination, the properties of the rank-one formula seem to make a smaller contribution to practical algorithms, although their elegance is undimi-
nished. The rank-one formula is not used very
much now, because it does not maintain positive
definiteness in the sequence of matrices $H^{(k)}$
($k = 1,2,3,\ldots$), and because special precautions
have to be included in case the denominator of
expression (3.9) is zero or is unacceptably small.
Suitable precautions are described by Murtagh and
Sargent (1970).

Dixon (1973) describes a technique for
obtaining quadratic termination without line searches in variable metric algorithms that maintain
positive definite second derivative approximations.
From the changes in gradient that occur when steps
are taken along the directions $\underset{\sim}{d}^{(k)}$ ($k = 1,2,3,\ldots$),
he predicts what the values of $\underset{\sim}{x}^{(k)}$ and $\underset{\sim}{g}^{(k)}$ would
be if exact line searches were made on every iter-
ation and $F(\underset{\sim}{x})$ were quadratic. These predicted
values are used in the calculation of the matrices
$H^{(k)}$ and the directions $\underset{\sim}{d}^{(k)}$ ($k = 1,2,3,\ldots$). Thus,
if $F(\underset{\sim}{x})$ is a quadratic function, $H^{(k)}$ and $\underset{\sim}{d}^{(k)}$ are
not affected by the actual choice of $\lambda^{(k)}$
($k = 1,2,3,\ldots$). After n iterations a step is

taken in the space of the variables to the point
which, according to the predictions, would have
been reached if exact line searches were made,
which completes the calculation in the quadratic
case. Otherwise, when $F(x)$ is a general function,
the process may be restarted from the point that
has given the least calculated value of $F(x)$.
This technique does not give substantial improvements to algorithms for unconstrained optimization.
However, if linear constraints are present, and if
they do not allow the value $\lambda = 1$ to be tried when
we use the function (2.4) to calculate $\lambda^{(k)}$, then
it may be advantageous to allow for the fact that
the line search is incomplete.

4. WOLFE'S ALGORITHM FOR MINIMIZING FUNCTIONS WITH FIRST DERIVATIVE DISCONTINUITIES

We noted in Section 1 that some techniques
for constrained optimization calculations require
the unconstrained minimization of a function with
first derivative discontinuities. Therefore we
would like to have an algorithm that not only
minimizes quadratic functions well, but also is
suitable for finding the least value of a piecewise linear function. A single algorithm that has
both these properties was developed recently by
Wolfe (1974a). It is described in this section.

It is a gradient algorithm, and, as in Section 2, we have to calculate the gradient of the
objective function $F(x)$ at a sequence of points
$x^{(k)}$ ($k = 1,2,3,...$). Therefore now we explain how
$g^{(k)}$ may be obtained when $x^{(k)}$ is a point where
the gradient is discontinuous. In order to avoid
difficulties due to cusps, we suppose that the
objective function has the form

$$F(x) = \max_{1 \leq i \leq m} f_i(x), \qquad (4.1)$$

where the functions $f_i(x)$ have continuous first
derivatives. Then the gradient is discontinuous
at $x^{(k)}$ only if there are at least two different
values of i that satisfy the equation

$$F(\underset{\sim}{x}^{(k)}) = f_i(\underset{\sim}{x}^{(k)}). \tag{4.2}$$

In this case we may set $\underset{\sim}{g}^{(k)}$ to any vector of the form

$$\underset{\sim}{g}^{(k)} = \underset{\sim}{\nabla} f_i(\underset{\sim}{x}^{(k)}), \qquad 1 \leqslant i \leqslant m, \tag{4.3}$$

where the integer i satisfies one or two conditions. Firstly we require equation (4.2) to hold. The second condition on i applies when $\underset{\sim}{x}^{(k)}$ is obtained by a line search from $\underset{\sim}{x}^{(k-1)}$ along $\underset{\sim}{d}^{(k-1)}$, the choice of λ being based on inequalities (3.7) and (3.8). In this case the definition (4.3) must also satisfy the inequality

$$(\underset{\sim}{g}^{(k)}, \underset{\sim}{d}^{(k-1)}) \geqslant c_1(\underset{\sim}{g}^{(k-1)}, \underset{\sim}{d}^{(k-1)}). \tag{4.4}$$

These conditions can always be obtained, if $F(\underset{\sim}{x})$ is bounded below.

In Wolfe's algorithm the search direction $\underset{\sim}{d}^{(k)}$ depends on the gradients $\{\underset{\sim}{g}^{(k_0)}, \underset{\sim}{g}^{(k_0+1)}, \ldots, \underset{\sim}{g}^{(k)}\}$, where k_0 is an integer that is set to one initially, and that is controlled by rules that are given later. In fact $\underset{\sim}{d}^{(k)}$ is defined by the equation

$$\underset{\sim}{d}^{(k)} = -Nr\{\underset{\sim}{g}^{(k_0)}, \underset{\sim}{g}^{(k_0+1)}, \ldots, \underset{\sim}{g}^{(k)}\}, \tag{4.5}$$

where $Nr\{\underset{\sim}{g}^{(k_0)}, \underset{\sim}{g}^{(k_0+1)}, \ldots, \underset{\sim}{g}^{(k)}\}$ is the "shortest" vector in the convex hull of the point set $\{\underset{\sim}{g}^{(i)}; i = k_0, k_0+1, \ldots, k\}$, the length of vectors being the Euclidean length. Thus the calculation of $\underset{\sim}{d}^{(k)}$ is a quadratic programming problem, whose second derivative matrix is simply the unit matrix.

Except for the final iteration, every one either uses $\underset{\sim}{d}^{(k)}$ as a search direction or sets $k_0 = k+1$. The search direction is used if and only if the inequality

$$\| \underset{\sim}{d}^{(k)} \| \geqslant \varepsilon_1 \tag{4.6}$$

is satisfied, where ε_1 is a small positive constant that is set by the user of the algorithm. In this case $x^{(k+1)}$ is the vector

$$x^{(k+1)} = x^{(k)} + \lambda^{(k)} d^{(k)}, \qquad (4.7)$$

where $\lambda^{(k)}$ is obtained from a line search calculation based on the function (2.4). However, if inequality (4.6) fails, then the vector $d^{(k)}$ is judged to be too small for a line search calculation. Then we let $x^{(k+1)} = x^{(k)}$ and $k_0 = k+1$, in order that the search direction of the next iteration is down the gradient of $F(x)$. The method of this paragraph is applied for $k = 1,2,3,\ldots$, until inequality (4.6) fails and the condition

$$\sum_{j=k_0}^{k-1} \| x^{(j+1)} - x^{(j)} \| \leq \varepsilon_2 \qquad (4.8)$$

is satisfied, where ε_2 is another small positive constant that is set by the user.

The algorithm is suitable for minimizing quadratic functions because, if each line search is exact, then the points $x^{(k)}$ and the search directions $d^{(k)}$ ($k = 1,2,3,\ldots$) are the same as those obtained by the conjugate gradient method (Fletcher and Reeves (1964)), until a value of k is reached that does not satisfy condition (4.6). This statement is proved by Wolfe (1974b). One method of proof is to show that, for the early values of k, the direction $d^{(k)}$ is the vector

$$d^{(k)} = \frac{-\sum_{j=1}^{k} g^{(j)}/\| g^{(j)} \|^2}{\sum_{j=1}^{k} 1/\| g^{(j)} \|^2} \qquad (4.9)$$

Thus $-d^{(k)}$ is in the interior of the convex hull of the set $\{g^{(1)}, g^{(2)}, \ldots, g^{(k)}\}$. It follows from the definition (4.5) that $d^{(k)}$ is orthogonal to the changes in gradient $\{g^{(i+1)}-g^{(i)}; i=1,2,\ldots,k-1\}$ so the conjugacy condition (2.7) is satisfied. Moreover $d^{(k)}$ is in the linear space spanned by the gradients $g^{(j)}$ ($j = 1,2,\ldots,k$). Thus Wolfe's algorithm is similar to the conjugate gradient method.

The behaviour of the algorithm when $F(x)$ is a piecewise linear function is also good, and to consider it we suppose that expression (4.1) is made up of the linear functions

$$f_i(\underset{\sim}{x}) = (\underset{\sim}{a_i}, \underset{\sim}{x}) + b_i, \qquad i = 1, 2, \ldots, m. \quad (4.10)$$

Then the right hand side of equation (4.3) is the vector a_i. Thus the set of vectors $\{g^{(k_0)}, g^{(k_0+1)}, \ldots, g^{(k)}\}$ is a subset of the vectors $\{a_i; i = 1, 2, \ldots, m\}$. We use the notation I_k for the set of integers i such that the sets $\{g^{(k_0)}, g^{(k_0+1)}, \ldots, g^{(k)}\}$ and $\{a_i; i \epsilon I_k\}$ are the same. Then, if condition (4.6) is satisfied, the search direction (4.5) is a descent direction at $\underset{\sim}{x}^{(k)}$ for the function

$$\Phi_k(\underset{\sim}{x}) = \max_{i \epsilon I_k} f_i(\underset{\sim}{x}), \qquad (4.11)$$

and $\Phi_k(x^{(k)} + \lambda d^{(k)})$ decreases strictly monotonically for $\lambda > 0$. We note that the definitions (4.1) and (4.11) provide the inequality

$$\Phi_k(\underset{\sim}{x}^{(k)} + \lambda d^{(k)}) \leqslant F(\underset{\sim}{x}^{(k)} + \lambda d^{(k)}), \quad (4.12)$$

and, because a value of i satisfies equations (4.2) and (4.3), the two sides of expression (4.12) are equal when $\lambda = 0$.

In fact, when $F(\underset{\sim}{x})$ is a piecewise linear function, it is often the case that $f_i(\underset{\sim}{x}^{(k)})$ is equal to $F(\underset{\sim}{x}^{(k)})$ for all values of i in I_k. If this happens on every iteration, then $F(\underset{\sim}{x})$ is minimized in at most m iterations, provided that the value of ϵ_1 in inequality (4.6) is so small that every non-zero vector $\underset{\sim}{d}^{(k)}$ is used as a search direction. This requirement can be satisfied, because the number of different values of $\|\underset{\sim}{d}^{(k)}\|$ that can occur is at most the number of different sets I_k that can be achieved, which is finite.

The following remarks prove that the least

value of $F(x)$ is found when $f_i(x^{(k)}) = F(x^{(k)})$ for all i and k. If $d^{(k)}$ is non-zero, then the left hand side of inequality (4.12) is linear and decreasing for $\lambda > 0$. Therefore, either we terminate because $F(x)$ is not bounded below, or inequality (3.7) holds, in which case $q^{(k+1)}$ is not in the set $\{a_i, i \varepsilon I_k\}$. Thus most iterations obtain I_{k+1} by adding one integer to I_k. Because I_k contains at most m elements, it follows that, either we find that $F(x)$ is not bounded below, or $d^{(k)}$ is zero for some $k \leqslant m$. If $d^{(k)}$ is zero, then there is no downhill search direction for the function $\Phi_k(x)$ at $x^{(k)}$. Since this function is convex, $\Phi_k(x^{(k)})$ is the least value of $\Phi_k(x)$. Now $\Phi_k(x) \leqslant F(x)$ for all x, and $\Phi_k(x^{(k)}) = F(x^{(k)})$. Therefore $x^{(k)}$ is the vector of variables that minimizes $F(x)$.

A promising feature of Wolfe's algorithm is that the very strong conditions demanded by the proof of the last paragraph are often obtained naturally. In particular, if $\lambda^{(k)}$ is positive and if $f_i(x^{(k)}) = F(x^{(k)})$ for all $i \varepsilon I_k$, then $f_i(x^{(k+1)}) = F(x^{(k+1)})$ for all $i \varepsilon I_k$ only if the scalar products $\{(d^{(k)}, a_i); i \varepsilon I_k\}$ are independent of i. These scalar products are independent of i when $-d^{(k)}$ is an interior point of the convex hull of the set $\{a_i; i \varepsilon I_k\}$, because the remarks following equation (4.9) show that $d^{(k)}$ is orthogonal to the differences $(a_i - a_j)$ for all i and j in I_k. In fact the equation

$$(d^{(k)}, a_i) = -\| d^{(k)} \|^2, \qquad i \varepsilon I_k, \qquad (4.13)$$

is satisfied.

Even when $-d^{(k)}$ is inside the convex hull of the point set $\{a_i; i \varepsilon I_k\}$, and when $f_i(x^{(k)}) = F(x^{(k)})$ for all $i \varepsilon I_k$, it may happen that $f_i(x^{(k+1)})$ is not equal to $F(x^{(k+1)})$, unless we choose $\lambda^{(k)}$ in equation (4.7) carefully. The definition (4.11) shows that we would like $\lambda^{(k)}$ to satisfy the equation

$$\Phi_k(x^{(k)} + \lambda^{(k)} d^{(k)}) = F(x^{(k)} + \lambda^{(k)} d^{(k)}), (4.14)$$

which imposes an upper bound on $\lambda^{(k)}$ in the usual
case when the left hand side of inequality (4.12)
decreases monotonically, and the right hand side
is bounded below. It is appropriate to let $\lambda^{(k)}$
equal this upper bound, for the function (2.4) has
a first derivative discontinuity at this point, so
inequality (3.7) can be satisfied if c_1 is sufficiently small. Specifically, if the function $f_j(x)$
in expression (4.1) causes $F(x^{(k)} + \lambda d^{(k)})$ to be
greater than $\Phi_k(x^{(k)} + \lambda d^{(k)})$ when λ is larger
than the upper bound, then j is not in I_k, and the
directional derivative $(d^{(k)}, a_j)$ is greater than
the right hand side of equation (4.13). If the
condition

$$(d^{(k)}, a_j) \geqslant -c_1 \| d^{(k)} \|^2 \qquad (4.15)$$

holds, then we let $\lambda^{(k)}$ equal the upper bound.
Thus the set I_{k+1} is obtained by adding the integer
j to I_k, and $g^{(k+1)}$ is the vector a_j.

If $\lambda^{(k)}$ is defined in this way on every
iteration, and if equation (4.13) holds for all
values of k, then we have a very fast algorithm
for minimizing a piecewise linear function. However, because the general linear programming problem can be expressed in this form, usually it is
necessary to reset k_o during the calculation. We
have shown that Wolfe's algorithm has features
that are suitable for minimizing piecewise linear
functions, in addition to the quadratic termination property.

It is an open question whether the method
obtained from equation (4.14) for choosing $\lambda^{(k)}$ is
the best one to use. We have suggested it,
because if instead $\lambda^{(k)}$ is calculated to minimize
the function (2.4), and if $F(x)$ is a piecewise
linear function of two variables whose contour
lines are concentric regular pentagons, then the
rate of convergence of the algorithm is only
linear.

5. ALGORITHMS THAT ARE SUITABLE WHEN n IS LARGE

Except for the conjugate gradient algorithm (Fletcher and Reeves (1964)), and for techniques that come from numerical linear algebra and linear programming to exploit sparsity in matrices, little work has been done on the development of optimization algorithms that are especially suitable when the number of variables is large. At this conference Martin Beale observed that "large problems are sparse", and usually we rely on this fact to obtain solutions. However the conjugate gradient algorithm is different.

This algorithm is well suited to large problems, because, as we noted in Section 2, it requires storage for only a few vectors, and it does not involve any matrices. Moreover, applying the formula (2.9) to obtain the search direction $d^{(k)}$ requires only of order n operations. However it usually requires more function evaluations than the variable metric algorithms mentioned in Section 2, and it does not take advantage of any sparsity in second derivative matrices. It is an excellent algorithm for the following types of minimization calculations.

The most important applications are covered by the obvious statement "The conjugate gradient algorithm should be used when it works well". This form of words is given to emphasize the fact that the most successful applications of the method have been found by experiment rather than by analysis. Indeed it is much easier to try the algorithm in practice, than to work out theoretically whether the objective function and the starting point $x^{(1)}$ are suitable. Some excellent successes have been obtained. For example only about thirty iterations are needed to find acceptably small values of functions of hundreds of variables that occur in the control of distillation columns. Equally good results have been obtained in minimization calculations that arise from numerical methods for solving nonlinear partial differential equations. This good behaviour occurs when $F(x)$ is approximately quadratic, and when the initial gradient $g^{(1)}$ can be expressed as a linear combination of a few eigenvectors of the second derivative matrix (see Stewart (1973) for instance). Equally good behaviour is obtained by the variable metric algorithms that employ formula (2.10), and they should be preferred if the amount of extra work per iteration is small in comparison with the time taken to calculate $F(x^{(k)})$ and $g^{(k)}$, because the variable metric algorithms have better convergence properties, and they do not require a restarting procedure.

Most versions of the conjugate gradient algorithm use the restarting procedure that sets $d^{(k)}$ to $-g^{(k)}$, instead of applying formula (2.9), about every n iterations. This device provides a superlinear rate of convergence, but it is a rather weak superlinear rate when n is large. Moreover, when n is large, one usually cannot afford to run the algorithm for n iterations, unless the calculation of $F(x^{(k)})$ and $g^{(k)}$ requires not more than about 0.1 second. In this case the second derivative matrix is probably sparse, so some of the other methods considered in this section may be better than the conjugate gradient algorithm.

It is easy to take advantage of sparsity when the user provides expressions for second derivatives. Then we only have to calculate and store the elements of $G(x)$ that may be non-zero, and we can obtain the sequence $x^{(k)}$ ($k = 1,2,3,\ldots$) by applying formula (2.14). If $\mu^{(k)}$ is chosen so that the matrix $[G^{(k)} + \mu^{(k)}I]$ is positive definite, then there are excellent methods for calculating the vector $[G^{(k)} + \mu^{(k)}I]^{-1}g^{(k)}$, that make use of the sparsity (for example see Reid (1972)). Thus the sparsity is very helpful to the numerical linear algebra, but it makes no other contribution to the current algorithms for minimization when second derivatives are available.

139

The situation is quite different when only first derivatives are calculated, because a good method for taking advantage of sparsity in variable metric algorithms has not yet been found. Some remarks on this question are made later. The best available technique is to estimate second derivatives by differencing first derivatives, forcing zeros that are known to occur. Then we apply the method outlined in the last paragraph. Three comments on this procedure may be helpful. The idea of Curtis, Powell and Reid (1974) may be applied to estimate all the elements of $G(x^{(k)})$ from only about $(r+1)$ gradient evaluations, where r is the greatest number of non-zero elements in a row of $G(x^{(k)})$. Small errors in the elements of $G(x^{(k)})$ are usually unimportant, and certainly they are more tolerable than similar errors in $g^{(k)}$, because we expect $G(x^{(k)})$ to tend to a positive definite matrix, while $g^{(k)}$ should tend to the zero vector. It may be more efficient to use one matrix $G(x^{(k)})$ for a number of iterations. If $[G^{(k)} + \mu^{(k)}I]$ is positive definite, then this technique retains a downhill search direction in equation (2.14). Also, if $\mu^{(k)}$ is unchanged, then the matrix factorization that is used to calculate the search direction can be carried forward to the next iteration. This device may also be useful when second derivatives are calculated explicitly.

In view of these helpful features when second derivatives are approximated by differences of first derivatives in large sparse problems, do we want variable metric algorithms also. I think we do for the following three reasons. A superlinear rate of convergence is obtained by variable metric algorithms that, towards the end of the calculation, require only one function and gradient evaluation per iteration (Dennis and Moré (1973)), but, if derivatives are estimated by differences, then each iteration requires more function evaluations. We are not very skilled at choosing automatically step-lengths for obtaining derivative estimates from differences. The third reason depends on the observation that for general calculations it is preferable to use a variable metric algorithm when there is no sparsity. If this is so, and if variable metric algorithms are about as efficient as their rivals on sparse problems, then using a variable metric algorithm all the time avoids the need to decide when to switch algorithms because of sparseness in the second derivative matrix. Here I have in mind a single variable metric algorithm that treats the second derivative matrix as full, unless it is given that certain elements of the second derivative matrix are zero or constant. A method of this type for solving nonlinear equations is described by Schubert (1970).

One method of taking advantage of sparsity in variable metric algorithms has been tried at the National Physical Laboratory (W. Murray, private communication). It is to calculate $L^{(k+1)}$ and $D^{(k+1)}$ from a formula that is analogous to equation (3.4), and then to set to zero any elements of $L^{(k+1)}$ that should be zero because of the known sparsity. The numerical results obtained by this procedure are not very good. Instead I think that new research in this area should try to assess only one or other of the following two advantages of sparsity, and then any benefits from the two sides can be merged later. One advantage is that known sparsity gives information about the second derivative matrix that may reduce the number of iterations and that may save function and gradient evaluations. The other advantage is that sparsity can help the solution of the linear equations that occur on each iteration. It would be interesting to discover the strength of the first advantage, even when the associated linear calculations are done inefficiently. We will probably find that it is well worthwhile to take advantage of sparsity on smaller problems, for example $n=10$, where the routine work per iteration is often negligible.

The following technique, which as far as I know has not been tried in practice, can be used to calculate a sequence of second derivative

141

approximations $B^{(k)}$ (k = 1,2,3,...), that preserves any known constant elements of the required second derivative matrix. As in Section 2, the unknown elements of $B^{(k+1)}$ depend on the matrix $B^{(k)}$, and on the vectors $\delta^{(k)}$ and $\gamma^{(k)}$ of equation (2.10). Specifically, to give $B^{(k+1)}$ a property that is present in the true second derivative matrix when $F(x)$ is quadratic, we require the equation

$$B^{(k+1)}\underset{\sim}{\delta}^{(k)} = \underset{\sim}{\gamma}^{(k)} \qquad (5.1)$$

to be satisfied. Also we require $B^{(k+1)}$ to be symmetric, and to have the known constant elements. Any freedom that remains is fixed by minimizing the expression

$$\| B^{(k+1)} - B^{(k)} \|_E^2 = \sum_{i=1}^{n} \sum_{j=1}^{n} (B_{ij}^{(k+1)} - B_{ij}^{(k)})^2. \qquad (5.2)$$

This is a quadratic programming problem, that has a large amount of structure that helps the calculation of $B^{(k+1)}$ (k = 1,2,3,...).

This technique is suggested, because it gives good results in the "full" case, and we expect it to be even better in the "sparse" case. Here the term "full" means that no elements of the second derivative matrix are known, and the term "sparse" means that some elements are known constants. Numerical results for the full case are reported by Powell (1970), but his formula for $B^{(k+1)}$ is not derived by the variational method of the last paragraph, although it is equivalent to it. The equivalence follows from Greenstadt's (1970) work.

We expect better results in the sparse case for the following reason. Suppose $F(x)$ is a quadratic function whose second derivative matrix is G. Then, if $B^{(k+1)}$ minimizes expression (5.2) subject to linear equality constraints that are satisfied by the matrix G, the equation

$$\| B^{(k+1)} - G \|_E^2 = \| B^{(k)} - G \|_E^2 - \| B^{(k+1)} - B^{(k)} \|_E^2 \qquad (5.3)$$

is satisfied. Now the linear constraints in the
sparse case include all the linear constraints of
the full case, and there are some extra ones also.
The extra constraints usually increase the least
feasible value of expression (5.2). Therefore it
follows from equation (5.3) that the sparseness
usually improves the accuracy of the approximation
$B^{(k+1)}$. The validity of this argument ought to be
tested by numerical computation.

A disadvantage of this approach, which is
not present in the variable metric algorithms of
Section 2, is that the matrix norm (5.2) introduces
the need to scale the variables carefully, before
the main calculation is begun. Further, we may
find that some of the matrices $B^{(k)}$ $(k = 1,2,3,...)$
are not positive definite. Therefore we would
like to find a variational method for calculating
$B^{(k+1)}$ that provides the nice properties of variable metric algorithms, and that allows elements
of $B^{(k+1)}$ to be specified. Goldfarb (1970) shows
how some variable metric formulae may be derived
by variational methods in the full case. However,
hardly any research has been done on methods that
use sparsity to save function and gradient evaluations.

6. THE CONTRIBUTION FROM THEORY TO THE DEVELOPMENT OF ALGORITHMS

In this section we consider the effect that
theoretical work has had on the design of practical
algorithms for unconstrained optimization, and we
note that the theory is not always helpful. Some
experiences are described, that relate to the
balance between theoretical analysis and the development of automatic methods by experimentation
with numerical test problems. Because more and
more effort is going into the theoretical side of
most branches of numerical analysis, and because
the number of numerical analysts that take an
interest in optimization is increasing, it is
hoped that these remarks will be of some value to
the direction of future research.

I have been particularly concerned with convergence theorems for variable metric algorithms of the type described in Section 2. Therefore most of my comments relate to the development of the theory of these algorithms. The most important one is that it is usual for knowledge obtained from test problems to be well in advance of knowledge obtained from theoretical analysis. Therefore sometimes algorithms should be published before they are understood properly, provided that a careful attempt is made to relate them to existing algorithms, and that numerical results show merit.

The prime example of the truth of this remark is the original variable metric algorithm (Davidon (1959)). When it was published it could be proved only that the algorithm works well when $F(x)$ is a quadratic function, but it is designed to minimize general functions. Therefore a theoretical fanatic might have tried to suppress its publication, despite the excellence of the numerical results. This remark is not entirely hypothetical, because in the field of rational approximation an algorithm, whose numerical results were excellent, was superseded by an inferior method, because the known theory of the inferior method was more comprehensive. Later the theoretical analysis of the better method was completed, and it proves that the inferior method really is inferior (Barrodale, Powell and Roberts (1972)). Experiences of this type, and fears that they may happen, have made many developers of algorithms feel that they should justify their methods by convergence theorems. To correct this view, I wish to point out that designing an algorithm so that convergence can be proved may detract from the usefulness of the algorithm.

A good example is provided by the extensions that have been made to the variable metric algorithm, to bring proofs of convergence within our present range of competence. Without these extensions the theoretical analysis is very difficult, because of the possibility that the search directions $d^{(k)}$ ($k = 1,2,3,...$) may tend to be orthogonal to the gradients $g^{(k)}$ ($k = 1,2,3,...$). Therefore one extension is to make the "angle test"

$$|(g^{(k)}, d^{(k)})| \geq \varepsilon \|g^{(k)}\| \|d^{(k)}\|, \qquad (6.1)$$

where ε is a small positive constant, whose value has to be chosen. If the test is satisfied by the search direction $d^{(k)}$ that is obtained naturally, then the search direction is used. Otherwise $d^{(k)}$ is modified to satisfy the angle test, for instance it would be sufficient to let $d^{(k)} = -g^{(k)}$. A different extension that is used sometimes is to define $H^{(k+1)}$ by equation (2.10), unless the condition number of $H^{(k+1)}$ would exceed another constant, c_3 say, in which case $H^{(k+1)}$ is modified so that the condition numbers of all the matrices $H^{(k)}$ ($k = 1,2,3,...$) are bounded above by c_3. The search direction $d^{(k)}$ is defined by equation (2.2) for all values of k. This extension is very similar to the one that uses the angle test, because it may be proved that inequality (6.1) is satisfied, where ε has the value $2\sqrt{c_3}/(1+c_3)$.

Both these extensions destroy a very valuable feature of the variable metric algorithm. It is that, except for the initial choice of the approximation to the second derivative matrix, the algorithm is invariant under linear transformations of the variables. However, the vector norms on the right hand side of inequality (6.1) are suitable for computation, only if the magnitudes of the different components of x are scaled so that they are all of about the same size. Therefore the extensions may prevent the user from letting the components of x be in units that are natural to his problem. Further, if in the search for the least value of x, some variables change by several orders of magnitude, then the prescaling demanded by the extensions may not be possible. In my opinion these disadvantages should be tolerated only if they are compensated by real gains. If we stop using the variable metric method now, then we may be making a mistake of the same kind as the one

145

mentioned earlier from the field of rational
approximation.

A similar cautionary experience occurred in
about 1968. Then it was known that convergence
could be proved if the matrix $H^{(k)}$ was reset to
the unit matrix every n iterations, and also numerical examples showed that resetting often saved
iterations and function values (McCormick and
Pearson (1969)). Therefore the reasons in favour
of resetting were quite strong, but some of us
were stubborn, because we enjoyed the elegance of
the variable metric algorithm. Soon afterwards we
began to use the BFS formula (2.12), instead of
the DFP formula, for calculating $H^{(k+1)}$. The
gains that were obtained by this change were greater than the gains obtained by resetting, so the
stubborn ones were right. However, if resetting
had become widespread in 1969, then we might still
be subject to its disadvantages.

The switch from the DFP formula to the BFS
formula was not immediate, because the DFP formula
was a fairly trustworthy friend. I thought that
the advantages of the BFS formula reported by
Broyden (1970) and Fletcher (1970) might be due to
their skills at Fortran programming. I will never
know if eventually I would have been convinced by
their numerical results, because I was converted
by Dixon's (1972) important theorem. It states
that, if a variable metric algorithm is applied,
and if all line searches are exact, then the calculated sequence of points $x^{(k)}$ $(k = 1,2,3,...)$ is
independent of the parameters $\beta^{(k)}$ $(k = 1,2,3,...)$
in equation (2.10), provided only that $F(x)$ has
continuous first derivatives, and that any ambigu-
ities in line searches are resolved in a consistent
way.

Dixon's result directed attention to inexact
line searches. Here, again, practical experience
is far ahead of theoretical analysis, although the
theory has made some valuable contributions. For
example in Section 3 we noted that a good final

rate of convergence can be obtained by setting $\lambda^{(k)} = 1$, and that good behaviour on quadratic functions does not require exact line searches. Another example is the following theorem, which I proved very recently (Powell (1975)).

Theorem If $F(x)$ is convex, if it has a continuous gradient, and if the level set $\{x; F(x) \leqslant F(x^{(1)})\}$ is bounded, then the variable metric algorithm that uses the BFS formula to calculate $B^{(k+1)}$ $(k = 1,2,3,...)$, and that calculates each $\lambda^{(k)}$ to satisfy conditions (3.7) and (3.8), makes $F(x^{(k)})$ $(k = 1,2,3,...)$ tend to the least value of $F(x)$.

I believe that this theorem is the least re-strictive known theorem for an unmodified variable metric algorithm. However it is far from explain-ing the good results of numerical calculations, because of the condition that $F(x)$ be a convex function. At least the theorem is useful in two ways. One is that, when minimizing a general function, it happens usually that the sequence $x^{(k)}$ $(k = 1,2,3,...)$ reaches a region where $F(x)$ is locally convex, and then, if condition (3.8) keeps the sequence within the region of local convexity, the theorem shows that convergence to a local mini-mum is obtained. The other way is that, if one believes that sometimes the variable metric algor-ithm does not work well, and one wishes to show failure by means of a numerical example, then the theorem shows that the search for a suitable example must go beyond convex functions $F(x)$.

Many theoretical papers have been published on variable metric algorithms, that make assump-tions that are much stronger than convexity of the objective function. Some assume that $F(x)$ is exactly quadratic, for example most of the known quadratic termination properties are given by Powell (1972). Others assume that $x^{(1)}$ is very close to the vector of variables that minimizes $F(x)$, in order to derive local convergence proper-ties. Excellent surveys of this local convergence analysis are given by Dennis and Moré (1974) and

147

by Stoer (1974), including some results that do not demand that $B^{(1)}$ is close to $G(\underset{\sim}{x}^{(1)})$.

In addition to the several contributions that theory has made to the development of numerical methods, there is another benefit, which is probably the most important of all. It is that theoretical work has identified some key ideas, which give the subject of unconstrained optimization a structure, while ten years ago we had a collection of disjoint algorithms. Therefore now the subject is taught at many universities, and it makes a pleasant lecture course. This may attract more people into optimization, which would be good, because there is still plenty of research to be done on general techniques, especially on methods without derivatives, on methods that take advantage of sparsity, on global convergence, and on algorithms for constrained optimization. Further, many of the papers at this conference show that skills in optimization are a strong help in the solution of some important practical problems.

In this section we have seen that the development of the subject of unconstrained optimization has depended strongly on practical computation. It is hoped that the given experiences and comments will help to maintain this dependence.

ACKNOWLEDGEMENT

I thank Dr. D. Goldfarb for studying a draft of this paper and for the improvements he made to the original manuscript.

REFERENCES

Bard, Y. (1968) "On a numerical instability of Davidon-like methods", *Maths of Comp.*, **22**, 665-666.

Barrodale, I., Powell, M.J.D. and Roberts, F.D.K. (1972) "The differential correction algorithm for rational ℓ_∞-approximation", *SIAM J. Numer. Anal.*, **9**, 493-504.

Bertsekas, D.P. (1974) "On penalty and multiplier methods for constrained minimization", presented at the SIGMAP Nonlinear Programming Symposium, University of Wisconsin.

Box, M.J. (1966) "A comparison of several current optimization methods, and the use of transformations in constrained problems", *Computer J.*, 9, 67-77.

Brent, R.P. (1973) "Algorithms for minimization without derivatives", Prentice-Hall Inc., Englewood Cliffs, N.J.

Brodlie, K.W., Gourlay, A.R. and Greenstadt, J. (1973) "Rank-one and rank-two corrections to positive definite matrices expressed in product form", *JIMA*, 11, 73-82.

Broyden, C.G. (1967) "Quasi-Newton methods and their application to function minimization", *Maths of Comp.*, 21, 368-381.

Broyden, C.G. (1970) "The convergence of a class of double-rank minimization algorithms", *JIMA*, 6, 222-231.

Broyden, C.G., Dennis, J.E. and Moré, J.J. (1973) "On the local and superlinear convergence of quasi-Newton methods", *JIMA*, 12, 223-245.

Curtis, A.R., Powell, M.J.D. and Reid, J.K. (1974) "On the estimation of sparse Jacobian matrices", *JIMA*, 13, 117-119.

Davidon, W.C. (1959) "Variable metric method for minimization", AEC Research and Development Report ANL-5990 (Rev.).

Davidon, W.C. (1968) "Variance algorithm for minimization", *Computer J.*, 10, 406-410.

Dennis, J.E. and Moré, J.J. (1973) "A characterization of super-linear convergence and its application to quasi-Newton methods", Report TR 73-157, Cornell University.

Dennis, J.E. and Moré, J.J. (1974) "Quasi-Newton methods, motivation and theory", Report TR 74-217, Cornell University.

Dixon, L.C.W. (1972) "Quasi-Newton algorithms generate identical points", *Math. Prog.*, 2, 383-387.

Dixon, L.C.W. (1973) "Conjugate directions without linear searches", *JIMA*, 11, 317-328.

Fletcher, R. (1970) "A new approach to variable metric algorithms", *Computer J.*, 13, 317-322.

Fletcher, R. (1973) "An exact penalty function for nonlinear programming with inequalities", *Math. Prog.*, 5, 129-150.

Fletcher, R. and Powell, M.J.D. (1963) "A rapidly convergent descent method for minimization", *Computer J.*, 6, 163-168.

Fletcher, R. and Reeves, C.M. (1974) "Function minimization by conjugate gradients", *Computer J.*, 7, 149-154.

Gill, P.E. and Murray, W. (1972) "Quasi-Newton methods for unconstrained optimization", *JIMA*, 9, 91-108.

Gill, P.E., Murray, W. and Pitfield, R.A. (1972) "The implementation of two revised quasi-Newton algorithms for unconstrained optimization", Report NAC 11, National Physical Laboratory.

Goldfarb, D. (1970) "A family of variable-metric methods derived by variational means", *Maths of Comp.*, 24, 23-26.

Goldfarb, D. (1973) "Factorized variable metric methods for unconstrained optimization", IBM Research Report RC 4415, IBM Research Center, Yorktown Heights.

Greenstadt, J. (1970) "Variations on variable-metric methods", *Maths of Comp.*, 24, 1-22.

Hebden, M.D. (1973) "An algorithm for minimization using exact second derivatives", Report T.P. 515, AERE, Harwell.

Huang, H.Y. (1970) "Unified approach to quadratically convergent algorithms for function minimization", *JOTA*, 5, 405-423.

Kowalik, J. and Osborne, M.R. (1968) "Methods for unconstrained optimization problems", Elsevier Publishing Co. Inc., New York.

McCormick, G.P. and Pearson, J.D. (1969) "Variable metric methods and unconstrained optimization", *in* "Optimization", *ed.* R. Fletcher, Academic Press, London.

Murtagh, B.A. and Sargent, R.W.H. (1970) "Computational experience with quadratically convergent minimization methods", *Computer J.*, **13**, 185-194.

Oren, S.S. (1974) "Self-scaling variable metric (SSVM) algorithms II: implementation and experiments", *Man. Sci.*, **20**, 863-874.

Pietrzykowski, T. (1969) "An exact potential method for constrained maxima", *SIAM J. Numer. Anal.*, **6**, 299-304.

Powell, M.J.D. (1964) "An efficient method for finding the minimum of a function of several variables without calculating derivatives", *Computer J.*, **7**, 155-162.

Powell, M.J.D. (1970) "A new algorithm for unconstrained optimization" *in* "Nonlinear programming", *eds.* J.B. Rosen, O.L. Mangasarian and K. Ritter, Academic Press, New York.

Powell, M.J.D. (1972) "Quadratic termination properties of minimization algorithms I. Statement and discussion of results", *JIMA*, **10**, 333-342.

Powell, M.J.D. (1974) "A view of minimization algorithms that do not require derivatives", Report C.S.S.9, AERE, Harwell.

Powell, M.J.D. (1975) "Some global convergence properties of a variable metric algorithm for minimization without line searches", Report CSS 15, AERE, Harwell.

Reid, J.K. (1972) "Two Fortran subroutines for direct solution of linear equations whose matrix is sparse, symmetric and positive definite", Report AERE-R.7119, Harwell.

Sargent, R.W.H. and Sebastian, D.J. (1972) "Numerical experience with algorithms for unconstrained minimization", *in* "Numerical methods for non-linear optimization", *ed.* F.A. Lootsma, Academic Press, London.

Schubert, L.K. (1970) "Modification of a quasi-Newton method for nonlinear equations with a sparse Jacobian", *Maths of Comp.*, 24, 27-30.

Stewart, G.W. (1967) "A modification of Davidon's minimization method to accept difference approximations of derivatives", *J. Assoc. Comput. Mach.*, 14, 72-83.

Stewart, G.W. (1973) "The convergence of the method of conjugate gradients at isolated extreme points of the spectrum", Computer Science Report, Carnegie-Mellon University.

Stoer, J. (1974) "On the convergence behaviour of some minimization algorithms", presented at the IFIP Congress, Stockholm.

Wolfe, P. (1967) "Another variable metric method", working paper.

Wolfe, P. (1969) "Convergence conditions for ascent methods", *SIAM Review*, 11, 226-235.

Wolfe, P. (1974a) "Note on a method of conjugate subgradients for minimizing nondifferentiable functions", *Math. Prog.*, 7, 380-383.

Wolfe, P. (1974b) "A method of conjugate subgradients for minimizing nondifferentiable functions", Report RC 4857, IBM Research Center.

Zoutendijk, G. (1960) "Methods of feasible directions", Elsevier Publishing Co., Amsterdam.

OPTIMIZATION OF FREQUENCY SELECTIVE ELECTRICAL NETWORKS

J.K. Fidler and R.E. Massara*

(University of Essex)

*now at University of Keele

SUMMARY

This paper will describe a novel approach to optimization in the context of the Computer-Aided Design of electrical filter networks.

In this study an error-space, based on the sum of squared residuals from electrical network functions, is shown to have certain favourable properties which have been exploited in the development of a simple but effective optimization strategy based on an extension of a Direct Search method and which has been termed the "Alternating Pattern Move" procedure.

In practice it is found that this technique provides excellent initial convergence but rather less impressive end-stage performance, and consequently a two-stage algorithm has been developed which uses the above method in conjunction with a modification of the Fletcher-Powell optimization method: this provides a consistently effective procedure.

Results will be presented to demonstrate some of the practical problems encountered in Computer-Aided Circuit Optimization and their solution by the methods evolved in this work.

153

This paper describes the development of optimization techniques specifically directed towards the minimization of error functions associated with the design of an important class of electrical networks.

There is a considerable requirement for electrical circuits having prescribed frequency selective characteristics, *i.e.*, circuits which transmit signals over some specified range of frequencies and reject signals not in this range. In some cases, analytic synthesis techniques exist whereby these electrical filtering networks may be designed to exhibit the transmission performance required; in other cases, such techniques do not exist and computer-aided design methods are then required. It is this area with which the present work is concerned.

The place of optimization methods in a computer-aided circuit design (CACD) scheme may be stated briefly. A specification relating to a particular performance criterion is generated and an error function is formulated by comparing the current value of this performance measure with that specified: an optimization strategy is then directed to the minimization of the error function by the adjustment of the network variables - usually component values. In this simple scheme, a central problem in the electrical engineering application of optimization may be identified: each formulation of the error function requires the evaluation of the chosen performance measure for the current variable set and usually involves at least one, and often many, network analyses. Network analysis is an inherently costly process and, accordingly, the aim of this study was to develop methods which reduce the number of function evaluations and hence network analyses necessary to achieve a desired solution.

The conventional approach to CACD consists of comparing the actual and specified magnitude (amplitude) and/or phase relationships between input and output signals of the network over some set of discrete signal frequencies. The error function is then typically formulated in minimax or least squares form. Fig. 1 illustrates such a case with A_{si} and A_i representing corresponding specified and actual amplitudes at the ith frequency point, ω_i rad/s. A sum of squares error function may then be formed as:

$$F = \sum_i (A_{si} - A_i)^2 \qquad (1)$$

These frequency points are chosen to correspond to regions of importance - thus, in the case illustrated, the network is required to transmit a particular band of frequencies and to reject frequencies outside this band.

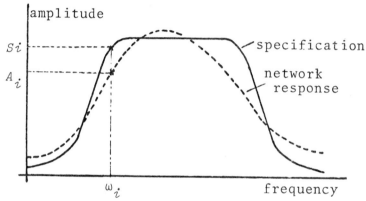

Fig. 1. Formulation of amplitude-matching error function

This type of approach, whilst often corresponding to the way in which specifications originate, has certain serious disadvantages, in particular the need to perform a network analysis at each of the frequency points for every error function evaluation. In addition, the gradient or Jacobian evaluations required by the more powerful optimization methods are difficult and costly to perform.

These more efficient optimization algorithms themselves give rise to further difficulties in the present context in that certain approximations involved are usually invalid at remote starting points. For example, the well known Fletcher-Powell algorithm (1963) requires that the functional surface be well approximated by a quadratic form whilst the least squares methods (for example, Murray (1972)) assumes that a predicted minimizing step is small. These assumptions are reasonable in the closing stages of the design process but are unjustified at points distant from the solution. The algorithms usually include devices to improve their early-game performance, typically biasing early searches towards the negative gradient direction, but these are only partially successful and there is a tendency for the methods to enter extended periods of slow convergence in valley-like conditions with consequent extravagant increases in computation cost.

In view of these problems, the present study considers an alternative performance measure formulation which reduces the number of network analyses required per error function evaluation. Optimization methods are then constructed as early-game procedures which do not rely on assumptions concerning the error-surface and which, in addition, exploit certain properties of the error function resulting from its origins in a physically realizable electrical network.

The dynamic behaviour of circuit components ensures that input and output time domain signals are related through a differential equation. For linear circuits, application of the Laplace transform method to such equations results in the concept of the network function relating, for example, input and output voltages as functions of s, the complex frequency variable

$$\frac{V_{out}(s)}{V_{in}(s)} = \frac{b_r s^r + b_{r-1} s^{r-1} + \ldots + b_j s^j + \ldots + b_1 s + b_0}{c_p s^p + c_{p-1} s^{p-1} + \ldots + c_k s^k + \ldots + c_1 s + c_0} \quad . \quad (2)$$

The amplitude and phase characteristics of the circuit are uniquely determined by this network function and, as an alternative approach, the network function has been used as the basis for optimization in the present study.

The numerator and denominator coefficients, the b_j and c_k in equation (2), are functions of the network elements; specifically, it may be shown (Bode (1945)) that each coefficient is no more than a first order function of a given element, *i.e.*, the coefficients are, in general, multilinear functions of the network elements.

Denoting a general numerator or denominator coefficient realized by the current element set and the corresponding specification coefficient as a_i, a_{si}, respectively, a sum-of-squared-residuals error function is formed as:

$$F = \sum_{i=1}^{m} (a_{si} - a_i)^2 \qquad (3)$$

- it is assumed that there are m such coefficients. Methods are available which permit all the coefficients to be evaluated, and hence F, in a single network analysis; this resolves the difficulty of excessive numbers of network analyses in the classical amplitude/phase matching approach.

The property of multilinearity allows each a_i to be written in the form:

$$a_i = \alpha_{ij} e_j + \beta_{ij} \qquad (4)$$

where e_j is the jth network variable and α_{ij}, β_{ij} are linear functions of the other network variables (either may be zero). By substituting (4) into (3) it will be noted that F is a quadratic in e_j and that its unique turning point (for given values of $e_i, i \neq j$), if it exists, is a minimum. It is then only necessary to evaluate the partial derivatives α_{ij} to locate the position of this minimum and, since a_i is linear in e_j, this may be done exactly by simple perturbation.

In addition to permitting simple one-step minimization of the error with respect to a given variable, the multilinearity property gives rise to certain favourable features in the coefficient-derived error function when compared with the corresponding amplitude/phase matching function including fewer stationary points and a reduced risk of convergence to a non-global minimum.

It has been stated that one aim of this work was to produce effective early-game optimization procedures using the network theoretic properties of the error function. It has also been shown that the multilinearity property provides an efficient means of minimizing the error with respect to the variables selected one at a time: this suggests that a univariate minimization-based algorithm be considered.

It is well known that the efficiency of univariate minimization schemes depends on the form of the error function. In the case of the coefficient-derived error function, the multilinearity of the coefficients leads to the general isovalue error contour form illustrated in Fig. 2 for a two variable case. Fig. 2(a) also shows the route taken by univariate minimization for various, relatively remote, starting points and suggests that such a scheme would provide rapid initial convergence in the case illustrated as well as starting point insensitivity. The low error contour form (Fig. 2(b)) indicates, however, that the method would provide slow, oscillatory, progress after this initial phase.

The present approach uses this method as the basis for a first-order pattern move scheme similar to the Direct Search method of Hooke and Jeeves (1961). In order to augment the curved valley negotiation capabilities of this algorithm, a second-order pattern move scheme was developed which steps along a quadratic fitted to three points in the error space. It was found most efficient to generate the three points with the

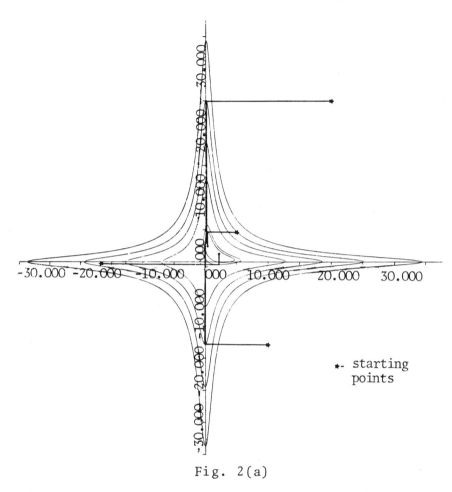

Fig. 2(a)

Fig. 2. Contours in a two-variable, coefficient-
derived error space.

(a) high error (5-1000); also shows first
iteration of univariate minimization
scheme for four starting points

159

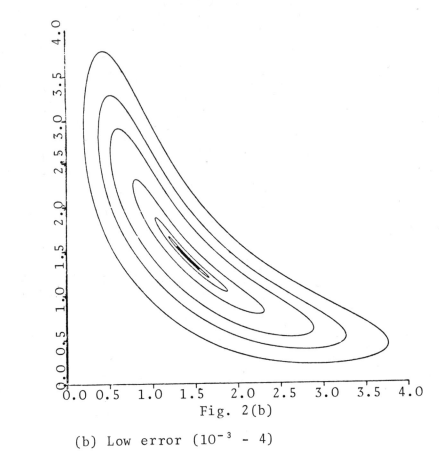

Fig. 2(b)

(b) Low error $(10^{-3} - 4)$

univariate minimization-based first-order pattern move and to alternate between first- and second-order moves. This alternating pattern move (APM) procedure is discussed in more detail elsewhere (Massara and Fidler (1974)).

Fig. 3 shows typical comparative convergence characteristics obtained with these algorithms: the problem was a nine variable filter design example. The curves show $\log_{10}$ of error plotted against number of function evaluations for (a) univariate minimization alone, (b) univariate minimization and first-order pattern move and (c) APM, and demonstrate the superiority of the latter method over the other algorithms. This scheme, then, was used as the first part of a two-stage procedure to be discussed next.

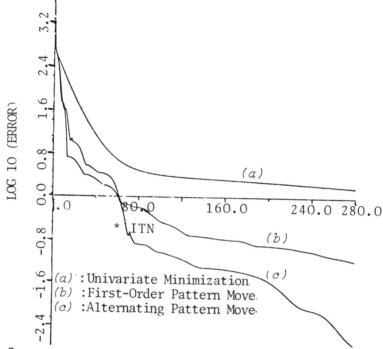

Fig. 3. Performance of direct search algorithms. Design of a passive, 8th-order Butterworth lowpass filter. Log_{10} of error *vs.* number of function evaluations

It was noted earlier that the APM algorithm was developed to provide rapid initial convergence. Fig. 4(b) demonstrates its behaviour on a six variable filter design problem: convergence is fast at first but slows as the solution is approached. Fig. 4(a) shows the behaviour of the Fletcher-Powell algorithm (1963) on the same problem (abscissae have been adjusted to permit direct cost comparison) and illustrates the initial superiority of the APM algorithm. Fig. 4(c) shows the results obtained by combining these two

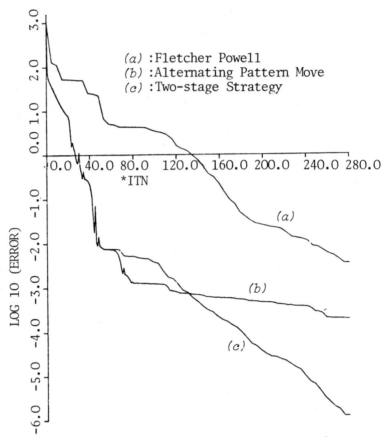

Fig. 4. Comparison of (b) APM, (a) Fletcher-Powell and (c) two-stage algorithms. Design of a passive 5th-order Chebyshev lowpass filter

algorithms so that the APM strategy is used for the initial stages of the design whilst control is transferred to the Fletcher-Powell algorithm when convergence becomes slow. It will be noted that the two-stage program exploits the best features of its constituents to optimum effect.

Since entry into the Fletcher-Powell algorithm is now effected at a point relatively close to the ultimate solution, it becomes reasonable to employ a more enlightened initial approximation to the inverse Hessian matrix at the solution than the simple unit matrix conventionally used and which provides an initial search in the negative gradient direction; a useful measure in the absence of information on the local nature of the error surface and at remote points, but inappropriate here. The procedure adopted consists of evaluating and inverting the Hessian corresponding to the current element set on entry into the algorithm.

Fig. 5 illustrates the order of improvement obtainable by this measure: the problem here was a four variable network design example. Curve (a) shows the behaviour of this modified Fletcher-Powell algorithm and (b) that of the basic algorithm.

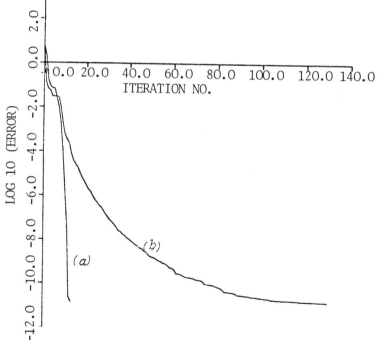

Fig. 5. Effect of using (a) inverse Hessian instead of (b) unit matrix on first entry into Fletcher-Powell subroutine. Design of 3rd-order Butterworth lowpass filter

163

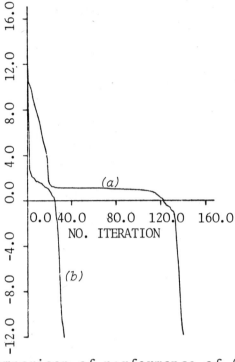

Fig. 6. Comparison of performance of (*a*) damped
least squares method and (*b*) APM-least
squares two-stage algorithm

Results obtained with a two-stage strategy
using the APM and Fletcher-Powell methods have been
discussed. More recently, the behaviour of a similar scheme has been studied with a damped least
squares algorithm (Murray (1972)) substituted for
the Fletcher-Powell procedure. As before, the
two-stage procedure is found to be generally
superior to the least squares algorithm used alone.
Fig. 6 illustrates this point with results obtained
on a problem based on a nine variable filter design
example. Curve (*a*) shows the convergence behaviour
of the damped least squares algorithm used alone
whilst (*b*) represents the behaviour of the two-
stage algorithm. As the error function is formulated as a sum of squared residuals, the least
squares approach is to be preferred and is found

to provide consistently faster convergence than the Fletcher-Powell based algorithm. The behaviour of the APM/Fletcher-Powell two-stage algorithm is nevertheless of interest since, it is suggested, a similar scheme could be constructed for more general function minimization.

As further examples, Figs. 7 and 8 represent the results of lossy filter design problems (eight and nine variables, respectively: component R_1 was fixed in both cases) obtained using the two-stage APM/Fletcher-Powell algorithm. CPU times, using a DEC PDP-10 time sharing system, were 178 and 206 seconds, respectively. Problems of similar complexity were found to require CPU times in the order of 10 to 50 seconds using the APM/least squares procedure. These run times should be regarded in the light of results obtained by amplitude/phase matching techniques where CPU times are typically in the order of many minutes, or in some cases, hours.

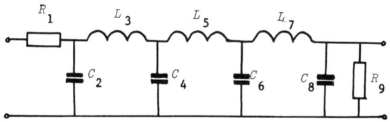

$R_1=1\Omega$ $C_4=2.3360F(2000)$ $L_7=1.5829H(100)$

$C_2=1.4850F(1000)$ $L_5=1.6427H(50)$ $C_8=1.5104F(\infty)$

$L_3=1.4641H(100)$ $C_6=2.1836F(2000)$ $R_9=3.1874\Omega$

(Figures in brackets are Q's defined at band-edge)

Fig. 7.   7th-order Chebyshev lowpass filter with non-uniformly lossy elements. (Bracketed figures define component loss levels)

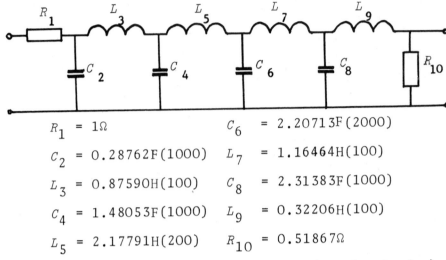

R_1 = 1Ω C_6 = 2.20713F(2000)

C_2 = 0.28762F(1000) L_7 = 1.16464H(100)

L_3 = 0.87590H(100) C_8 = 2.31383F(1000)

C_4 = 1.48053F(1000) L_9 = 0.32206H(100)

L_5 = 2.17791H(200) R_{10} = 0.51867Ω

(Figures in brackets are Q's defined at band-edge)

Fig. 8. Lossy 8th-order Butterworth lowpass filter

CONCLUSIONS

The need to minimize the number of error function evaluations (and hence network analyses) in the electrical engineering application of optimization has been discussed. In order to achieve this, a particular error function formulation was considered and properties of this function have been exploited in the development of the pattern move algorithm based on simple, one-step, univariate minimization. This algorithm has been used in conjunction with a modified Fletcher-Powell algorithm, widely held to be the most powerful, general, function minimization method available, and with a damped least squares algorithm, probably the most powerful method for appropriately formulated error functions: the resulting two-stage procedures are found to offer consistent savings over the latter strategies used alone. These techniques have been found, in practice, to provide rapid and efficient optimization of a wide variety of electrical networks.

REFERENCES

Bode, H.W. (1945) "Network Analysis and Feedback Amplifier Design", Van Nostrand.

Fletcher, R. and Powell, M.J.D. (1963) "A rapidly convergent descent method for minimization", *Comput. J.*, **6**, 163-168.

Hooke, R. and Jeeves, T.A. (1961) "Direct search solution of numerical and statistical problems", *J. Assoc. Comput. Mach.*, **8**, 212-219.

Massara, R.E. and Fidler, J.K. (1974) "Computer Optimization of Frequency Selective Networks", *in* "Proceedings 1974 European Conference on Circuit Theory and Design", IEE Conference Publication no. 116.

Murray, W. *Ed.* (1972) "Numerical Methods for Unconstrained Optimization", Academic Press.

APPENDIX

All the circuit design examples cited in this paper are based on the design of electrical lowpass ladder networks having the general structure illustrated in Fig. 9. The order of a network of this form is n, where n is the number of reactive elements (capacitors, C, and inductors, L). In some of the examples (e.g., Figs. 7 and 8) lossy networks are considered: here, the design process is required to encompass the parasitic dissipative losses associated with real, as opposed to ideal, components.

To achieve this, the reactive elements depicted in Fig. 9 are replaced by more realistic models in which the loss is represented by an additional resistor arranged as shown in Fig. 10. The values of these loss components are related to the associated reactances as:

$$R_L = d_L L, \quad R_c = \frac{1}{d_c C}$$

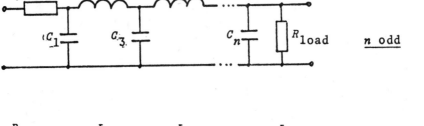

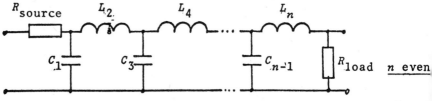

Fig. 9   General form of lowpass ladder networks

Fig. 10   Lossy inductors and capacitors

where d_L and d_C are the "loss factors" of the components L and C, respectively. Under the assumptions of the problem, the "quality factor", or Q, of the components is the reciprocal of these loss factors, *i.e.*,

$$Q_L = \frac{1}{d_L}, \qquad Q_C = \frac{1}{d_C}$$

- this figure appears in brackets after the relevant values quoted in Figs. 7 and 8. As an example, inductor L_3 in Fig. 8 has $Q = 100$: d_L is then 10^{-2} and the value of the associated series resistor, R_L, is $0.876 \times 10^{-2}\Omega$.

For the lowpass networks of Fig. 9, the voltage transfer function (2) can be expressed in the form:

$$\frac{V_{out}}{V_{in}} = \frac{1}{\sum_{i=0}^{n} a_n s^n}$$

and a particular response is specified by the values of the $n+1$ coefficients a_i.

For the contours plotted in Fig. 2, the network has $n=2$: the terminating resistors, R_{source} and R_{load}, are fixed at 1Ω so that a two-variable problem results. The specified coefficient values, a_{si}, are those yielding the 2nd-order "Butterworth" response (Weinberg (1962)) for which

$$a_{s0} = 2, \qquad a_{s1} = 2(2)^{\frac{1}{2}}, \qquad a_{s2} = 2 .$$

In this relatively simple case, the expressions for the coefficients realized by the network may be easily found explicitly as

$$a_0 = 2, \qquad a_1 = C_1 + L_2, \qquad a_2 = C_1 L_2$$

and the error function is formed by comparing the a_{si} and the a_i as in equation (3).

For higher order networks, it is difficult and inefficient to evaluate the realized coefficients in explicit form and a network analysis procedure is used to compute the numerical values of a_i, which are then compared with the a_{si}, for a current set of element values.

Figs. 3, 6 and 8 relate to 8th-order Butterworth specifications, Fig. 5 relates to a 3rd-order Butterworth specification whilst Figs. 4 and 7 relate to 5th- and 7th-order Chebyshev specifications, respectively. The numerical values of the specified coefficients in these cases are given in the reference.

REFERENCE

Weinberg, L. (1962) "Network Analysis and Synthesis", McGraw-Hill, Chapter 11.

The Choice of Design Parameters for Overhead Line Vibration Dampers

D.A.D. Cooke and M.D. Rowbottom*

(Central Electricity Research Laboratories)

*now at CEGB Northeast Region, Harrowgate

SUMMARY

This work arose out of a problem with certain wind induced oscillations of overhead, electrical power transmission lines. A study was undertaken to evaluate the feasibility of using mechanical dampers to control the oscillations. The most general damper arrangement that could be envisaged involved 12 parameters, and it was necessary to determine the most effective values of these parameters to use in a practical design. For each choice of the parameters, the logarithmic decrements of the relevant modes of oscillation can be determined from the complex roots of a frequency equation. The problem reduces to those of (a) finding the roots of the frequency equation and (b) choosing the parameters so as to maximize all the logarithmic decrements. The paper describes how the simplex method was applied to both these problems. The method worked well for finding the roots of the frequency equation, but a number of difficulties arose in optimizing the parameters. These difficulties are discussed, and the *ad hoc* way in which they were overcome is described.

LIST OF SYMBOLS

f	complex function
f_1, f_2	functions in f
i	$\sqrt{-1}$

x,y real and imaginary parts of a root of eqn. (1).

z non-dimensional argument of f

z',z'' real and imaginary parts of z

z_1,z_2 roots of eqn. (1).

$\lambda,\lambda_1,\lambda_2$ fraction of span to damper location

p_1 non-dimensional stiffness

p_2 non-dimensional damping

p_5 non-dimensional mass

1. INTRODUCTION

Ice and snow accretion can occur on overhead power transmission lines under blizzard conditions, and the effects of a cross wind on the resulting non-circular cross sections can give rise to a type of oscillation known as full-span galloping. The electrical conductors in a span between successive towers oscillate vertically at frequencies of the order of $\frac{1}{6}$ Hz, often with the whole span moving in phase. Typical span lengths are 350 m, and mid-span amplitudes of 5 m or more have been experienced in the UK. The oscillations can cause electrical faults on the line, they can accelerate the mechanical wear of the conductors and fittings and are a source of some operational difficulty in the control of the grid system.

Aerodynamic dampers have been developed to give reduced oscillation amplitudes (Hunt and Richards (1969)), but it is not certain that these dampers can give sufficient control over the oscillations. There is still a need for an improved method for reducing or eliminating the oscillations.

With this thought in mind, Cooke and Rowbottom (1974) compared the relative effectiveness of aerodynamic and mechanical dampers. They considered primarily the lowest frequency mode of the span, for which the largest oscillation amplitudes occur

172

for any given wind-speed. No specific mechanical dampers were considered; it was assumed that the effect of the dampers would be to give a logarithmic decrement to the mode considered, leaving the mode shape unchanged. On this basis they were able to show that mechanical dampers would be valuable if they could provide large damping levels (logarithmic decrements of 0.5 or greater).

Having established the level at which the dampers would become useful, the next step is to see if this level can be achieved for any notional design of damper. If this can be done the final step is to engineer the damper into existence.

This paper is concerned with the work that was undertaken on the first of these steps, the examination of the effects of mechanical dampers on the damping level of the line span. Only passing reference will be made to the problems of producing a finished damper.

2. THE DAMPERS CONSIDERED

As a first approximation the conductors were modelled by a horizontal stretched string. It was felt that treating the conductors as catenaries would complicate the problem without adding to the understanding of the effects of the dampers. At the start of this work it was believed (Den Hartog (1956)) that the dampers would need to have very large masses. The only way in which these masses could be supported would be by placing them at the ends of the spans, and the most general damper of this type that could be envisaged is shown in Fig. 1. This involves twelve parameters, five parameters for each damper, and the two damper locations in the span. In addition to this arrangement, the two types of dampers suspended from the lines as shown in Fig. 2 were also considered. The type *B* in-span damper should be easier to engineer than the type *A*.

173

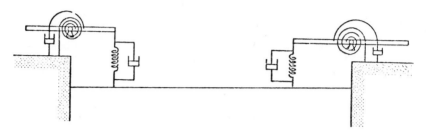

Fig. 1.   General damper configuration

Type *A*

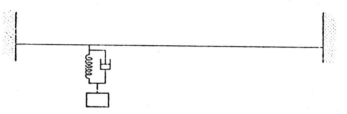

Type *B*

Fig. 2.   In-span dampers

For these mechanical dampers it is fairly straightforward to extend the work of Allnutt and Rowbottom (1974) to show that the non-dimensional complex frequency of free vibrations (z) has to satisfy an equation of the form

$$f(z) = \sin(z) + f_1(z) \sin(\lambda_1 z) \sin(\{1-\lambda_1\}z)$$

$$+ f_2(z) \sin(\lambda_2 z) \sin(\{1-\lambda_2\}z) + \{f_1(z)f_2(z)$$

$$\sin(\lambda_1 z) \sin(\lambda_2 z) \sin([1-\lambda_1-\lambda_2] z)\} = 0 \quad (1)$$

where f_1 and f_2 are quotients of polynomial functions of z whose coefficients depend on the non-dimensional parameters of the two dampers.

The solutions have the form $z' + iz''$, where z' is the non-dimensionalized frequency of oscillation, and $2\pi z''/z'$ is the logarithmic decrement associated with this frequency.

The function f clearly has no finite number of roots. If the work is to proceed, it is necessary to be able to evaluate the roots that are of interest. In this case these are usually the first few roots with positive real parts. (If $x + iy$ is a root of f then so is $-x + iy$ for all the functions f_1 and f_2 considered here). In particular it was decided to concentrate on the lowest frequency mode for the reasons mentioned before.

Some thought was given to the problem of finding the roots, and a number of methods were tested. The simplex method of Nelder and Mead (1965) was used to minimize $|f|$ with respect to the two real variables z' and z''. This converged on to the roots satisfactorily, and was used in the early work when the nature of the solutions of the equation was not known. Later in the work it was found that the Newton-Raphson method using numerically determined derivatives converged on to the roots more quickly, and this was used in the majority of the work.

3. EARLY RESULTS

It was initially thought (wrongly as it transpired) that the particular damper shown in Fig. 3 would be the most effective, and the initial work concentrated on this damper type. Only one damper per span was considered, and for a damper of fixed mass operating at a fixed location there are only two parameters to be chosen (corresponding to the spring rate and the damper's constant of proportionality). In the first instance, the optimum values of the parameters were found by obtaining the roots of equation (1) for a range of parameters, and choosing further ranges for examination based on the results from the previous range. This process was very tedious and slow, several computer runs being required to find the near optimum parameters. On the other hand, the results obtained gave valuable information on the nature of the roots of equation (1), and the optimum values obtained agreed well with those obtained later by the numerical optimization scheme.

Fig. 3. Particular end-of-span damper

In parallel with this early numerical work, the problem was attacked analytically. On the assumption that the roots of equation (1) were differentiable with respect to the parameters of the problem, it was shown that for a single damper at any position along the line, none of the dampers could give more damping than a suitably chosen viscous damper attached between that point and the ground.

This result was quickly overturned by the numerical work mentioned above, which showed that the optimum parameters gave larger damping values than the predicted maximum. A detailed examination of the numerical results showed that the variation of the roots with some of the parameters was very rapid, particularly over certain ranges near the optima. It could not be established numerically that there was a discontinuity in gradient, but the results obtained and the failure of the theory together suggested that this was the case.

This early work left us with two conclusions. The work would never be completed using the manual

optimization process, and the roots of equation
(1) had apparent discontinuities in gradient.
Taken together, the obvious conclusion was to use
an automatic numerical optimization process, and a
process not requiring the derivatives. The simplex method of Nelder and Mead (1965) was selected.

4. THE USE OF THE SIMPLEX METHOD

The first problem was to choose a function
to be minimized. The general dampers shown in
Fig. 1 introduce roots in addition to the stretched
string roots, and the work described in the previous section showed that for the maximum logarithmic decrements, these extra roots would be close
to the fundamental mode that we had chosen to concentrate on. We suspected intuitively, and some
numerical work mentioned by Myerscough (1973) confirmed, that the nature of the oscillations is
such that when the span has two close natural frequencies, one lightly damped and the other heavily
damped, the lightly damped mode only is excited.
It is clear that our optimum damper will be the
one that gives all the natural frequencies near
the fundamental more or less equal logarithmic
decrements.

We chose to calculate the logarithmic decrements of the modes for a given set of parameters,
to choose the smallest one and to pass the inverse
of this to the optimization routine as the "cost"
for this choice of parameters.

In practice the main difficulty arose with
keeping track of the roots, for two separate reasons. Firstly the proximity of the roots presented some problems to the root finding routine.
This was overcome by finding the first root (z_1)
of $f(z)$, the second root (z_2) of $f(z)/(z-z_1)$ and
so on. Occasionally roots other than the ones
desired were found but it was relatively easy to
restrict the area of search to overcome this.

The second cause of difficulty was not so

easily overcome. If the parameters are too far from the optimum, either because of a bad initial estimate, or because of a subsequent step by the optimizing routine, the roots may move away from the region of interest and become confused with other roots of the function. Once this has occurred it is exceedingly difficult to devise an automatic procedure to sort out which root is which. The situation is compounded by the physically different dampers that arise from the general dampers if the spring or damper parameters become very large or very small. It proved impossible to overcome the difficulty for the general problem shown in Fig. 1.

Attention was then concentrated on the case of one general damper per span. In this slightly simpler situation some progress was made. It was found that the presence of the two springs and the two dampers makes the problem rather ill-conditioned. For a damper of a given natural frequency there is apparently no really optimum division of the stiffness between the two springs. A similar situation applies in the case of the dampers.

In view of this we decided to study one damper per span, this damper to have not more than one spring and one damper. This left the 8 dampers described in Table I to be considered, together with the two in-span dampers of Fig. 2.

Even in these cases occasional loss of contact with the roots was experienced, but because of the simpler nature of the problems now being tackled, it was possible to modify the cost function to give a high value if the required number of roots could not be found in the vicinity of the fundamental.

Table I

Dampers with not more than 2 variables

Type	Central Spring	Central Dashpot	Linear Spring	Linear Dashpot
C	*R*	*O*	*O*	*F*
D	*O*	*F*	*F*	*O*
E	*F*	*O*	*O*	*F*
F	*O*	*O*	*F*	*F*
G	*F*	*F*	*R*	*O*
L	*R*	*O*	*F*	*F*
N	*O*	*O*	*O*	*F*
O	*O*	*F*	*R*	*O*

Note *R* denotes rigid
     O denotes free
     F denotes finite value

5. RESULTS

Having knocked the original problem about because of the difficulties described in the previous section, the application of the simplex method to the revised problem was gratifyingly successful. The optimum values were usually found inside 100 steps, and always inside 200, the number of steps depending largely on the initial estimates for the parameters. The chosen cost function worked well. As might have been anticipated, those dampers having an extra degree of freedom resulted in two modes of oscillation with close natural frequencies and identical logarithmic decrements.

Each of the dampers was considered for a range of locations in the span and, where appropriate, for a range of masses. On performance, the dampers fell into two classes - those like the viscous damper attached to ground mentioned in Section 3, and those like the in-span damper *A*.

179

This latter class gives higher logarithmic decrements than the former. The results for the two
types are given in Figs. 4 and 5.

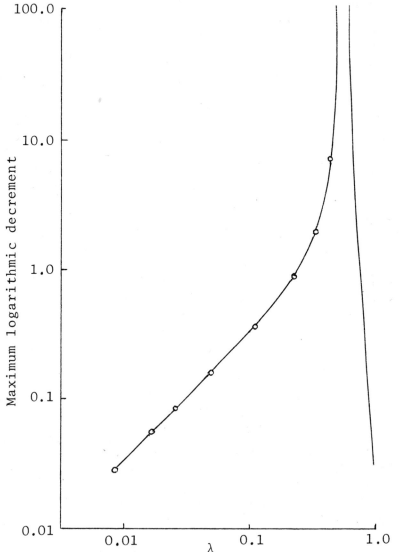

Fig. 4. Results for viscous damper (Types **C**,*E*,*G*,
L,*N*,*O*)

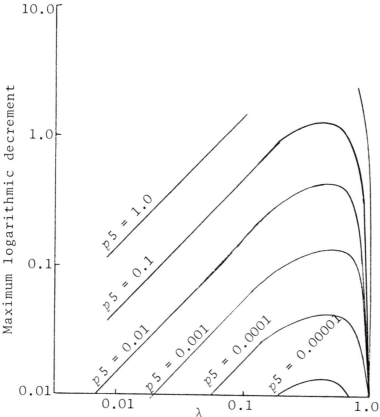

Fig. 5. Results for tuned damper (Types A,D,F)

The use of the automatic optimization routine threw up one surprise result. The two in-span dampers were found to give different maximum logarithmic decrements for the same mass, the type B damper giving lower values. In an attempt to understand this, some work was done on a damper of this type having the mass and the moment of the inertia of the beam about its centre of mass as independent variables. The unconstrained optimization moved to make the moment of inertia zero with an improvement in damping up to the type A levels. When this was investigated we found that a damper with the mass concentrated at the end of

181

the beam (and hence with zero moment of inertia about the centre of mass) has identical behaviour to the type A damper of the same mass.

DISCUSSION

The general problem considered at the outset proved too difficult to handle, mainly because of the difficulty of devising a satisfactory automatic cost function. It is possible that a suitable function could have been derived given time, but we decided to consider a larger number of more simple designs on the grounds that these could be handled easily by the simplex method, and that they were more likely to result in realistic engineering designs.

We did briefly consider a constrained optimization problem, but we felt that a great deal of investigation of the behaviour of the roots of equation (1) would have been necessary to define the constraints sufficiently well. Our preliminary investigation in this area showed that the optimum values of the parameters vary widely with the mass and location of the dampers, and in a nonlinear fashion. Again we judged that the approach of considering several simple dampers was a more certain line of attack.

While the optimum values of logarithmic decrement are high enough to be worth pursuing, the dampers turn out to be very sensitive to their parameters, as illustrated in Fig. 6. It may well turn out that the high levels obtainable in theory cannot be realized in a practical design.

We found the situation on the in-span dampers much better than we expected from Den Hartog (1956). The results of Fig. 5 are predicting logarithmic decrements of 1.5 for dampers having masses of the order of 50 kg - certainly far removed from many tons. They are still subject to the extreme sensitivity mentioned above.

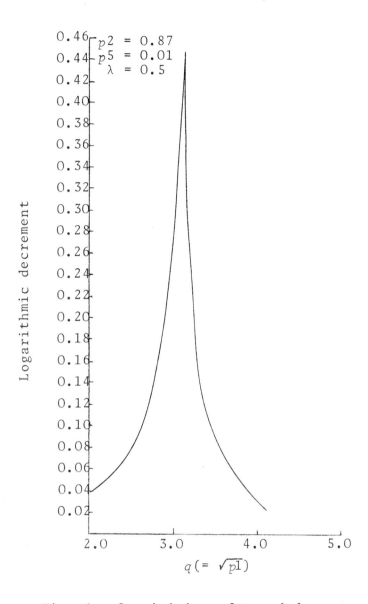

Fig. 6. Sensitivity of tuned damper

ACKNOWLEDGEMENT

This work was carried out at the Central Electricity Research Laboratories and this paper is published by permission of the Central Electricity Generating Board.

REFERENCES

Allnutt, J.G. and Rowbottom, M.D. (1974) "Damping of aeolian vibration on overhead lines by vibration dampers", *Proc. I.E.E.*, **121**, No. 10, 1175-1179.

Cooke, D.A.D. and Rowbottom, M.D. (1974) "Effects of mechanical and aerodynamic damping on the galloping of overhead lines", *Proc. I.E.E.*, **121**, No. 8, 845-848.

Den Hartog, J.P. (1956) "Mechanical Vibration", (fourth edition), McGraw-Hill.

Hunt, J.C.R. and Richards, D.J.W. (1969) "Overhead line oscillations and the effect of aerodynamic dampers", *Proc. I.E.E.*, **116**, No. 11, 1869-1874.

Myerscough, C.J. (1973) "A simple model of the growth of wind induced oscillations in overhead lines", *J. Sound and Vib.*, **28**, 699-713.

Nelder, J.A. and Mead, R. (1965) "A simplex method for function minimization", *Computer J.*, **7**, 308-313.

Progress in the Development of a Modularized Package of Algorithms for Optimization Problems

K. Brown

(University of Minnesota)

M. Minkoff

*(Argonne National Laboratory and
.Northern Illinois University)*

K. Hillstrom, L. Nazareth, J. Pool and B. Smith

(Argonne National Laboratory)

SUMMARY

Motivated by the success of EISPACK and
FUNPACK and by the increasing demand at Argonne
National Laboratory for a comprehensive and coher-
ent collection of optimization software, we initiated the MINPACK Project to examine the organization of a package of algorithms for nonlinear optimization problems and for systems of nonlinear
equations. Experience with EISPACK and FUNPACK
and early experiences with MINPACK have demonstrated the advantages of organizing packages with an
emphasis on modularity. Advantages include facil-
itation of software maintenance and development
and provision of a flexible tool both for solving
optimization problems and for developing and improving algorithms. Although both EISPACK and FUN-
PACK were based upon a modularization approach,
the development of MINPACK involves resolution of
questions at a higher level of difficulty. However, preliminary studies indicate both the feasi-

ibility and the desirability of a modularization approach to MINPACK. The experience gained in the evolution from a collection of conventionally organized software for optimization problems to a prototypical version of MINPACK is reported together with the anticipated directions of this research venture.

1. INTRODUCTION

Today there are only the beginnings of an organized body of knowledge addressing the problems of producing and utilizing quality mathematical software. The need for quality mathematical software in the computing field is made evident by the demand for packages such as EISPACK and FUNPACK and libraries such as NAG and IMSL. The cost, both in time and in manpower, of producing EISPACK and FUNPACK demonstrates the necessity of a methodology of mathematical software engineering. Consequently, the Mathematical Software Research Program at Argonne National Laboratory seeks a scientific basis for such a methodology by exploring the generation and utilization of mathematical software.

One aspect of this program is the exploration of systematized collections of numerical software addressed to specific classes of mathematical problems. Our current efforts, EISPACK, FUNPACK, and MINPACK, each represent increasing levels of complexity. The complexity of a collection is measured here, for example, in terms of the nature of interactions among components of the collection both among themselves and with their computing environment. The experience gained in each case contributes to the solution of the problems of systematizing a collection at a higher level of complexity. Indeed, this test of transferability from one major problem area to another and from one level of complexity to a higher level is one criterion on which to judge whether individual experiences with systematization represent a principle or an *ad hoc* construct.

The origin of the MINPACK Project at Argonne can be traced to an effort to evaluate and upgrade the nonlinear optimizers and system solvers available in our Central Computing Facility's numerical software library. The expansion of this effort into the current MINPACK Project followed the simultaneous recognition that:

a) increased emphasis on the analysis of data from scientific and engineering experiments and on the modelling of engineering and socio-economic systems required the availability of a more comprehensive collection of optimization routines;

b) although the development of algorithms for optimization problems is a very active field, several basic algorithms exist which outperform most other algorithms currently available to the scientist or engineer; and

c) the nonlinear optimization problem provides a challenging test of the applicability of insights into systematization gained in the EISPACK and FUNPACK Projects (Boyle (1972), Cody (1975), Smith *et al.* (1974) and Smith *et al.* (1974)).

Consequently, the MINPACK Project was initiated as a research venture to examine the organization of a collection of algorithms for nonlinear optimization. The package elements currently under development will form the basis for MINPACK I, Edition 1, a package of modularly structured algorithms addressing the following problems:

a) general nonlinear optimization - minimize an objective function, $F(X)$, which is a user-provided function of the n-component real vector X;

b) nonlinear least-squares optimization - minimize an objective function, $F(X)$, which is a user-provided function formed as a sum of squares of m nonlinear functions of the n-component real vector X;

c) solution of nonlinear systems - find an n-component real vector X which is simultaneously a zero for m user-provided nonlinear functions of X.

187

The MINPACK effort described in this paper serves the dual goals of identifying basic principles of mathematical software engineering and of providing a systematized collection of software for optimization. The testing criteria used to select algorithms for inclusion in Edition 1 of MINPACK I are discussed in Section 2. In Section 3, we discuss the advantages for modularization, experience gained in this direction, and the drivers and modules currently developed or under development. The impact of extensions in the direction of unifying the algorithms and modules into polyalgorithms and their associated control programs is the subject of Section 4. Extensions necessary to address constrained optimization problems are briefly considered in Section 5. Modules are discussed further in Section 6. Future goals of the project are outlined in Section 7.

2. ALGORITHM SELECTION PROCEDURE FOR MINPACK I

Twenty computer algorithms, developed by prominent researchers in the field, have been considered for inclusion in the first edition of MINPACK I. The routines demonstrating the greatest reliability and efficiency are now being processed into package members. Some routines and their sources are listed in the Appendix. The aim of this section is to provide the motivation for, and a description of, the testing program developed to evaluate and compare these algorithms.

Tests commonly used to evaluate and compare optimization algorithms have the following characteristics:

a) a well-known difficult test problem, such as Rosenbrock's descending parabolic valley, is selected as a test problem;

b) .the same initial estimate is used in all tests;

c) the computational effort required to find the extrema is measured in terms of iterations or equivalent function evaluations.

188

Utilizing the essential ingredients of these procedures, we have enlarged upon their scope, while attempting to avoid the following two classes of problems.

First, the employment of a test problem with a single starting point, designed to initiate a search through a certain difficult topographical feature, may not realistically test an algorithm's ability to locate the minimum sought. The subsequent search path depends on the precise location of the starting point. It may be that a particular starting point may be to the advantage of one algorithm but not to another. For example, the starting point in Rosenbrock's problem was chosen to bias the search down a descending parabolic valley. However, in practice, the algorithm may well be called upon to find a minimum at the bottom of a descending curved valley but with starting points located throughout the topography. Thus, an innovation in testing is to repeat the same experiment using several different randomly chosen starting points. The precise number used depends upon the criticality of the choice in a particular problem.

Second, computational effort measured in terms of iterations or calls to user supplied subroutines is subject to large discrepancies when comparing algorithms. Iterations generally vary greatly in scope from algorithm to algorithm. The scope and number of user supplied subroutines may also be algorithm dependent, a fact evident when comparing derivative-free with derivative routines and least-squares optimizers with general optimizers. In the former, the amount of computation per call differs, entailing function evaluation alone on the one hand and function and gradient evaluation on the other. In the latter, the number of calls, in addition to the amount of computation per call differs since least-squares optimizers require calls to function vector and Jacobian evaluation routines whereas general optimizers call only the objective function evaluator. Fin-

189

ally, reliance on function evaluations alone as a measure ignores the often considerable overhead computational effort required in obtaining a solution.

The techniques employed in evaluating MINPACK candidates alleviate, to a large extent, these limitations (Hillstrom (1974, 1975)). The first objection is addressed by utilizing a set of twenty randomly generated starting points for each optimization problem instead of only one fixed point. Thus, the effect of biased starts is reduced, the initial estimates simulate those encountered in practice, and a survey is obtained of the algorithm's ability to converge along various paths. Moreover, if an algorithm has been tailored to solve a particular problem with a particular starting point, the random starts will aid in revealing its limited applicability. The second objection is addressed by utilizing measures of both direct and indirect computational effort. The direct computational effort, *i.e.*, the effort expended in the user supplied subroutines, is measured using a modification of the Horner unit assignment technique (Traub (1964)). The Horner unit technique measures the computational effort per function call of general optimizers by assigning one unit to the effort required to calculate an objective function of N variables and adding N units if the N component gradient (first partial derivative) vector is calculated. This assumes that the gradient calculation is N-fold more expensive than the function calculation. However, estimates obtained in this manner may be grossly inaccurate as, for example, when complicated expressions are common to the function and the gradient; then the objective function computation may dominate with relatively little extra effort required to obtain the gradient. The unit assignment technique adopted here is a generalization of the Horner unit scheme permitting greater accuracy in the estimates. The technique is a generalization since direct computation units (DCU) are assigned to all the elements of user supplied optimization software, *i.e.*, the objective function, gradient

function vector and Jacobian matrix. The assign-
ments are made in proportion to the number of operands and operations in the arithmetic assignment
statements of the element with functions such as
EXP, SIN, SQRT etc., which may be required by an
element, accounted for by the addition of units
proportional to the number of their operations and
operands. The gross inaccuracies of the Horner
assignment scheme are reduced since expressions
common to one or more of the above mentioned elements are represented by an appropriate unit assignment upon the first occurrence and are subsequently accounted for as an operand. The proportional representation of the computational effort
required by these elements creates a medium facilitating comparison of derivative with derivative
free optimizers and general with least-squares
optimizers. The DCU measurement technique may be
objected to because it ignores control statements
and differences in execution times of various operation. However, the impact on computational
effort of control statements and the differences
in execution times is difficult to ascertain
since it is obscured by the parallel processing
capacity of current hardware. The point is that,
although the technique avoids the gross inaccuracies of the Horner scheme, it is still only designed to obtain estimates of and comparisons between
the order of computational effort required by the
user supplied software. The indirect or overhead
computational effort is measured in terms of the
number of DCU units per unit of elapsed time. A
large number of units per unit of time indicates
relatively low overhead effort while a small number indicates dominating overhead.

In addition, the following measures of performance are recorded. Reliability is measured by
categorizing the results of each set of twenty
cases of a problem as following: occurrences of
global convergence, occurrences of local convergence, failures to converge within a specified number of function evaluations, and failures to converge for other reasons. Various statistical

191

averages are calculated over the successes for each test problem:

a) the average of DCU units required to obtain a solution;

b) the average vector error; and

c) the average efficiency rates.

The major deficiency of these testing procedures is their problem orientation, which appears to be a fundamental limitation. In addition, the majority of the test problems currently used have been designed to probe for weaknesses in optimization algorithms. However, this "artificial" nature of test problems can be avoided by collecting and employing test problems arising naturally in applications.

A major difficulty encountered in constructing a self-contained computer algorithm is comparison of the contribution of a specific algorithm component to the over-all performance of the algorithm. This difficulty is considerably eased when the computer algorithms are modularized to permit experimentation with different techniques. For example, two general optimizers, which use different Hessian updates, can have their Hessian modules interchanged. A modularized implementation permits the essential areas in which they differ to be isolated and facilitates the implementation of other basic steps in an identical way. Moreover, a more uniform evaluation and comparison of the algorithms is made possible. For example, individual components, e.g. line searches, may be compared not only through interchanges as indicated above but also by stand-alone tests using standard input data for the module.

Finally, it is planned that tests will be run in which minimizations to intermediate accuracies are recorded. In this way, initial, intermediate, and final rates of progress can be monitored and the effects of various starting and convergence criteria can be measured.

3. STRUCTURE OF MINPACK I

The package of algorithms constituting Edition 1 MINPACK I addresses the class of problems specified in Section 2. Initially the candidates tested for MINPACK were self contained in the sense that they did not share subroutines or modules. Indeed, in this form they are still currently being utilized by collaborators in other Argonne scientific and engineering groups to determine user reactions. Experience from the EISPACK and FUNPACK Projects suggested that it was well worth investigating the feasibility of reorganizing the software in a form that emphasized modularity.

However, unconstrained optimization problems and algorithms have some quite distinctive features. For example, only limited information about a function at any point can economically be made available. Algorithms differ in the way they use this information to perform the basic steps of optimization. For example, they differ in the use of:

a) the information obtained from a sequence of successive iterates to deduce further information about the function (e.g., deductions based upon quadratic approximation);

b) the information provided or inferred to develop subsequent search directions; and

c) the search direction to develop subsequent iterates.

Consequently,we faced the question: Does the blending of these and other basic steps into an unconstrained optimization algorithm require the implementation of each step to be so closely interrelated to others as to make the algorithm unique? In order to answer this question a study was conducted which revealed that the routines did, in fact, generally consist of distinct logical components often functionally similar to, if not identical to, corresponding components in other routines.

We, therefore, decided to modularize most of the routines in the first edition of MINPACK I into these components. Our objective was to gain the following advantages of modularized packages found in the EISPACK and FUNPACK Projects:

a) algorithm development, which frequently involves experimentation with only one of the constituent components, is facilitated in that components are readily interchanged and the effects on performance measured;

b) code redundancies are eliminated since common components are shared;

c) library maintainence is simplified since algorithm updates often involve only the modification and replacement of one of the constituent components;

d) algorithm structure and logical flow are improved, since components originally addressed from various places in the algorithm are treated as subprograms;

e) a discipline for analyzing the structure of optimization algorithms is imposed; and

f) within an algorithm, each module used may be exercised independently; therefore, checking the validity of an implementation is facilitated.

Our concept of what constitutes a module and a preliminary categorization evolved through the process of modularizing the algorithms selected for MINPACK I. Modularizing an algorithm involves the following basic steps. First, a module is specified by the identification of a section of the over-all algorithm devoted to performing a well-defined task. Examples of well-defined tasks are initializing matrices, calculating search direction vectors, conducting minimal line searches, and updating hessian or hessian inverse matrices. Second, a determination is made of the module interface, *i.e.*, identification of the information required by the module to perform its specified task and of the information returned by the module

194

necessary for execution of a typical algorithm
utilizing the module. Third, the module is designed, *i.e.*, an implementation of the task and interface is constructed with special emphasis on robustness such as the trapping of underflows, over-
flows, and divide checks whenever feasible. Fourth,
the driver or connecting code is developed. Fifth,
the modularized version is validated against the
original algorithm.

The modularized collection serves as the
basis for an on-going process of experimentation
with and refinement of the package. Thus, for
example, three of the modularized general optimizers originally invoked different implementations
of functionally similar line searches. These optimizers were altered by replacing the three search
modules by one common search module. The package
was thus simplified; moreover, tests demonstrated
that the performance of two routines was preserved
and the over-all performance of the third routine
was greatly enhanced. Numerical evidence for these
assertions is reported (in Hillstrom (1974)).
Further experiments of this nature, e.g., combining Powell's DOGLEG strategy with the updating
strategy developed (in Davidon (1975)) are being
undertaken.

The collection of modules that constitute
the prototype MINPACK I are documented briefly
(Hillstrom (1974)). The following observations
and the discussions in Sections 4 and 5 will be
particularly important in defining future steps in
the project.

The tasks performed by some modules potenti-
ally have a broad range of applicability and represent self-contained tasks, e.g., a line search,
estimation of a Jacobian matrix by differences, or
a Hessian update. Certain other modules, in particular those arising from hybrid algorithms, may
not be self-contained. They essentially derive
meaning from the context within which they are
used. An example is the module which calculates a

195

special correction vector designed to maintain
sufficient linear independence in successive corrections in Powell's hybrid method GQSYM.

The module designs and interfaces reflect
the requirements of the current set of algorithms.
The needs of algorithms to be added in the future
(especially the constrained algorithms to be discussed in Section 5) and of the polyalgorithm (to
be discussed in Section 4) will require a significant expansion of the interface. We anticipate
the need to include more flags and return additional data to give a more thorough record of the
task accomplished. The major deficiency of modularization appears to be the increase in elapsed
time arising from the transfer of information between algorithm driver and modules. Tests comparing modularized routines with their self-contained
counterparts indicate, for small problems (N L.T.
5), a significant increase in elapsed time for the
execution of the modularized version. However,
these increases arise from the transfer of information via calling sequences. If the calling sequences are eliminated by the use of common storage
the modularizations are as efficient in overhead
time as the self-contained versions. Thus, it may
be argued that the deficiency is not a problem of
modularization, but of the way in which information is transferred. These observations also reflect the fact that many of the test functions used
are easy to evaluate whereas in many real life situations the function evaluation is the most costly
item. In terms of the number of DCU's required,
the modularized and unmodularized versions do not
significantly differ.

4. POLYALGORITHM APPROACH TO OPTIMIZATION PROBLEMS

The collection of algorithms in Edition 1
of MINPACK I form the nucleus of MINPACK; however,
certain extensions are necessary to develop a more
effective problem solving tool. The following
user requirements, algorithm deficiences, and pro-

blem characteristics exemplify needs which may call
for the introduction of additional features.

The efficiency and robustness of many local
methods is intimately related to how close the
starting point is to the solution. Often good
starting points are not available and the user
can, at best, supply bounds delimiting a region of
search. With this in mind, Rice and Aird (1975)
have recently proposed an algorithm for systematically distributing a specified number of points
within a hypercube containing the region of search.
The points obtained by this starting point generator may be used in various ways. For example,
Aird (1973) in his least-squares polyalgorithm
carries out a few iterations of an efficient local
method from each such point, using the best point
generated over-all as the starting approximation
from which the local method may be continued to
obtain the accuracy desired. Alternatively, as is
often done, the user may wish to repeat the complete search from each starting point in the hope
of obtaining a global solution.

Rather than making a commitment to a particular strategy, many implementations use hybrid
techniques, for example, choosing between a
Levenberg-Marquardt correction, the steepest descent vector, or a linear combination of the two.
These may be interpreted as making adaptive choices
between different models of the function based
upon information gathered during the search. Polyalgorithms represent a further development in this
direction. As defined by Rice and Rosen (1975),
Rice (1968), a polyalgorithm is "a synthesis of
a group of numerical methods and a logical structure into an integrated procedure for solving a
specific type of mathematical problem". For
example, the least squares polyalgorithm developed
by Aird (1973) utilizes a hierarchy of methods.
A fast local optimizer forms the top level of the
hierarchy. Below it in the hierarchy, "alternative search methods", *c.f.* (Gill and Murray (1974))
are utilized which are more robust but less

efficient. Assisted by user-supplied parameters, a control program monitors and evaluates progress and guides the selection of methods. Parameters currently used in MINPACK I for evaluating algorithms, e.g., average rate of convergence over a number of iterations, could be monitored by this control program and guide its decisions.

Current implementations have often paid inadequate attention to the choice of suitable termination criteria. Two difficulties may be encountered. The first difficulty is that a poor choice of the termination criteria results in an iteration being falsely declared to be convergent. This may occur, for example, with the resolution ridge problem discussed by Brent (1973). The second difficulty is the inability to recognize when rounding errors predominate or to recognize ill-conditioning. This can result in an excessive number of iterations with little improvement. Further work in this area along the lines laid out by Lyness(1969) for quadrature routines is much needed.

Problems arising in applications are often poorly scaled in their initial formulation. While users should be encouraged to scale their problems properly, this cannot always be done, and some automatic scaling technique should be implemented.

Currently, our studies related to a possible future polyalgorithm version of MINPACK are proceeding along two lines.

First, we recognize that the above remarks indicate the need for developing additional modules. A starting point generator is a useful pre-optimizer module. A post-optimizer module incorporating features which permit an analysis of the obtained solution in relation to selected termination criteria is also required. For example, the accuracy testing routine of Aird, which performs a sampling around a point declared to be optimal, and the techniques suggested by Brent (1973) are useful ideas to avoid false convergence. Also algorithms not currently in the MINPACK collection may be required, e.g., the more robust but less efficient end of the hierarchy of a polyalgorithm may utilize the simplex method, Spendley, Hext, and Himsworth (1962); Nelder and Mead (1965); and Parkinson and Hutchinson (1971). We have also experimented with hybrid techniques for nonlinear least squares, Nazareth (1975).

Second, we observe that the development of the control program for a polyalgorithm has many implications. Consequently, a polyalgorithm for a *limited* problem area is currently the focus of experiments. The above-mentioned modules and algorithms must be made available to the control program. Initial experience indicates that this has implications for the way modules are structured, (for example, providing the capability to restart certain modules, in particular, linear searches, from the point where they left off.) If this polyalgorithm proves to be promising, this approach will be extended to the other problem areas addressed by MINPACK.

5. EXTENSION TO CONSTRAINED OPTIMIZATION PROBLEMS

Although Edition 1 of MINPACK I will be limited to solving systems of nonlinear equations, nonlinear least-squares data fitting problems, and unconstrained optimization problems, it will also serve as a basis for expanding the current project to constrained optimization problems. To avoid major changes to MINPACK during the incorporation of capabilities for constrained optimization, a preliminary study has been initiated to assess the impact of modularization on this extension and conversely, the impact of this extension upon modularization. The comments made here are, of course, quite preliminary and deal with only a few

199

of the constrained methods we hope to study.

In examining the extension of Edition 1 of MINPACK I, we first consider approaches to constrained optimization which directly utilize unconstrained methods. In these cases, the unconstrained software of MINPACK I can be used with perhaps minor modifications. For example, the augmented unconstrained transformations of Rockafeller (1974), Fletcher (1973), and Mangasarian (1974) are possible candidates for directly utilizing the unconstrained software of MINPACK I.

On the other hand, certain algorithms will require entirely new modules to deal with well-defined steps which are unique to constrained optimization. In particular, modules will be needed to check feasibility and identify violated constraints, compute projections of search directions, and conduct convergence testing for a candidate for a constrained solution.

In some cases it will be desirable to provide modules which are alternatives to the corresponding unconstrained module. This will be useful in situations where a particular subproblem arising from a constrained algorithm has special properties which can be exploited. For example, the one-dimensional search in an interior penalty method involves a constrained search with a singularity at the boundary. Because of this singularity a polynomial interpolatory scheme, while acceptable for an unconstrained search, should not be used (see Ryan (1974)). Instead, an interpolatory method designed to handle such singularities should be considered as in Murray (1969) and Fletcher and McCann (1969).

Finally, we consider the effect of the extension to constrained algorithms on modularization. The modularization approach to constrained optimization is illustrated by the subroutine structure of several well-known codes, e.g., Buckley (1973), Mylander, Holmes and McCormick (1974), and

Rosen and Wagner (1975) . In examining several
of these codes, we find that the interaction with
the user is, of course, more involved than in unconstrained techniques. Thus, as we extend MINPACK
to constrained optimization we expect, as in the
case of the polyalgorithm, to encounter new sophi-
stications which will help to refine our concept
of modularization.

6. DISCUSSION OF MODULES

Experience gained from the studies described
in the previous sections indicates that three
broad categories of modules may be identified.
These are determined by their primary usage.

a) ANCILLARY MODULES have their primary use within
the context of manipulating an optimization problem
or aiding its solution. Examples are modules to
generate starting points, to perform post-optimal
analysis or to perform automatic rescaling. Such
modules provide the user with a more effective
problem solving aid and will also be of use within
a polyalgorithm.

b) LOGICAL MODULES have their primary use within
the context of development and experimentation
with algorithms. They correspond to a well-defined
optimization subtask, e.g., a derivative free
linear search or a hessian update, and are considered to form a logical and self-contained building block of current and possibly of future algorithms. Logical modules are thus mainly intended
for use by an algorithm developer or within a polyalgorithm.

c) CODE MODULES correspond to the usual notion of
a subroutine and are provided to improve structure,
to eliminate redundancies, and to facilitate validation, maintenance and further development of the
software package. Consequently, there are many
examples of code modules. An inner product routine may be isolated to facilitate its replacement
over the complete package. A specific subroutine
may be introduced to avoid unnecessary duplication
of code. A component of a hybrid algorithm may be

specifically tailored to the over-all design of
this algorithm; therefore, it does not fall into
category *(b)*, but nevertheless is isolated since
this improves structure, and, for example, permits
future replacement by an improved version.

The categories of modules discussed above
are not necessarily disjoint. For example, a code
module may be upgraded to an ancillary or logical
module. Within the basic steps of module development - specification, interface, and design - outlined in Section 3, it is clear that modules in
categories *(a)* and *(b)* require careful attention
to all three aspects particularly to specification
and interface if they are to be made available to
the sophisticated user in much the same way as the
algorithm drivers will be made available to the
everyday user. Similarly careful attention must
be paid to the design stage of code modules in
order to obtain robustness. However, specification
and interface requirements of code modules, although
important, are not quite as pressing since usage
of code modules is internal to the package.

That there are many benefits to be gained
from providing well structured algorithms we take
to be self evident. However, we have not yet
settled the question as to whether certain additional facilities in the form of modules of type
(a) and *(b)* should also be made generally available.
This will require further feedback from users and
algorithm developers.

7. FUTURE PLANS FOR MINPACK

Some aspects of the future plans for the
MINPACK Project have been implicit in the previous
sections. The purpose of this section is to del-
ineate the plans of the MINPACK Project as they
are currently conceived. The primary objective of
the MINPACK Project is to gain understanding of
the problems of systematizing collections of numerical software. Consequently, the MINPACK Project is a research venture rather than a software
development task. To achieve the research goals,

however, it is necessary to develop software just
as the experimental and theoretical physicist
relies upon data from scattering experiments to
plan new experiments and to test new theories. As
in the case of the EISPACK and FUNPACK Projects,
the software developed is sought by the scientist
and engineer for use in computational modelling.
It is, however, essential that pressure from potential users of MINPACK should not be allowed to
deflect the principal thrust of our research,
namely, to understand the interrelations among
components of a complex collection of numerical
software. In common with many major, complex research ventures EISPACK, FUNPACK and MINPACK have
each experienced one major project "restart".
Based on our experience with EISPACK and FUNPACK,
we realize that the plans described here represent
honest intentions rather than firm commitments and,
furthermore, that the time schedules are only
approximate. With these disclaimers, the following represents a summary of the plans for the MIN-
PACK Project. The three plateaux of the MINPACK
Project are referred to as MINPACK I, MINPACK II
and MINPACK III.

Briefly, MINPACK I is conceived as a modularized collection of optimizers (both unconstrained
and constrained) and system solvers available on a
restricted range of computers (perhaps only IBM
370 and CDC) for experimental use. Edition 1 of
MINPACK I will be restricted to unconstrained optimization problems; however, subsequent editions
are expected to feature capabilities for solving
constrained optimization problems. It is currently
anticipated that Edition 1 of MINPACK I will be
released to NATS Test Sites for exploration of the
problems of software validation in the fall of
1975. This step, however, points out an essential
weakness in the testing procedures employed previously in the MINPACK Project. While these procedures have been adequate for evaluation and comparison of algorithms, they are totally inadequate
for rigorous validation of software. A key to the
success of the EISPACK and FUNPACK Projects has

been the availability of techniques to detect minute algorithmic and implementation errors. We are
planning an extensive investigation into the applicability of techniques based on performance pro-
files, which have already been developed by Lyness
and Kaganove (1972, 1975) to evaluate automatic
quadrature routines. If adequate validation techniques can be developed and if procedures for exe-
cuting these tests at the NATS Test Sites can be
formulated and implemented, then Edition 1 of MIN-
PACK I will be released for experimental use after
satisfactory completion of tests.

Through the addition of further modules,
algorithms and control programs, it is intended
that the core of algorithms comprising MINPACK I
will gradually evolve towards polyalgorithms for
general optimization, least squares optimization
and solution of systems of equations. The fact
that MINPACK I has been modularized simplifies
this task considerably. However, the exact definition of the logical structure required by the
polyalgorithm is anticipated to accelerate the
progress in systematically developing module specifications, interfaces and designs. Achievement
of polyalgorithms satisfying the performance characteristics outlined for MINPACK I will represent
completion of MINPACK II.

MINPACK III will represent an attempt at a
level of quality comparable to EISPACK and FUNPACK.
Because of the nature of the optimization problem,
the precise criteria by which high quality software is judged cannot be the same as those used in
FUNPACK and EISPACK. This is because it is always
possible to provide an objective function and
starting vector combination which will defeat any
specified software. One of the challenges in this
research venture is to devise alternate generally
acceptable criteria. This is a major hurdle and
it would be premature to attempt to estimate a
timescale for completing this. However, if and
when this is done, we shall be ready to produce
MINPACK III.

8. CONCLUDING REMARKS

We hope that the users of optimization software present at this conference will communicate their needs and their criticism of our interpretations of the users' needs to us. Many of the problems posed by these needs which have been addressed in the previous sections still await satisfactory solutions. Therefore, we hope that these problems will stimulate researchers in the field of optimization present at this conference to find solutions suitable for implementation and experimentation within the context of the MINPACK venture.

ACKNOWLEDGEMENTS

The authors wish to acknowledge the insights they have gained from the critiques of George D. Byrne, William J. Cody, and James N. Lyness.

APPENDIX

A partial list of algorithms (drivers) currently considered for inclusion in the first edition of MINPACK I is as follows.

System Solvers

SBROWN - a modification of a routine for solving simultaneous nonlinear equations written by K.M. Brown (1973) which employs a derivative free quadratically convergent Newton-like method based upon Gaussian elimination.

SNGINT - a modularized modification of the nonlinear simultaneous system solving Harwell library routine NSO1A written by M.J.D. Powell (1970 *a* and *b*) which combines a derivative free quasi-Newton and a steepest-descent method.

SAMULL - a modification of a routine which finds all the zeros of one analytic function of a single complex variable written by K.M. Brown which is based upon Muller's method with deflation.

General Optimization

GQSYM - a modification of the function minimization Harwell library routine VAO6A written by M.J.D. Powell (1970 *c*) which combines quasi-Newton and steepest-descent methods where the quasi-Newton technique is based on a symmetrized update (M.J.D. Powell (1970 *d*)).

GCFLRV - a modularized modification of the function minimization Harwell library routine VAO8A written by R. Fletcher (1972 *a*) which uses the Fletcher-Reeves version of the conjugate gradient technique.

GQBFGS - a modularized modification of the original Davidon-Fletcher-Powell quasi-Newton function minimization algorithm (1963), but with a BFGS update (1972 *b*).

GQSWCH - a modularized modification of the function minimization Harwell library routines VAO9 and VA1OA written by R. Fletcher (1972 *b*) which uses a quasi-Newton algorithm based on a switching strategy employing the DFP and BFGS updates.

GQOCON - a modularized modification of an optimally conditioned search free optimization algorithm written by W.C. Davidon (1975).

GCPRAX - a modularized modification of M.J.D. Powell's conjugate direction method (1964) as extended by R.P. Brent (1973).

Least-Squares Optimizers

LMGENV - a modularized modification of the least-squares optimization Harwell library routine VAO5A written by M.J.D. Powell which combines derivative free Levenberg-Marquardt and steepest-descent methods where the Levenberg-Marquardt equations are solved by a generalized inverse method.

LMCHOL - a modularized modification of the least-squares optimization Harwell library routine VAO8A by R. Fletcher (1970) based on a modified Levenberg-Marquardt algorithm where the Levenberg-Marquardt equations are solved by Cholesky decomposition.

REFERENCES

Aird, T.J. (1973) "Computational Solution of Global Nonlinear Least-Squares Problems", PhD Thesis, Purdue University.

Boyle, J.M., Cody, W.J., Cowell, W.R., Garbow, B. S., Ikebe, Y., Moler, C.B. and Smith, B.T. (1972) "NATS, a collaborative effort to certify and disseminate mathematical software", *in* "Proceedings 1972 National ACM Conference", Vol. II, 630-635.

Brent, R.P. (1973) "Algorithms for Minimization without Derivatives", Prentice Hall, Englewood Cliffs, New Jersey, 116-167.

Brown, K.M. (1973) "Computer Oriented Algorithms for solving systems of simultaneous nonlinear algebraic equations", *in* "Numerical Solution of Systems of Nonlinear Equations", *Eds.* G.D. Byrne and C.A. Hall, Academic Press, New York, 281-348.

Buckley, A. (1973) "An Alternate Implementation of Goldfarb's Minimization Algorithm", AERE Harwell Report, T.P. 544.

Cody, W.J. (1975) "The FUNPACK Package of Special Function Subroutines", *Trans. on Mathematical Software,* 1.

Davidon, W.C. (1975) "An optimally conditioned optimization algorithm without linear searches", *Math. Prog.*, **9**, 1-30.

Fletcher, R. (1970) "A modified Marquardt subroutine for nonlinear least-squares", Report No. R-6799, AERE, Harwell, England.

Fletcher, R. (1972*a*) "A FORTRAN subroutine for minimization by the method of conjugate gradients", Report No. R-7073, AERE, Harwell, England.

Fletcher, R. (1972*b*) "FORTRAN subroutines for minimization of quasi-Newton Methods", Report No. R-7125, AERE, Harwell, England.

Fletcher, R. (1973) "An ideal penalty function for constrained optimization", UKAEA Research Group Report CSS2.

Fletcher, R. and McCann, A.P. (1969) "Acceleration techniques for nonlinear programming", *in* "Optimization", *Ed.* R. Fletcher, Academic Press.

Fletcher, R. and Powell, M.J.D. (1963) "A rapidly convergent descent method for minimization", *Computer J.*, **6**, 163.

Gill, P.E. and Murray, W. (1974) "Numerical Methods for Constrained Optimization", Academic Press.

Hillstrom, K.E. (1974)"MINPACK I - A study in the modularization of a package of computer algorithms for the unconstrained nonlinear optimization problem", Technical Memorandum 252, Applied Mathematics Division, Argonne National Laboratory.

Hillstrom, K.E. (1975) "A simulation test approach to the evaluation and comparison of unconstrained nonlinear optimization algorithms", *in preparation*.

Lyness, J.N. (1969) "The effect of inadequate convergence criteria in automatic routines", *Computer J.*, **12**, 279-281.

Lyness, J.N. (1972) "Guidelines for Automatic Quadrature Routines", *in* "Proceedings of the IFIP Congress 1971", North Holland, Amsterdam 1351-1355.

Lyness, J.H. and Kaganov, J.J. (1976) "Comments on the Nature of automatic quadrature routines", *ACM Trans. of Maths. Software,*2, No. 1, 65-81.

Mangasarian, O.L. (1974) "Unconstrained methods in optimization", University of Wisconsin Computer Sciences Report 224.

Murray, W. (1969) "Constrained optimization", Report Ma79, National Physical Laboratory, Teddington, England.

Mylander, W.C., Holmes, R.L. and McCormick, G.P. (1974) "A Guide to SUMT-Version 4", Research Analysis Corporation.

Nazareth, J.L. (1965) "A Hybrid Least Squares Method", Tech. Memo 254, Applied Mathematics Division, Argonne National Laboratory.

Nelder, J.A. and Mead R. (1965) "A simplex method for function minimization, *Computer.J.,*7, 308-313.

Parkinson, J.M. and Hutchinson, D. (1971) "An investigation into the efficiency of variants of the simplex method"*in* "Numerical Methods for Nonlinear Optimization", *Ed.* F.A. Lootsma, Academic Press, London 115-147.

Powell, M.J.D. (1964) "An efficient method for finding the minimum of a function of several variables without calculating derivatives", *Computer J.,* 7, 155.

Powell, M.J.D. (1970*a*) "A hybrid method of nonlinear equations", *in* "Numerical Methods for Nonlinear Algebraic Equations", *Ed.* P. Rabinowitz, Gordon and Breach, London, 87-114.

Powell, M.J.D. (1970*b*) "A FORTRAN subroutine for solving systems of nonlinear algebraic equations", *in* "Numerical Methods for Nonlinear Algebraic Equations" *Ed.* P. Rabinowitz, Gordon and Breach, London,115-161

Powell, M.J.D. (1970*c*) "A FORTRAN subroutine for unconstrained minimization requiring first derivatives of the objective function", Report No. R-6469, AERE, Harwell, England.

Powell, M.J.D. (1970*d*) "A new algorithm for unconstrained optimization", Report No. T.P.-393, AERE, Harwell, England.

Rice, J.R. (1968) "On the construction of polyalgorithms for automatic numerical analysis", *in* "Interactive Systems for Experimental Applied Mathematics", *Eds*. M. Klever and J. Reinfelds, Academic Press, 301-313.

Rice, J.R. and Aird, T.J. (1975) "Systematic search in high dimensional sets", *to appear*.

Rice, J.R. and Rosen, S. (1966) "NAPSS, A numerical analysis problem solving system", *in* "Proceedings of the ACM National Conference at Los Angeles", ACM Publication P-66, 51-66.

Rockafellar, R.T. (1974) "Augumented Lagrange Multiplier functions and duality in nonconvex programming", *SIAM J. Control,* 12(2), 268-285.

Rosen, J.B. and Wagner, S. (1975) "GPM Nonlinear Programming Package Instruction Manual", University of Minnesota Computer Information Sciences Tech. Report 75-14.

Ryan, D.M. (1974) "Penalty and barrier functions", UKAEA Research Group Report, CSS4.

Smith, B.T., Boyle, J.M. and Cody, W.J. (1974) "The NATS approach to quality software, Software for Numerical Mathematics" *in* "Proceedings of the IMA Conference on Software for Numerical Mathematics, Loughborough University, April 1973" *Ed.* D.J. Evans, Academic Press, 393-405.

Smith, B.T., Boyle, J.M., Garbow, B.S., Ikebe, Y., Klema, V.C. and Moler, C.B. (1974) "Matrix Eigensystem Routines - EISPACK Guide", Lecture Notes in Computer Science, 6, Springer-Verlag, Heidelberg.

Spendley, W., Hext, G.R. and Himsworth, F.R. (1962) "Sequential application of simplex designs in optimization and evolutionary operations", *Technometrics,* 4, 441.

Traub, J.F. (1964) "Iterative Methods for the Solution of Equations", Prentice Hall, Englewood Cliffs, New Jersey.

Note added in proof: The references to Buckley (1973) and Ryan (1974) have recently appeared as follows.

Buckley, A. (1975) "An alternate implementation of Goldfarb's minimization algorithm", *Math. Programming* **8**, 207-231.

Ryan, D. M. (1974) "Penalty and barrier functions", in "Numerical Methods for Constrained Optimization", *Ed.* P. E. Gill and W. Murray, Academic Press, New York, 175-190.

DISCUSSION SESSION ON UNCONSTRAINED OPTIMIZATION

P.E. Gill: Mr. Powell, you mentioned that to prove convergence with the variable metric method, two conditions such as (3.7) and (3.8) are required (see p. 129-130).

I would like to draw your attention to an NPL report* in which a step length rule based upon equations (3.7) and (3.8) has <u>already</u> been analyzed and implemented.

In the NPL suite of algorithms, safeguarded cubic or quadratic interpolation is used until (3.7) is satisfied. If this step length does not satisfy (3.8) it is reduced by a suitable fixed multiple. We have found in practice that if $c_2 \lll c_1$ <u>the first step length need never be reduced in order to satisfy (3.8)</u>.

If an initial step of unity is used to commence the cubic or quadratic interpolation we have found that, with a value of c_1 = 0.9, an average of 1.1 function evaluations are required per iteration.

The parameter c_1 is set by the user and it is possible to vary the accuracy of the linear search depending upon the type and difficulty of the problem being solved.

M.J.D. Powell: I agree that with Fletcher's implementation most iterations work with $\lambda = (0.1)j$, $j=0$ and the step is very rarely reduced.

L.C.W. Dixon: In our implementations we have chosen not to use (3.7), but have chosen instead yet another of Wolfe's conditions

$$F(\underset{\sim}{x}^{(k)} + \lambda^{(k)} \underset{\sim}{q}^{(k)}) \geqslant F(\underset{\sim}{x}^{(k)}) + c_3 \lambda^{(k)} (\underset{\sim}{q}^{(k)T} \underset{\sim}{d}^{(k)}) \tag{3.7b}$$

where $c_3 \geqslant c_2$ and usually $c_3 = 1 - c_2$ and $c_2 = 0.001$.

---

*Gill, P.E. and Murray, W. (1974) "Safeguarded step length algorithms for optimization using descent methods," Nat. Phys. Lab. NAC 37.

We have chosen this alternative for two reasons.
First it avoids the necessity to calculate
$g(\underset{\sim}{x}^{(k)}+\lambda^{(k)}\underset{\sim}{d}^{(k)})$ at points that are going to be
rejected which seems an unnecessary expense and
secondly it is obvious that there are $\lambda^{(k)}$ that
satisfy (3.7b) and (3.8) but we have never been
convinced there must be a solution to (3.7) and
(3.8). If

$$F(\underset{\sim}{x}^{(k)}+\lambda^{(k)}\underset{\sim}{d}^{(k)}) \geqslant F(\underset{\sim}{x}^{(k)})+(1-c_2)\lambda^{(k)}(g^{(k)T}\underset{\sim}{d}^{(k)})$$

then the function has decreased significantly
better than the linear prediction, would you advocate continuing the linear search in these circumstances?

M.J.D. Powell: It is quite straightforward to prove
that $\lambda^{(k)}$ exist which satisfy (3.7) and (3.8).

If the function decrease is better than linear you may well have entered a nonconvex region
and the value of $d^T y$ may lead to non-positive definite matrices $H^{(k)}$. It therefore seems better
to continue the line search by seeking a larger
value of $\lambda^{(k)}$ which will satisfy (3.7) and (3.8).

P.E. Gill: With reference to your comments below
equation (6.1), we introduce the bound on the
condition number for reasons of numerical expedi-
ency. It is no good having an excellent proof of
convergence for an algorithm and then performing
the calculation with a Hessian matrix that is so
ill-conditioned that there are no significant figures accurate in the resultant direction of search.
At least the bound on the condition number guarantees you will get some measure of accuracy in the
direction of search.

M.J.D. Powell: You must agree surely that one of
the nice properties of the variable metric method
is that it is truly invariant under scaling of the
variables. The introduction of such a bound des-
troys this property.

P.E. Gill: But the invariancy is only true if you are working in infinite precision.

M.J.D. Powell: It is possible, by changing the scale of the variables, to violate completely the bound on the condition number and yet to obtain an accurate direction even when working to an accuracy of 3 decimal places.

W. Murray: If you have a set of linear equations with a condition number of 10^8 then you cannot in general solve them on a machine with 3 figure precision.

A. Curtis: I agree with Powell that a bad condition number is not a sufficient condition for a bad solution.

W. Murray: I agree with Mr. Curtis but none-the-less, if in ninety-nine cases out of one hundred an ill-conditioned system leads to an inaccurate search direction, one can improve the general efficiency of an algorithm by anticipating this. The adverse effect of using an inaccurate search direction is often discovered only after several wasted function evaluations. In the rare case when the search direction would have worked the effect on the search direction obtained by altering the system to comply with a very large bound on the condition number would very rarely incur any inefficiency and would not in any case cause the algorithm to fail.

D. Goldfarb: May I return to the discussion on the linear search. Did you say that you increase λ if condition (3.7) is not satisfied?

M.J.D. Powell: You increase λ when (3.7) is not satisfied provided that condition (3.8) holds. If you have negative curvature then you must continue to increase λ until the convex region is re-entered. As the function will be decreasing fast while the negative curvature prevails, this is a good technique.

Could I now ask Mr. Massara a question?   In
your paper you compare algorithms on the basis of
iterations. The variable metric DFP approach uses
derivatives but the pattern search does not.
Could you tell us how you took account of this in
making the comparison?

R.E. Massara: The DFP and pattern search iterations are more directly comparable than might
appear to be the case: it is necessary to evaluate the partial derivatives of the coefficients
to implement the pattern search scheme (see (4),
p.157) as it is to implement the DFP gradient
evaluations. Because of the linear dependence
of the coefficients on the variables, it turns
out that if we estimate the coefficient derivatives by finite differencing, the results are
exact. We therefore use the same method for both
optimization strategies.
      In any case, the curves in Fig. 4 (p.162)
are scaled so that the abscissae effectively measure equal computational cost. This was arranged
for by allowing the basic DFP and APM- DFP two-
stage strategies to run for the same time as had
the APM strategy previously. The DFP iteration
counts were then scaled prior to plotting so that
all curves occupied the equivalent abscissa length
of 280 APM iterations. Hence the coordinate direction of labelled "ITN" actually measures approximate CPU time and, since the storage requirements
of the three algorithms do not differ significantly, approximate computational cost. A corresponding approach was used in Fig. 6 (p.164). This is
the only basis on which meaningful engineering
comparisons can be made.

Anon: Dr. Pool, one of the essential features in
writing such a package is presumably to have
built up a large number of satisfied customers in
earlier projects, but I gather from what you said
that you do not intend initially to circulate your
programs to your customers for testing. Would
you care to comment?

J.C.T. Pool: Within our laboratory we have many
user divisions and material that is only available
in the laboratory can therefore still be tested on
many large projects.

We plan to make candidate routines for MIN-
PACK available for outside testing; however, we
must have sufficient control to assure the feedback
necessary for improving these candidates.

You will appreciate that this is a completely new area for us. Indeed it was chosen for that
very reason. We wished to investigate whether the
ideas that had been successfully developed during
the FUNPACK and EISPACK projects could be successfully transferred to a completely new area.

---

*Our acknowledgements are due to I. Dawson whose
tape recordings of the session form the basis of
the above.

Methods for Constrained Optimization

Walter Murray

(National Physical Laboratory, Teddington)

SUMMARY

The paper reviews the basic nature of constrained problems and attempts to give a geometric insight into the workings of various algorithms designed to solve them. There are several different types of constrained problems. These are categorized and an indication given as to which algorithms are suitable for which problem. It is hoped the paper will also act as an outline of the IMA/NPL conference on "Numerical Methods for Constrained Optimization" and will, therefore, be an aid to anyone wishing to study the proceedings of that conference.

1. INTRODUCTION

In this paper we shall consider the general constrained problem:

P $\qquad$ minimize $\{F(x)\}$
$\qquad\qquad\quad x$

subject to $\qquad x \in R \subset E^n$.

Thus in this problem, in contrast with the unconstrained case, possible solutions are restricted to some subregion R of n-dimensional space, normally termed *the feasible region*. If $x \in R$ then x is said to be feasible and conversely if $x \notin R$ it is said to be *non-feasible*. Usually R is defined by a set of linear and nonlinear functions termed *constraints*, *i.e.*, $x \in R$ if

$$c_i(x) \geqslant 0 \qquad i = 1, 2, \ldots, m.$$

217

If subroutines are available to compute $c_i(x)$ for some given x then it is easy to check that this point is feasible. Most algorithms would require this facility, but strictly all that is necessary is some means of identifying feasible points.

A large number of algorithms have been designed to solve the general constrained problem. In part this is due to the varying character of practical applications but it is also a reflection of the rapid development of research in this area and of inadequate means of easily determining the best algorithms. It is not possible, therefore, to review all algorithms in a single paper even superficially; instead a general description of *classes* of algorithms will be given leaving the reader to seek more detailed information elsewhere. This task has recently been made easier by the availability of the proceedings of a previous IMA conference on optimization (Gill and Murray (1974)), which dealt specifically with constrained problems. This paper could be viewed as a synopsis of those proceedings.

For the understanding of algorithms it is necessary to have some appreciation of the features of a problem that have an important bearing on its solution. In the unconstrained case it is generally known that if $F(x)$ has continuous first derivatives, the gradient is zero only at local minima and other stationary points. The gradient vector, can, therefore, nearly always be used to generate a direction along which $F(x)$ decreases. There is obvious merit when minimizing a function to generate a sequence of estimates for which the function is monotonically decreasing. The theoretical basis of algorithms for constrained optimization is more complex and most algorithms generate a sequence of estimates for which neither $F(x)$ nor any other related function is monotonically decreasing. It is, therefore, not possible to decide which of two points is a better estimate of the solution. Each algorithm will require some consistent means of deciding which point to choose but it could be that different algorithms take opposing decisions since there is no universal agreement on which is the better point. In aiming to give some insight into the problem, we shall simply quote results without proof, giving some explanation of their origin. The reader may then be able to follow a more rigorous development elsewhere.

In two dimensions a typical problem, say

$$\min\{F(x)\}$$

subject to
$$c_1(x) \geqslant 0$$
$$c_2(x) \geqslant 0, \quad (x) = (x_1, x_2)$$

could have the following pictorial representation.

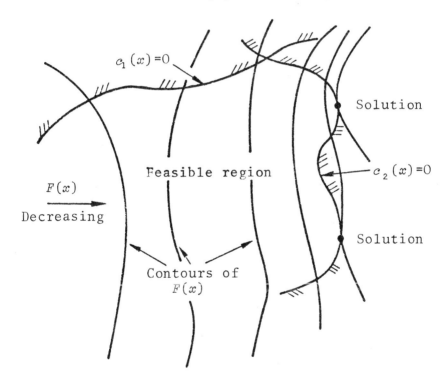

Fig. 1.

Simply by inspecting Fig. 1 a number of observations can be made.

(i) There could be more than one solution (by solution we mean a local minimum with respect to points in R even if the *unconstrained* problem has a single global minimum).

(ii) The curvature of both $F(x)$ *and* the constraint functions play a rôle in defining the solution.

(iii) There could be redundancy in the statement of the problem since in the example illustrated in Fig. 1 the solutions are unaltered by removing $c_1(x) \geqslant 0$ from the set of constraints. The constraint $c_2(x)$ is said to be *active* or binding at the solution.

Clearly algorithms for constrained optimization should have a facility for determining the constraints relevant to the problem. If the constraints active at the solution are known we can replace the original problem with a *reduced* problem involving only equality constraints, *i.e.*,

$\min\{F(x)\}$

$c_1(x) \geqslant 0$

$c_2(x) \geqslant 0$

.
.
.

$c_m(x) \geqslant 0$

$\min\{F(x)\}$,

$\hat{c}_1(x) = 0$

.
.

$\hat{c}_t(x) = 0$

$t \leqslant m$,

where $\hat{c}_1(x) = c_{i_1}(x)$ etc. and $i_1 \ldots i_t$ are the suffices of the active constraints. Equality constraints are usually easier to deal with than inequality constraints. If from physical considerations a constraint is known to be active then it is better to include this in the problem as an equality.

An exception to this rule is when the method used is based on barrier functions (see section 6), since such algorithms are unable to cater for equality constraints. It is however possible to adopt the "reverse" transformation if it is known on which side of the constraint the function is decreasing at the solution.

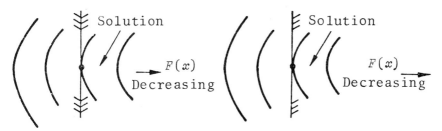

Equality constraint   Inequality constraint
Fig. 2.

Fig. 2 illustrates this with one constraint in two dimensions. In general

$$\min\{F(x)\} \qquad\qquad \min\{F(x)\}$$
$$c_i(x) = 0 \quad\implies\quad (-1)^{j_i}c_i(x) \geqslant 0,$$
$$i = 1,\ldots,m$$

where $j_i = 1$ or 2 depending on whether we want $-c_i(x) \geqslant 0$ or $c_i(x) \geqslant 0$ in order to force the constraint to be binding. It may be thought that by making this transformation we have discarded useful information but the procedure for determining the active set can be modified accordingly.

An equivalent problem to P, which emphasizes the importance of constraint curvature, is the following

$$\min\{F(x)\} \qquad\qquad \min\{x_{n+1}\}$$
$$c(x) \geqslant 0 \quad\implies\quad c(x) \geqslant 0$$
$$x \in E^n \qquad\qquad x_{n+1} - F(x) \geqslant 0 \qquad x \in E^{n+1}.$$

Since the objective function is now linear the curvature of $F(x)$ can be included only by considering constraint curvature.

It would not normally be advantageous to make this transformation although some algorithms have been devised that assume $F(x)$ is linear. In fact an algorithm ought to be invariant (generate the same sequence of estimates) under this transformation.

2. LINEAR CONSTRAINTS

An important class of constrained problems are those in which the constraints are all linear and these are referred to as *linearly constrained problems* (LCP). Advantage can be taken of this linearity to devise special algorithms that are very efficient. In fact this is one class of constrained problems where the algorithms are almost as refined as in the unconstrained case. Study of linearly constrained problems has merit since problems of this type occur in practice. The techniques developed for LCP are still applicable to problems with a mixture of linear and nonlinear constraints. Moreover it is possible to transform the nonlinear constrained problem into the solution of a sequence of LCP.

We shall discuss only the case where all the constraints are inequalities but the extension to the case with a mixture of equalities and inequalities is straightforward. The problem of concern is, therefore,

$$\min\{F(x)\}$$

subject to $\qquad c(x) = A^T x - b \geqslant 0,$

where A is an $n \times m$ matrix and b is an $m \times 1$ vector. At the solution only a subset of these constraints will be active and these will be denoted by

$$\hat{A}^T x = \hat{b},$$

where the matrix $\hat{A}^T$ consists of $t \leqslant m$ rows of the matrix A^T and $\hat{b}$ contains the corresponding elements of b. Fig. 3 illustrates a typical two-dimensional problem.

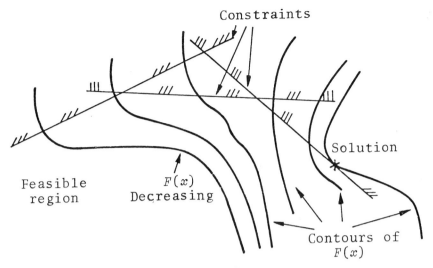

Constraints

Solution

Feasible
region

$F(x)$
Decreasing

Contours of
$F(x)$

Fig. 3.

It is essential to know which properties characterize the solution $\overset{*}{x}$. In the unconstrained case
if $F(x) \in c^1$ we know that $g(\overset{*}{x}) = 0$, where $g(x)$ is
the gradient vector of $F(x)$. Fig. 4 illustrates
that this is not the case for the constrained
problem. Consider the simple case of minimizing a
two-dimensional function subject to a single active
constraint. We shall assume $F(x) \in c^2$ (second
derivatives continuous).

Fig. 4(ii) depicts the variation in $F(x)$
along the constraint. By inspection, one characteristic of the solution is that the component of
the gradient in the plane of the constraint is
zero, $i.e.$, if z is a vector such that $z^T a = 0$,
$\| z \| \neq 0$ then $z^T g(\overset{*}{x}) = 0$. In fact in this plane we
are simply determining an unconstrained minimum.
In the plane of the constraints we also have posi-
tive curvature which implies $z^T \overset{*}{G} z \geqslant 0$, where $\overset{*}{G}$ is
the Hessian matrix of $F(x)$ evaluated at $\overset{*}{x}$. Consider
now a three-dimensional example with two active
constraints.

223

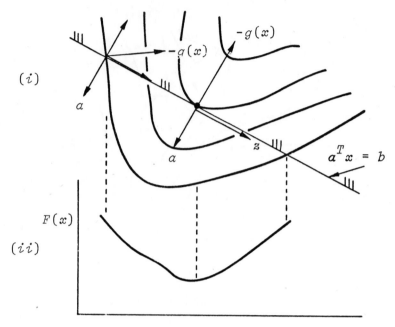

(i)

$-g(x)$

$-g(x)$

a

a

z

$a^T x = b$

(ii)

$F(x)$

Fig. 4.

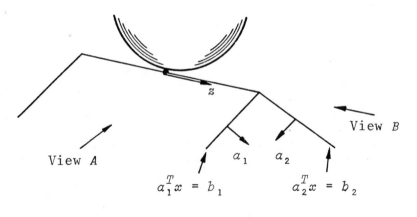

View B

View A

z

a_1

a_2

$a_1^T x = b_1$

$a_2^T x = b_2$

Fig. 5.

Fig. 5 depicts two constraints forming a ridge (like a roof, with the feasible region inside). The surface of the ball represents one of a family of concentric spherical contours whose value decreases as the radius of the contour decreases. If we imagine a ball held just touching this ridge then the point of contact is the solution.

The vector z lying along the ridge is obviously orthogonal to the constraint normals a_1 and a_2, *i.e.*, $a_1^T z = a_2^T z = 0$. Since z is tangential to the sphere it is orthogonal to the radius at that point which is along the vector $-g(\overset{*}{x})$ hence $z^T g(\overset{*}{x}) = 0$. Figs. 6 and 7 depict the variation in $F(x)$ along z and the view from B respectively.

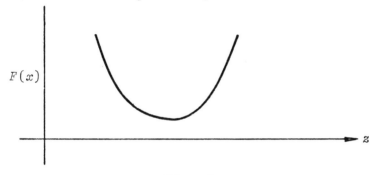

Fig. 6.

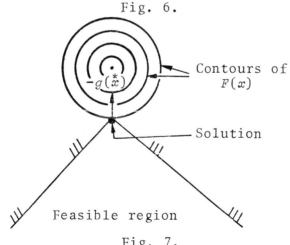

Contours of $F(x)$

Solution

Feasible region

Fig. 7.

Although there is positive curvature both along z and in the plane (view from B) orthogonal to z the latter is not necessary in order to define $\overset{*}{x}$. For example, we could alter the function so as to deform the contours in the plane orthogonal to z and this is depicted in Fig. 8.

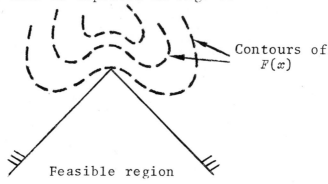

Contours of $F(x)$

Feasible region

Fig. 8.

The solution remains the same but $F(x)$ now has positive curvature only in the z direction.

Sufficient conditions for the solution of LCP

Let the active constraints at the solution be

$$\hat{A}^T_x = \hat{b}$$

and assume $\hat{A}$ has full column rank t. Define z to be an $n \times (n-t)$ matrix of full column rank such that

$$\hat{A}^T z = 0. \qquad (2.1)$$

If the following four conditions hold:

(i) $\hat{A}^T\overset{*}{x} = \hat{b}$,

(ii) $\bar{A}^T\overset{*}{x} \geqslant \bar{b}$, (redundant constraints)

(iii) $z^T g(\overset{*}{x}) = 0$,

(iv) $z^T \overset{*}{G} z$ is positive definite,

then $\overset{*}{x}$ is a strong local minimum of the LCP.

226

It has been assumed that we know which of the constraints are active at the solution. It is possible, as Fig. 9 illustrates, for a point to satisfy the four conditions just stated without it being a strong local minimum because it does not lie on the correct subset of constraints. What is required is some test which will indicate whether

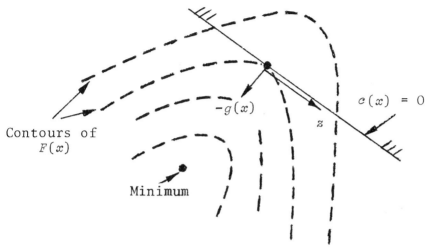

Contours of $F(x)$

$-g(x)$

$c(x) = 0$

z

Minimum

Fig. 9.

we have the correct active set. Since the test is crucial to the whole of constrained optimization and is simple to derive it is worthwhile to state it in full. Obviously $\overset{*}{x}$ will not be the solution if there exists an arbitrary small perturbation δ for which $F(\overset{*}{x} + \delta) < F(\overset{*}{x})$ and $\overset{*}{x} + \delta$ is feasible. Consider a perturbation in $\overset{*}{x}$, say δ, which stays on all the constraints active at $\overset{*}{x}$ except the jth. Then

$$\hat{A}^T(\overset{*}{x} + \delta) = \hat{b} + \alpha e_j \qquad (2.2)$$

and since we wish the perturbation to be feasible then $\alpha > 0$. It follows from (2.2) that

$$\hat{A}^T\delta = \alpha e_j.$$

227

The vector δ can be written as

$$\delta = \hat{A}w + \not{z}v, \qquad (2.3)$$

where w is given by

$$w = \alpha(\hat{A}^T\hat{A})^{-1}e_j,$$

(premultiply by $\hat{A}^T$ and recall $\hat{A}^T\not{z} = 0$) and v is an arbitrary $(n-t) \times 1$ vector. Substituting for w in (2.3) gives

$$\delta = \alpha\hat{A}(\hat{A}^T\hat{A})^{-1}e_j + \not{z}v. \qquad (2.4)$$

Now $F(\overset{*}{x} + \delta) < F(\overset{*}{x})$ for arbitrary small α only if the gradient along δ is negative, *i.e.*, if

$$\overset{*}{g}^T\delta = \alpha\overset{*}{g}^T\hat{A}(\hat{A}^T\hat{A})^{-1}e_j + \overset{*}{g}^T\not{z}v, \qquad (2.5)$$

is negative where $\overset{*}{g} = g(\overset{*}{x})$.

Define $\lambda = (\hat{A}^T\hat{A})^{-1}\hat{A}^T\overset{*}{g}$ and recall that $\not{z}^T\overset{*}{g} = 0$ then (2.5) reduces to

$$g^T\delta = \alpha\lambda_j.$$

It follows that if $\lambda_j > 0$, $j = 1,\ldots t$, there is no arbitrary small perturbation δ for which $F(\overset{*}{x}+\delta) < F(\overset{*}{x})$ and $\overset{*}{x}+\delta$ is a feasible point lying on all but one of the constraints active at $\overset{*}{x}$. It is a simple exercize to show that this result implies there is no arbitrary small perturbation δ for which $F(\overset{*}{x}+\delta) < F(\overset{*}{x})$ and $\overset{*}{x}+\delta$ is feasible. Thus we augment the four conditions already given with

(*v*) $\lambda_i > 0$ $i = 1,\ldots,t$, where

$$\lambda = (\hat{A}^T\hat{A})^{-1}\hat{A}^T g(\overset{*}{x}).$$

3. ALGORITHMS FOR LCP

All the algorithms for unconstrained problems can be generalized to the linearly constrained problem. If this is done carefully almost all the properties of the unconstrained algorithm can be retained. Obviously some algorithms will be easier to adapt than others and the adaptation will alter their relative merits. For instance, whereas conjugate-gradient procedures are only required to store two or three n-vectors, in unconstrained problems, quasi-Newton methods are required to store an $n \times n$ symmetric matrix. However, in constrained problems this symmetric matrix may be of dimension much smaller than n. Moreover *both* methods now require the storage of the matrix of constraint coefficients and factorizations based on these coefficients. The relative storage requirement of the two methods has, therefore, drastically altered and what was a major advantage is now of minor significance only. There is a similar erosion in the advantages of quasi-Newton methods over Newton-type methods (methods where a direct finite-difference approximation to the Hessian matrix is made at each iteration). In the unconstrained case Newton-type methods require n additional gradient evaluations at each iteration. For $n > 10$, this extra computation is usually sufficient to counter the savings made in the reduced number of function evaluations and iterations required by the method. If, on the other hand, there are linear constraints on the problem, the algorithm can be arranged so that the number of additional gradients required at a given iteration is $n-t$, where t is the number of constraints active. Even if t is not large at the solution compared with n, the average t over all iterations can usually be made significant. With this and other linearly constrained algorithms it may even be advantageous to use them to solve problems whose solution is known not to lie on any constraints.

Active set strategy

The basis of many algorithms for linearly constrained problems is to assume that some set of the constraints active at $x^{(k)}$, the kth iterate, will also be active at $x^{(k+1)}$. We define

$$x^{(k+1)} = x^{(k)} + \alpha^{(k)} p^{(k)},$$

where $p^{(k)}$ is called the search direction and $\alpha^{(k)}$

is the step-length. If $\hat{A}^{(k)T}x^{(k)} = \hat{b}^{(k)}$ is the required active set then the search direction $p^{(k)}$ is chosen to satisfy

$$\hat{A}^{(k)T}p^{(k)} = 0. \qquad (3.1)$$

Clearly

$$\hat{A}^{(k)T}x^{(k+1)} = \hat{A}^{(k)}x^{(k)} = b^{(k)}.$$

If there are less than $n-1$ active constraints there is a choice of directions which will satisfy (3.1). The search direction is made unique by using the remaining degrees of freedom to reduce $F(x)$. Quite how this is done will depend on the unconstrained analogue being used. In Fig. 10 a three-dimensional problem is depicted in which one constraint is active.

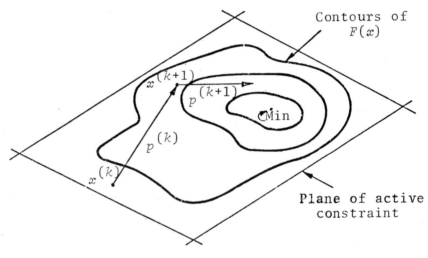

Fig. 10.

Two additional features need to be taken into account.

(1) The step taken along $p^{(k)}$ may have to be limited in order to ensure that $x^{(k+1)}$ is feasible. If the step is limited there will be an additional constraint active at $x^{(k+1)}$.

(2)   Not only may we add constraints to the active
        set, but we may wish to move off one of those
        currently active.

The direction of search (or step) at each iteration
of an algorithm for LCP's is usually determined by
solving a special quadratic program in which the
objective function is the current quadratic approximation to $F(x)$ and the constraints are some subset
of the original constraints. These quadratic programs (QP) are often easy to solve since the active
constraints are similar (perhaps even identical)
from one iteration to the next.

4. NONLINEAR CONSTRAINTS

        The problem we shall consider in this section
is

P1 $\min\{F(x)\}$

                $c_i(x) \geqslant 0$ $i = 1,\ldots,m.$

        We shall assume that all the constraints are
nonlinear. If linear constraints do occur it is
nearly always worthwhile to treat them separately.
For example if the problem with just nonlinear constraints is transformed to one of minimizing some
unconstrained function then the problem with mixed
nonlinear and linear constraints should be transformed to a linearly constrained problem. In some
of the algorithms we shall discuss, a linearization
of the constraint function is made, and the incorporation of linear constraints is straightforward.

        We state without proof *sufficient* conditions
for $\overset{*}{x}$ to be a strong local minimum of P1. Let
$\hat{c}(x)$ be some subset of t constraints and $\hat{A}^T$ the
Jacobian of $\hat{c}$ evaluated at $\overset{*}{x}$ and it will be assumed
that $\hat{A}$ is of full rank. The Lagrange multipliers
λ are given by

$$\lambda = (\hat{A}^T\hat{A})^{-1}\hat{A}^T g(\overset{*}{x}).   \qquad (4.1)$$

        Let Z be the matrix defined by (2.1) and let
W be the Hessian matrix of the *Lagrangian function*

231

$$L(x,\lambda) = F(x) - \lambda^T \hat{c}(x) \qquad (4.2)$$

evaluated at $\overset{*}{x}$.

If the following conditions hold, then $\overset{*}{x}$ is a strong local minimum of P1.

(i) $\hat{c}(\overset{*}{x}) = 0$

(ii) $\bar{c}(\overset{*}{x}) > 0$ (remaining constraints)

(iii) $z^T g(\overset{*}{x}) = 0$

(iv) $z^T W z$ is positive definite

(v) $\lambda_i > 0$, $i = 1,\ldots,t$.

These conditions are almost identical to those given below (2.1) for the LCP. The exception is condition (iv) which reduces to the condition given previously if the constraints are linear since the Hessian matrices of the constraint functions are then null. In the introduction we emphasized the importance of constraint curvature in defining the solution and it is in (iv) only that the relevance of curvature is reflected. Fig. 11, which depicts a two-dimensional problem with a linear objective function, illustrates the need for condition (iv).

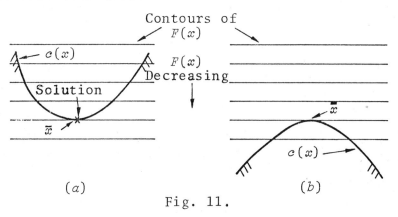

Fig. 11.

All the conditions are satisfied at $\bar{x}$ in (a) but in (b) all but (iv) are satisfied.

Many algorithms require the sufficient conditions (i) to (v) to hold if they are to be successful. This is especially true of algorithms which make use of the Lagrangian function. In some problems it may be that $z^T W z$ is only positive semi-definite at the solution. While this can pose difficulties, it does not usually have a catastrophic effect. A more serious situation arises if (iii) does not hold since the Lagrange multipliers do not then exist. A well known example of such a problem which has a perfectly well defined strong local minimum is the following

$$\min\{-x_1\} \quad \text{subject to} \quad c_1(x) = x_1 \geqslant 0$$
$$c_2(x) = x_2 \geqslant 0$$
$$c_3(x) = (1-x_1)^3 - x_2 \geqslant 0$$

This problem is illustrated in Fig. 12.

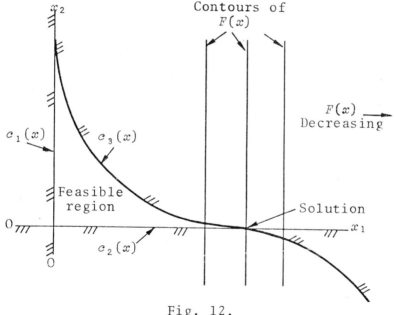

Fig. 12.

The solution is at $x_1 = 1$, $x_2 = 0$ and the two active constraints are $x_2 = 0$ and $(1-x_1)^3 - x_2 = 0$. The Jacobian of these constraints at $(1,0)$ is

$$\begin{bmatrix} 0 & 1 \\ 0 & -1 \end{bmatrix} ,$$

which is clearly rank deficient. Moreover $z = [\begin{smallmatrix} 1 \\ 0 \end{smallmatrix}]$ and $z^T g(\overset{*}{x}) \neq 0$ since $g(x) = [\begin{smallmatrix} -1 \\ 0 \end{smallmatrix}]$. It is likely that practical problems for which condition (iii) does not hold, occur infrequently. If we perturb a problem for which condition (iii) does not hold at the solution into one for which it does, then one or more of the Lagrange multipliers will be very large. Obviously, if a method fails on the origina problem it is still likely to fail or perform poorl on the perturbed problem, even though all the con- ditions (i) to (v) now hold.

5. DISCONTINUOUS DERIVATIVES

Occasional discontinuities in the derivatives of $F(x)$ and $c(x)$ along the solution path (especiall; if these do not occur in the neighbourhood of the minimum) are unlikely to seriously impede most algorithms that assume continuity. Problems do, however, occur in which the discontinuities defeat the sophisticated algorithms. (It is important to distinguish such problems from those for which analytical derivatives, although not available as a subroutine, do exist.) The only methods availabl for these difficult discontinuous cases are those based on trial and error. Since they are crude methods they are also relatively simple to implemen being natural extensions of unconstrained procedure: For instance, a univariate search procedure for con strained problems is identical to that for uncon- strained problems except that the search is cur- tailed prematurely if a constraint is violated. In fact, a linear-search procedure should have some facility for limiting the step size even for an unconstrained problem. Other techniques appli- cable to this class of problem are: (i) regular search over some grid again considering feasible points only, (ii) random search and (iii) algorithm: such as Complex (a generalization of the Simplex method for unconstrained minimization) which are based on geometric principles. There are few

guarantees with any of these methods and it is
easy to demonstrate their inefficiency on problems
with continuous derivatives.

6. PENALTY AND BARRIER FUNCTIONS

An obviously beneficial approach to constrained problems *if it can be achieved* is to transform them into an equivalent unconstrained problem.
By "equivalent" we mean that the solution of the
two problems are identical. Penalty and Barrier
functions were an early attempt to achieve this
objective by transforming the constrained problem
into that of solving a *sequence* of related unconstrained problems. A typical Penalty function is
given by

$$P(x,r) = F(x) + r^{-1} \sum_{i \in I} c_i^2(x), \qquad (6.1)$$

where I is the index set of constraints active at
$\overset{*}{x}$ and r is called the penalty parameter. Under
certain conditions

$$\lim_{r \to 0} \overset{*}{x}(r) = \overset{*}{x} \quad ,$$

where $\overset{*}{x}(r)$ is the minimum of $P(x,r)$. Since the
set of active constraints at the solution is usually
not known before we determine $\overset{*}{x}$, the summation in
(6.1) is made over those constraints for which
$c_i(x) < \delta$, where δ is a small positive scalar, and
x is the current estimate of $\overset{*}{x}$.

A typical Barrier function is given by

$$B(x,r) = F(x) + r \sum_{i=1}^{m} \{- \ln c_i(x)\}, \qquad (6.2)$$

If the initial estimate to $\overset{*}{x}$ is interior to
the feasible region an unconstrained algorithm
applied to $B(x,r)$ can be made to avoid generating
a non-feasible point since $B(x,r)$ is infinite
along the boundary of the feasible region. Again
under certain conditions (one of which is that the
constraints are all inequalities)

235

$$\lim_{r\to 0} \overset{*}{x}(r) = \overset{*}{x}.$$

Fig. 13 illustrates the two approaches when applied to the problem

$$\min (x-2)^2$$

$$1 - x \geqslant 0.$$

The solution is $x = 1$.

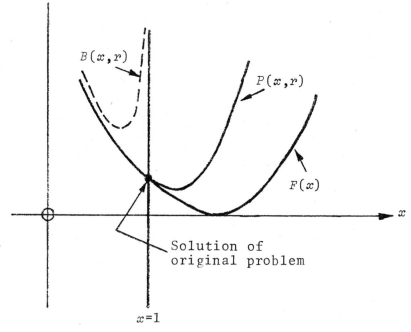

$x=1$

Fig. 13.

Taking $r = 10^{-4}$ the Penalty approach gives
$P(x,10^{-4}) = (x-2)^2+10^4(1-x)^2$ and $\overset{*}{x}(10^{-4}) \approx 1.0001$,
whereas
$B(x,10^{-4}) = (x-2)^2-10^{-4}\ln(1-x)$ and $\overset{*}{x}(10^{-4}) \approx 0.99995$.

It may be thought that we could simply set $r = 10^{-10}$ and minimize a single function. Unfortunately the numerical difficulties of minimizing

$B(x,r)$ and $P(x,r)$ increase as r decreases, consequently the usual practice is to minimize a sequence
of functions (say for $r = 1,0.1,0.001$, etc.) and to
extrapolate along the trajectory $x(r)$ to $\overset{*}{x}$. The
motivation for this strategy is that by using the
information gained in previous minimizations the
smaller the value of r the closer is the initial
estimate to $\overset{*}{x}(r)$; this reduces the likelihood that
the routine will get stuck at a point removed from
the solution. As practical tools for the solution
of constrained problems Penalty and Barrier functions are declining in importance. Efforts to
improve them have, however, led to important new
algorithms.

7. METHODS BASED ON AUGMENTING THE LAGRANGIAN FUNCTION

In this section we shall for simplicity mainly
consider problems with equality constraints only.
The extension of the algorithms to the inequality
case is straightforward theoretically but for some
it can lead to numerical difficulties.

Consider the *Lagrangian function* defined by
(4.2)

$$L(x,\lambda) = F(x) - \lambda^T \hat{c}(x), \qquad (7.1)$$

where λ is the vector of Lagrange multipliers
defined as

$$\lambda = (\hat{A}^T \hat{A})^{-1} \hat{A}^T g(\overset{*}{x})$$

and $\hat{A}^T$ is the Jacobian of $\hat{c}(x)$ evaluated at $\overset{*}{x}$.

The condition $z^T g(\overset{*}{x})$ implies that $g(\overset{*}{x})$ can be
expressed as a linear combination of the columns of
A, *i.e.*,

$$g(\overset{*}{x}) = \hat{A} w ,$$

where w is a $t \times 1$ vector. Premultiplying this
equation by $\hat{A}^T$ and rearranging gives

$$w = (\hat{A}^T \hat{A})^{-1} \hat{A}^T g(\overset{*}{x}) \equiv \lambda .$$

Differentiating (7.1) gives

$$\frac{\partial L}{\partial x}(x,\lambda) = g(x) - \hat{A}(x)\lambda \ ,$$

hence the gradient of the Lagrangian function evaluated at $\overset{*}{x}$ is zero. This is a necessary condition for $\overset{*}{x}$ to be a minimum of $L(x,\lambda)$ with respect to x. If, therefore, the curvature of $L(x,\lambda)$ is positive at $\overset{*}{x}$ and if λ is known, $\overset{*}{x}$ could be determined by minimizing $L(x,\lambda)$. Unfortunately, neither of these two conditions is likely to hold. However, $\overset{*}{x}$ is certainly a stationary point of $L(x,\lambda)$ and it follows from condition (iv) section 4 that this function has positive curvature in the subspace of E^n spanned by the columns of $\mathcal{Z}$. We therefore seek to augment the Lagrangian function with a multiple of some function which has the following properties:

(a) the function has a zero gradient at $\overset{*}{x}$

(b) it has zero curvature along the directions $\mathcal{Z}_i$, $i = 1,2,\ldots,n-t$ at $\overset{*}{x}$.

(c) it has positive curvature at $\overset{*}{x}$ along all directions orthogonal to $\mathcal{Z}$.

Clearly the augmented function will have a minimum at $\overset{*}{x}$ provided the multiple of the function added to the Lagrangian function is large enough.

Under certain assumptions one function satisfying (a), (b) and (c) is $\hat{c}^T\hat{c} \equiv \phi(x)$, say. We see that

$$\frac{\partial \phi(x)}{\partial x}\bigg|_{x=\overset{*}{x}} = \hat{A}\hat{c}\bigg|_{x=\overset{*}{x}} = 0$$

and

$$G_c \equiv \left[\frac{\partial^2 \phi(x)}{\partial x_i \partial x_j}\right]_{x=\overset{*}{x}} = \hat{A}\hat{A}^T \ ,$$

so that $\mathcal{Z}^T G_c \mathcal{Z}$ is zero and $\hat{A}^T G_c \hat{A} = (\hat{A}^T\hat{A})^2$ which is clearly positive definite if $\hat{A}$ is of full column rank. The augmented Lagrangian is then

$$L(x,\lambda,\rho) = F(x) - \lambda^T\hat{c} + \rho\hat{c}^T\hat{c} \ ,$$

where ρ is some sufficiently large (but finite) positive scalar.

Consider the technique applied to the simple problem

$$\min\{x_1^2 - x_2^2\}$$

subject to $\qquad\qquad x_2 = -1$.

The solution is obviously $x_1 = 0$, $x_2 = -1$. At the solution $\hat{A}^T = [0,1]$ hence

$$\lambda = ([0,1]\begin{bmatrix}0\\1\end{bmatrix})^{-1}[0,1]\begin{bmatrix}0\\2\end{bmatrix} = 2.$$

Equation (7.1) becomes

$$L(x,\lambda) = x_1^2 - x_2^2 - 2(x_2 + 1).$$

Differentiating gives

$$\frac{L(x,\lambda)}{\partial x} = \begin{bmatrix}2x_1\\-2x_2 & -2\end{bmatrix} = 0, \text{ if } x_1 = 0, x_2 = -1.$$

This is obviously only a stationary point since the Hessian of $L(x,\lambda)$ for all x is

$$\begin{bmatrix}2 & 0\\0 & -2\end{bmatrix}.$$

Now $\hat{A}\hat{A}^T = \begin{bmatrix}0 & 0\\0 & 1\end{bmatrix}$ for all x, and if ρ is 4 the Hessian of $L(x,\lambda,\rho)$ is

$$\begin{bmatrix}2 & 0\\0 & 2\end{bmatrix}.$$

Figs. 14 and 15 depict the contours of $L(x,-2)$ and $L(x,-2,4)$, respectively.

However, for $\rho < 2$ the augmented Lagrangian function does not have a minimum at $(0,-1)$. We could avoid such an eventuality by setting ρ at a large value but for some problems this would result in the Hessian matrix of $L(x,\lambda,\rho)$ being ill-conditioned.

We have already noted that in general the Lagrange multipliers are unknown, and the same is clearly true for the optimum choice of ρ. Consi-

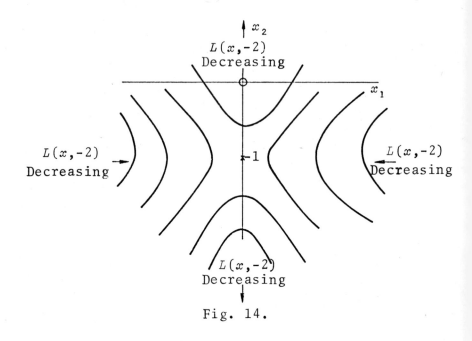

Fig. 14.

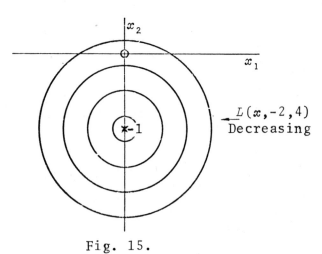

Fig. 15.

deration of these difficulties leads to the following three categories of algorithms. It is
important to appreciate that the closer an approximation we have to $\overset{*}{x}$ the better we can approximate
λ and ρ.

Algorithm (1a)

At some given initial point $x^{(1)}$ estimate λ
by $\lambda^{(1)}$ and ρ by $\rho^{(1)}$. Using $x^{(1)}$ as the initial
point minimize the function

$$L(x,\lambda^{(1)},\rho^{(1)}) = F(x) + \lambda^{(1)T}\hat{c} + \rho^{(1)}\hat{c}^{T}\hat{c}.$$

Denote the solution by $x^{(2)}$ and at this point re-
estimate λ by $\lambda^{(2)}$ and ρ by $\rho^{(2)}$ then repeat the
process.

Algorithm (1b)

In determining $x^{(2)}$ starting at $x^{(1)}$ in (1a),
a new estimate of $\overset{*}{x}$ is obtained at each iteration
of the unconstrained algorithm. Instead of keeping
the estimates to λ and ρ constant until $x^{(2)}$ is
found we adjust them at each iteration of the unconstrained algorithm. This implies that the function being minimized may differ at each iteration
but it is hoped as we approach the solution $\overset{*}{x}$ that
any changes to $\lambda^{(k)}$ and $\rho^{(k)}$ will be small.

Algorithm (1c)

Replace λ in the augmented Lagrangian function by $\lambda(x)$ where

$$\lim_{x \to \overset{*}{x}} \lambda(x) = \lambda.$$

One such function is

$$\lambda(x) = (\hat{A}^{T}(x)\hat{A}(x))^{-1}\hat{A}^{T}(x)g(x).$$

The algorithm is, therefore: choose an initial
point $x^{(1)}$, estimate $\rho^{(1)}$ and minimize the function

$$L(x,\lambda(x),\rho^{(1)}) = F(x) - \hat{c}^{T}\lambda(x) + \rho^{(1)}\hat{c}^{T}\hat{c}.$$

If during the course of the minimization $\rho^{(1)}$ is thought an unsuitable approximation to ρ it is altered. The disadvantage of this last algorithm is that the definition of the function being minimized involves the *gradients* of $F(x)$ and $\hat{c}(x)$.

There are functions other than $\hat{c}^T\hat{c}$ which satisfy the properties (a), (b) and (c) and these may be preferred but such a change would not alter the character of the algorithms.

8. LAGRANGIAN FUNCTIONS AND CONSTRAINT LINEARIZATIONS

An alternative to transforming the problem into an unconstrained problem is to transform it into a linearly constrained problem. One inducement to attempt this is that the function being minimized need no longer have a positive-definite Hessian matrix at $\overset{*}{x}$. In fact by judicious choice of the linear constraints the directions of negative curvature of $L(x,\lambda)$ at $\overset{*}{x}$ can be made infeasible

By examining the sufficient conditions for the solution of a LCP given in section 2 we see that the matrix of constraint coefficients needs to be chosen so that the associated $\mathcal{Z}$ matrix is such that the matrix

$$\mathcal{Z}^T W \mathcal{Z}$$

is positive-definite. One such matrix of constraint coefficients is $\hat{A}$, the Jacobian of $\hat{c}(x)$ evaluated at $\overset{*}{x}$. The linear constraints required, therefore, are linear approximations to $\hat{c}(x)$ at $\overset{*}{x}$, given by

$$\hat{c}(x) + \hat{A}^T(x - \overset{*}{x}) = 0. \qquad (8.1)$$

Fig. 16 illustrates the application of this idea to a two-dimensional problem with one nonlinear constraint.

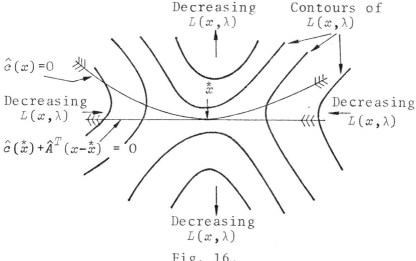

Fig. 16.

We have, therefore, replaced the original problem by

$$\min\{L(x,\lambda)\} \qquad (8.2)$$

subject to $\hat{A}^T(x-\overset{*}{x}) = -\hat{c}(\overset{*}{x})$,

where $\hat{A}$ is the Jacobian of $\hat{c}(x)$ evaluated at $\overset{*}{x}$. This of course does not get us very far since we are required to know $\overset{*}{x}$ and λ. In addition (8.2) is unsatisfactory because the Lagrange multipliers of this transformed problem are zero (since the gradient of $L(x,\lambda)$ is zero at $\overset{*}{x}$). With equality constraints this is not a difficulty since the Lagrange multipliers of the LCP do not contribute to the efficiency of the algorithm. With inequalities, however, $\overset{*}{x}$ may only be a *weak* local minimum of the LCP and there may be other equally valid solutions which are not solutions of the original problem. The following problem illustrates this point

$$\min\{x_1^2 - x_2^2\} \quad \text{subject to} \quad -x_1^2 - x_2^2 + 1 \geqslant 0$$

The solution is given by $\overset{*}{x} = (0,-1)$ or $(0,1)$. Taking the first solution λ is -1 and $L(x,\lambda)$ is $2x_1^2 - 1$.

The linear approximation to $\hat{c}(x) \geqslant 0$ at $\overset{*}{x}$ is

$$x_2 \geqslant -1.$$

This constraint is satisfied by $\overset{*}{x} = (0,-1)$ with $L(\overset{*}{x},\lambda) = -1$ however $x = (0,0)$ also satisfies the constraint and gives $L(x,\lambda) = -1$ and if this or a similar point was ever reached in the linearly constrained algorithm the algorithm would stop. The difficulty can be circumvented by minimizing

$$F = L(x,\lambda) + x^T\hat{A}\lambda;$$

clearly we have $\partial F/\partial x = g - \hat{A}\lambda + \hat{A}\lambda = g$ and the Lagrange multipliers of the linearly constrained problem are identical to those of the original problem. It can easily be verified that the required solution of the linearly constrained problem is unaltered by the addition of the linear function.

The reader may wonder why we do not simply $\min(F(x))$ subject to the linearization of the constraints, since this would certainly work for the example just quoted. However, if we consider the problem in three dimensions

$$\min\{x_1^2 - x_2^2 - x_3^2\}$$
$$-x_1^2 - x_2^2 - x_3^2 + 1 \geqslant 0.$$

The linearization of the constraint at the solution gives

$$\sqrt{2}\,x_2 + \sqrt{2}\,x_3 \leqslant 2$$

which can always be satisfied by setting $x_2 = -x_3$. We are now free to set $x_1 = 0$ and x_3 to any value say 100!

Returning to the Lagrangian formulation, we have transformed our original problem to

$$\min\{F = L(x,\lambda) + x^T\hat{A}\lambda\}$$

subject to $\quad \hat{A}^T(x - \overset{*}{x}) = -\hat{c}(\overset{*}{x})$

and this can be used as a basis for generating the following categories of algorithms.

Algorithm (2a)

Estimate λ by $\lambda^{(1)}$ and $\overset{*}{x}$ by $x^{(1)}$. Evaluate $\hat{A}^{(1)}$ the Jacobian of $\hat{c}(x)$ at $x^{(1)}$, and solve the problem

$$\min\{F^{(1)} = F(x) - \lambda^{(1)T}\hat{c}(x) + x^T\hat{A}^{(1)}\lambda^{(1)}\}$$

subject to $\hat{A}^{(1)T}(x-x^{(1)}) = -\hat{c}(x^{(1)})$. (8.3)

Let the solution be $x^{(2)}$ and repeat the process, using $x^{(2)}$ in place of $x^{(1)}$ and $\lambda^{(2)}$ (the estimate of λ at $x^{(2)}$) in place of $\lambda^{(1)}$ etc.

Algorithm (2b)

Instead of solving (8.3) to determine $x^{(2)}$ we simply perform a single iteration of a linear constraint algorithm (or perhaps several) and then re-estimate λ and recompute $\hat{A}(x)$ etc. This does not require any additional work since these quantities would be obtained during the execution of the linear constraint algorithm.

Each iteration of a linear constraint algorithm requires solving a quadratic program (QP) so algorithms of type (2b) also reduce to solving a sequence of QP's. The difference is that the constraints as well as the objective function change from iteration to iteration. Usually the QPs are of a special type that can be solved in single step.

One could, as with algorithm (1c), replace λ by $\lambda(x)$ but it would still be necessary to solve either a sequence of problems or to alter the linear constraints every iteration. One advantage of such an approach when solving inequality problems is that it would be less prone to numerical difficulties caused by derivative discontinuities in the term $\rho\hat{c}(x)^T\hat{c}(x)$ because of the need to alter the active set. Discontinuities would still exist but the change in value would in general be much smaller and also diminish in magnitude the closer they occurred to the solution. For example in (1c) changing the active set from $\hat{c}_i(x)$,

$i = 1, 2, \ldots, t$ to $\hat{c}_i(x)$, $i = 1, 2, \ldots, t-1$ results in a discrete change in the Hessian matrix of $\rho \hat{a}_t \hat{a}_t^T$, where $\hat{a}_t = \nabla_x \hat{c}_t(x)$, no matter how close we are to the solution. (One could take direct account of this by doing a special rank-one modification to the Hessian matrix but this alters the algorithm from simply applying an unadulterated unconstrained procedure to a transformed problem.)

It is necessary to modify algorithms of type (2a) still further because the linearly constrained problem is not guaranteed to have a bounded solution. It is usually unnecessary to modify algorithms of type (2b) because this point is normally catered for in the QP subproblem. One obvious way of ensuring a bounded solution is to augment the problem with constraints of the form

$$| x_i - x_i^{(1)} | \leqslant \beta_i,$$

where β_i is positive scalar, say 1. Alternatively the term $\gamma(x - x^{(1)})^T (x - x^{(1)})$ could be added to the objective function, however, the positive scalar γ must be chosen to be larger than some unknown threshold value. One adverse effect of these changes is that the Lagrange multipliers of the subproblem may no longer be related to those of the original problem.

A useful modification of algorithms of type (2b) that I have found worthwhile is to alter the constraints to be of the form

$$\hat{A}^{(k)T}(x - x^{(k)}) = -D^{(k)}\hat{c}(x^{(k)}),$$

where $D^{(k)}$ is a diagonal matrix whose ith diagonal element $d^{(k)}$ is such that $0 \leqslant d_i^{(k)} \leqslant 1$. This modification is particularly relevant when solving inequality problems as Fig. 17 illustrates. The use of I for $D^{(k)}$ can be unsatisfactory on two counts: not only may the true solution be non-feasible, but there may be *no* point which is feasible with respect to both the linear constraints and the original nonlinear constraints. This is essential to methods in this class which construct

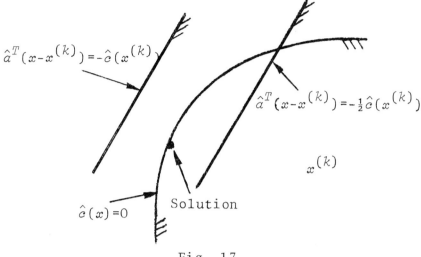

Fig. 17.

a sequence of estimates that are always feasible to the original problem.

An important requirement, whose significance is rarely appreciated, is that the sequence $x^{(1)}, x^{(2)}, \ldots$, should be such that

$$\lim_{k \to \infty} \frac{\hat{A}^{(k)T}(x^{(k+1)} - x^{(k)})}{\| x^{(k+1)} - x^{(k)} \|} \neq 0; \qquad (8.4)$$

this can be assured by judicious choice of $D^{(k)}$. Since

$$\lim_{k \to \infty} x^{(k)} \text{ is } \overset{*}{x}, \lim_{k \to \infty} g^{(k)} \text{ is } \hat{A}_\lambda^T$$

then (8.4) implies that

$$\lim_{k \to \infty} \frac{g^{(k)T} p^{(k)}}{\| p^{(k)} \|} \neq 0,$$

where $p^{(k)} = x^{(k+1)} - x^{(k)}$. This ensures that *linear* approximations always have relevance, since the first term in the Taylor expansion about $x^{(k)}$ is guaranteed not to vanish. Moreover, we have seen that $\nabla_x F$ is $g(x)$, so that the above result can be

247

written

$$\lim_{k \to \infty} \frac{p^{(k)T} \nabla_x F}{\| p^{(k)} \|} \neq 0.$$

This means that if the $p^{(k)}$ are considered as search directions, they will not tend to become orthogonal to the gradient of the function being minimized. Consequently the changes in F which arise from the step along $p^{(k)}$ will remain significant as the minimum is approached.

9. REDUCED GRADIENT METHOD AND GRADIENT PROJECTION METHOD

Although originally these were distinct methods, their developments have made them fundamentally equivalent. This has resulted in some confusion in the literature: they are sometimes treated separately and work on one category of algorithm is not quoted or seen to relate to the other category. The original variation arose because, given a steepest-descent direction in E^n, there are various ways of constructing a descent direction in some subspace of E^n. To simplify the description consider the case with just linear constraints. (This is not an over-simplification since nonlinear constraints are approximated by linear functions.) If we approximate $F(x)$ at $x^{(k)}$ by the linear function in x, $F(x^{(k)}) + g^{(k)T}(x-x^{(k)})$ then the approximate problem is linear. If p denotes $x - x^{(k)}$ and if we assume that $x^{(k)}$ lies on the linear constraints, the approximate problem is the linear program

$$\min_{p} \{ F(x^{(k)}) + g^{(k)T} p \} \qquad (9.1)$$
$$\hat{A}^{(k)T} p = 0$$

If we define the solution to be $p^{(k)}$ then the next approximation to $\overset{*}{x}$, $x^{(k+1)}$, is given by $x^{(k)} + \alpha p^{(k)}$ where α is a scalar chosen so that $F(x^{(k+1)}) < F(x^{(k)})$. As we saw in the discussion of the Active Constraint Strategy in section 3, the problem stated in (9.1) is not in general well-posed if the number of constraints is less than $n-1$. This is because there is more than one direction which is orthogonal to the constraint normals and along which the linear objective function is always decreasing. To resolve this non-uniqueness it is necessary to place additional restrictions on p, for example $p^T p = 1$. There are many similar restrictions on p that result in $p^{(k)}$ having the property

$$F(x^{(k)} + \alpha p^{(k)}) < F(x^{(k)}),$$

provided α is chosen sufficiently small and positive and $x^{(k)}$ is not the solution of the original problem. It is the variation in how (9.1) is altered to make it a well posed problem that produces the difference between gradient projection and various gradient reduction schemes. The terms projected gradient and gradient reduction arise because the solution of (9.1) augmented by additional restrictions on p is of the form

$$p^{(k)} = -Wg,$$

where W is an $n \times n$ symmetric matrix of rank $n-t$ such that

$$\hat{A}^T W = 0.$$

The search direction $p^{(k)}$ is, therefore, a projection of the gradient into the space spanned by the columns of $\hat{A}$. Moreover the gradient along $p^{(k)}$ (which lies in a reduced space) can not be greater (and will nearly always be smaller) in magnitude than $\| g^{(k)} \|_2$.

For the solution of unconstrained problems use of a *linear* approximation to $F(x)$ produces the method of steepest descent. These methods are, therefore, generalizations of the steepest descent method and consequently have the same poor rate of convergence. Use of a *quadratic* approximation to the objective function $F(x)$ transforms (9.1) to the following quadratic programming problem

$$\operatorname{Min}_{p}\{F(x^{(k)}) + g^{(k)T}p + \tfrac{1}{2}p^{T}Bp\} \tag{9.2}$$

subject to $\hat{A}^{(k)T}p = 0$,

where B is either the Hessian matrix of $F(x)$ at $x^{(k)}$ or some approximation to it. Provided $Z^{T}BZ$ (Z as defined earlier in the text) is positive definite then (9.2) is a well-posed problem. Consequently, there is no need to add additional restrictions to (9.2) (although when $x^{(k)}$ is a poor approximation this may be desirable for other considerations) and variations between different reduction and projection schemes vanish.

The general approach of the method is illustrated in Fig. 18.

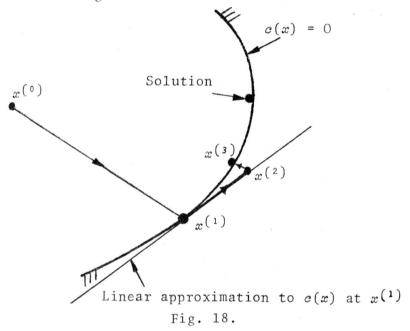

Fig. 18.

The principle being adhered to is try and remain feasible while reducing $F(x)$. Occasionally, perhaps even frequently, an iterate will be non-feasible so a special step to regain feasibility needs to be made. This is illustrated in Fig. 18

when moving from $x^{(2)}$ to $x^{(3)}$. The step is usually
determined by considering a search direction which
will reduce the term

$$\sum_{i \in I} c_i^2(x),$$

where I is the set of indices of violated constraints.

When nonlinear constraints are present the
inadequacy of considering a quadratic approximation
to $F(x)$ only has already been discussed both in the
introduction and the previous section. If we
modify (8.2) so that the objective function is a
quadratic approximation to the Lagrangian function
the algorithm falls into category (2b) discussed
in the previous section with the special proviso
of a step to regain feasibility. It is possible
to modify the quadratic program still further so
as either to avoid going infeasible or reduce the
probability. One way is to add a positive quantity
to the right hand side of the constraints in (8.2)
and this has the effect of rotating the solution
towards the feasible region. Alternatively we
could add a penalty term to the objective function
which would limit the degree of infeasibility.

ACKNOWLEDGEMENTS

I would like to thank my colleagues at the
NPL - Dave Martin, Jeff Hayes and Philip Gill for
their careful reading of the original manuscript
and suggesting a number of improvements.

REFERENCE

Gill, P.E. and Murray, W. (1974) *Eds*. "Numerical
Methods for Constrained Optimization", Academic
Press.

Nonlinear Optimization in Industry and the Development of Optimization Programs

F.A. Lootsma

(University of Technology, Delft, The Netherlands)

SUMMARY

This paper classifies the applications of nonlinear optimization in an industrial environment, and it analyses the absence of optimization programs in scientific software libraries. Particular attention is given to the question of how to build up an optimization library for technological problems. Finally, a possible strategy for the comparison of algorithms is briefly outlined. The paper emphasizes the significance of the journals and of a central clearing house to manage the reviewing of new and existing algorithms.

1. INTRODUCTION

Although it is dangerous to generalize from a limited number of observations, we feel that the applications of nonlinear optimization in industry fall into three distinct categories.

a) Technological applications. There is a variety of design problems in research, development and engineering departments (Bracken and McCormick (1968)). These problems have a highly nonlinear nature and can be formulated with a relatively small number of variables (two to twenty). Larger models (fifty to hundred variables) are frequently studied in electricity companies to solve dispatching problems. A significant proportion (70%) of the problems reduces to the minimization of a sum of squares arising from curve fitting. The number of research members and engineers concerned with a particular problem is relatively small (two to five), thus reducing the difficulties of communication and coordination.

b) *Business applications*. In this area, nonlinear optimization is mostly a refinement of a linear-programming study. Applications of linear programming are mainly found in the area of production planning (long term or medium term capacity planning, medium term or short term production allocation and production scheduling). The size of the problems tends to be very large, both in the number of variables and constraints (hundreds, or even thousands) and in the number of people involved. Mostly, linear programming can only be carried out after a favourable high-level management decision has been taken. In addition, data collection and the implementation of a solution calculated by linear programming techniques depend heavily on the continuous support by management. In many of these problems one finds a number of nonlinearities (due to economies of scale), although it is often extremely difficult to precisely determine these nonlinearities in a given practical situation. If they cannot be neglected, they can often be handled via local linearization and repeated application of linear programming (method of approximation programming), or via separable programming (Beale (1968)).

c) *Discretized problems*. There is a widespread interest in the attempts to solve problems arising from the discretization of continuous problems by means of nonlinear optimization techniques. This is a feasible approach to solve optimal control problems, for instance (Tabak and Kuo (1971)). The discretized problems tend to have hundreds of variables, each connected with a grid point (specification point) of the discretization. This is in sharp contrast with model formulation in the first-named category of technological applications where the majority of the variables have a distinct physical meaning (temperature, pressure, dimensions of devices, ...). There is a similar distinction with the second category where the majority of the variables are related to quantities of certain items (batch sizes, for instance) so that they have a managerial significance.

In this paper, we shall mainly be concerned with problems of the first category. The size of the problems, and the vagueness of the nonlinearities seem to indicate that it is preferable to consider the second category merely as an extension of linear-programming applications. For the time being, we hesitate to deal with the third category. We feel that our experience with large nonlinear problems is still too limited to draw any valid conclusions.

2. STUMBLING BLOCKS FOR APPLIED NONLINEAR OPTIMIZATION

It is our impression, from frequent contacts with users of optimization techniques and with optimization specialists, that the following features have a significant impact on the application of nonlinear optimization.

a) *Difficulties encountered in formulating a model.* It seems to be a common experience that the constraints can readily be formulated in many practical situations. A major stumbling block appears to be the identification of the objectives and the subsequent formulation of an objective function. Finally, satisfying the constraints is frequently the predominant concern of the user; optimization of the objective function does not have the same significance.

b) *Lack of expertise in nonlinear optimization.* Possible users do not easily find qualified help (neither in mathematics departments nor in computing centres) to support model formulation and model testing. Hence, many users also ignore the optimization software which is available in computing centres. They tend to develop their own specific optimization techniques for their particular problems. Alternatively, they resort to simple techniques (direct search methods) which are easy to program.

c) Career paths in optimization. Sooner or later, optimization specialists in industry have to face the choice between further specialization in a research environment (possibly a university) or increasing managerial responsibilities in a staff position. Anyhow, the period of active work in applied optimization is rather short (roughly five years). This rapid turnover seems to explain why successful optimization activities in a company may suddenly come to a halt.

d) Absence of optimization software. It is fair to say that computing departments often supply only a small number of optimization routines covering a limited range of applications, notwithstanding the abundant variety of published algorithms. More significant, however, seems to be the reluctance of the computer manufacturers to incorporate nonlinear optimization routines into their scientific software packages. A considerable amount of effort has gone into the development of linear programming codes (with some facilities for integer programming). Nonlinear optimization, however, is hardly represented in these packages. For the ultimate users this is a stumbling block preventing them to benefit from the recent activities in this area.

The difficulties in model formulation, the lack of communication between users and specialists, and the significance of career prospects are so notorious in operations research that we shall not further consider these aspects here. The fourth factor, however, the absence of optimization programs in software libraries, will be discussed more specifically. The dissemination of knowledge and experience in optimization is a major task of the audience in this conference. Hence, the present paper is devoted to the particular difficulties arising as soon as one has to evaluate the progress in optimization and to select a number of algorithms for incorporation in a software library.

It is not our intention to say that the availability of appropriate software is the crucial factor in applied nonlinear optimization. Even the best software library is useless if there is no communication between the users and the specialists responsible for the library.

3. EVALUATION CRITERIA FOR NONLINEAR OPTIMIZATION PROGRAMS

Today, there is a confusing variety of algorithms to solve general nonlinear optimization problems. Important, thoroughly tested algorithms are the reduced gradient method (Abadie and Carpentier (1969), and Lasdon (1970)), the gradient projection method (Rosen and Kreuser (1972)), and the penalty-function methods to solve constrained problems via unconstrained optimization (Fiacco and McCormick (1968), and the author (1972a)). Many variants of these methods have been programmed, comparisons between programmed algorithms have been made (Colville (1968), and Staha (1973)) and some programs are published or available on request. Nevertheless, as soon as one starts to think about the desirable properties of a general program for nonlinear optimization, several questions arise which are almost never discussed in the literature.

An important problem for the computer manager responsible for a software library is the choice of the programmed algorithms to be included in the library. The variety of algorithms proposed in the last ten or fifteen years indicates that we miss the unique, powerful method for solving any problem in this field. It is undesirable, however, that the library should contain all the proposed methods. The documentation would grow beyond reasonable limits. Moreover, one would have the intolerable task of finding the most appropriate method for each optimization problem

which is presented. Thus, one has to choose a
restricted set of algorithms, and the choice must
be such that one can roughly indicate, for each
particular problem, which of the selected methods
is the most suitable one to use.

The following list of criteria, to be used
in reviewing nonlinear optimization programs, was
drawn up during discussions between R.J. Goffin
(Information Systems and Automation Department,
N.V. Philips, Eindhoven), J. de Jong (University
of Technology, Eindhoven), C. Schweigmann (at that
time member of Research and Central Engineering,
AKZO, Arnhem) and the author (at that time member
of the Information Systems and Automation Department, N.V. Philips, Eindhoven). Following a working paper by C. Schweigmann, and a report by R.L.
Staha and D.M. Himmelblau (1973), we formulated
the following criteria as a possible guideline for
a broad comparison of programmed algorithms.

a) Generality. It is important to realize for
which type of problems the program has been designed, and under which conditions the underlying
algorithm converges to the desired solution. It
is also worthwhile to know whether the program is
applicable to special problems such as unconstrained problems, linearly constrained problems, problems with nonlinear constraints, least squares
problems, quadratic programming problems, and
other problems with a special structure.

b) Efficiency. The relative performance (in terms
of function evaluations, equivalent function evaluations, and/or computation time) of the program
with respect to other programs (under the same
termination conditions) is not only an important
criterion to decide whether the program should be
in the library, it is also the predominant concern
of many optimization specialists. However, there
is an important difference in motivation. Specialists are very much devoted to the design of algorithms which are faster than some of the existing
ones. The relative performance enables a computer
manager to <u>judge</u> and to <u>decide</u> whether the pro-

gram deserves the troubles of incorporation in the
library, given the software which is already available. Similarly, it is important to know <u>how</u>
<u>efficient</u> the program is for special problems.
Then, a computer manager may <u>decide</u> to drop some
of the programs which can only be used for those
special problems.

c) Reliability. In order to establish whether a
programmed algorithm calculates the desired solution with the required accuracy, a considerable
amount of numerical experience (possibly on a
variety of computers with different compilers)
should be available. Moreover, the underlying
algorithm should have a sound mathematical basis:
established convergence for well-behaved problems,
estimates of the rate of convergence, etc.

d) Capacity. The maximum size of the problems
that can generally be solved by the program, as
well as the storage requirements, are important
criteria to determine the range of possible applications for the program.

e) Simplicity of use. It is difficult to formulate any objective criterion to decide whether a
program is a simple, effective tool in the hands
of unsophisticated users. Nevertheless, simplicity of use may be decisive for the successful utilization of a program. Crucial questions are the
following. Does the program provide facilities to
ease problem preparation by the user (such as numerical differentiation to check analytical derivatives or to avoid supplying analytical derivatives)? Does the program have extensive output
provisions warning the user against possible input
errors and enabling him to follow the course of
the computations? Is the algorithm conceptually
simple, thus enabling both the user and the specialist to discuss the calculated results? How is
the quality of the documentation (and particularly
the quality of the user's manual)? Does the program already have a widespread circulation or is
it still a tool in the hands of the program designer only?

f) Simplicity of program. Error checking, maintenance and transfer to other computers are greatly
simplified if the program has a clear (possibly
modular) structure. If the underlying algorithm
is an extension of simpler algorithms, one has a
natural basis for a transparent structure corresponding to the construction of the algorithm. It
is desirable, of course, that the program should
be written in a higher-level language (Fortran IV,
Algol 60). Finally, the length of the program
(the number of statements, the number of lines
of coding) is an important feature of a program,
indicating a considerable amount of work to obtain
a workable tool in a variety of circumstances.

Obviously, it is out of the question that
there should be a unique criterion to review nonlinear optimization programs. However, are the
above criteria relevant and applicable? And, more
pragmatically, who is supposed to carry out a comparison guided by these criteria? The sections to
follow present a preliminary and tentative answer
to these questions.

4. COMPARING AND TESTING NONLINEAR OPTIMIZATION PROGRAMS

A well-known phenomenon in research and development laboratories is the popularity of the
direct search methods for unconstrained minimization, notably the methods of Hooke and Jeeves
(1961) and Nelder and Mead (1965). Some of the
reasons can easily be found in the list of criteria of the previous section. First, these methods
are extremely simple and easy to understand.
Second, a Direct Search program can be short
(about 50 lines of coding in Algol 60). Third,
many users in research and development laboratories do not require a high accuracy (a relative
accuracy of 1% is often sufficient). In addition,
if they have problems with a small number of variables (between 2 and 10) some Direct Search methods seem to be reliable and even rather efficient.
Hence, they are used for many purposes: for uncon-

strained minimization, for least squares problems, and for constrained minimization via simple transformations (to handle simple constraints like nonnegative or bounded variables) or via penalty functions.

The literature on unconstrained minimization is mainly concerned with gradient methods, particularly with the variable-metric or quasi-Newton methods (mathematically attractive, but conceptually not very simple). Whenever a comparison of variable-metric methods is made, it is almost invariably a mutual comparison (Himmelblau (1971) and the present author (1972b) made an attempt to escape from this practice), and not a comparison with other gradient methods or with non-gradient methods. The reasons are understandable. Any programmed algorithm contains a number of tolerances or threshold values to control some iterative subprocesses (like the linear search). This will greatly complicate the comparison of a variable-metric method (essentially an algorithm to generate successive search directions) with any method which is conceptually different. Another complication is that nonlinear optimization algorithms involve user-provided functions and possibly their first-order and second-order derivatives. The efficiency of a programmed algorithm depends materially on the manner in which the provision of the functions and the derivatives has been organized.

Because of the prevailing attention which is given to variable-metric methods, the significance of numerical differentiation is often overlooked. If a user is able and willing to supply the first derivatives of the problem functions, then the modified Newton-Raphson algorithm with numerical approximations to the second derivatives (obtained by differencing the first derivatives) appears to be almost consistently more efficient (F.A. Lootsma (1975) and R.J. Goffin, unpublished results). However, the amount of information to be supplied by the user is the same (the first-order derivatives only).

In the area of constrained nonlinear optimization, one finds a large number of competing algorithms and approaches. The reduced gradient method has been cheered as being faster and more accurate than any of the other methods. The penalty-function approach, conceptually simpler (particularly for nonlinear constraints) and permitting a transparent program structure since it is fully based on unconstrained minimization, has been considered as relatively slow and inefficient. Recent work by R.L. Staha and D.M. Himmelblau (1973), however, indicates that there are competitive penalty-function techniques (optimization by least squares). Moreover, we observed that current penalty-function programs with several facilities for unconstrained minimization of the penalty function tend to take 600-900 lines of coding (in Fortran and Algol 60), whereas reduced gradient programs may easily end up with 2000-3000 lines (J. Abadie and L.S. Lasdon, personal communication).

The computer technology and the software development in the last ten years have shown a rapid growth. In addition, there was a significant reorganization in the European computer industry. A consequence of the technological progress and the reorganization will be that many computing centres have to face a larger or smaller conversion to new computers. Keeping this in mind, a computer manager will be inclined to emphasize a structured style of programming facilitating both the maintenance and the conversion. A simple, effective tool to reorganize Algol 60 programs for these purposes appears to be the removal of labels and goto statements. This will mostly require a thorough, disciplined rethinking of the program, but it is rewarding.

We have presented these examples in order to sketch the significance of the criteria listed in the previous section. We emphasized the importance of program organization and programming style, and we also pointed at some of the inherent difficulties (termination criteria of iterative subprocesses, organization of user-supplied routines) to be encountered as soon as one tries to establish the relative performance of algorithms. It will be clear that a comparison of programmed algorithms, with a view on these criteria, does not only require actual running of the programs on a variety of test problems. A significant impression of the design can be obtained by inspection of the program listing. This will be the starting point of the recommendations in the next section.

It is our conviction, that the relative performance of optimization algorithms cannot be precisely defined and measured. First, it is computer dependent. Second, modern computers with multi-processing and time-sharing facilities do not always inform the user of the precise execution time for the total job (an important performance criterion), or for specified parts of the job. Lastly, there are too many arbitrary steps in the program, such as the setting of threshold values to terminate the linear search. Efficiency (relative performance) is just one item in the list of criteria, and we feel that it will be sufficient to establish <u>tendencies</u> or <u>orders of magnitude</u> in the relative performance. The decision whether a programmed algorithm should be included in a software library should accordingly be based on a large amount of information, ranging from generality and efficiency to programming style and simplicity of use.

5. MANAGING THE COMPARISON OF NONLINEAR OPTIMIZATION PROGRAMS

Is it necessary to carry out a broad comparison of optimization programs in a central institute with highly qualified optimization specialists? This seems to be a question which pops up in many discussions centring around the issue. Generally speaking, the task of the institute or clearing house would be threefold: collection of information about available programs, collection of test problems, and comparison (certification, validation) of the programs.

If one agrees that a central institute for the collection and validation has to be established, it should not be based on the voluntary cooperation of some optimization specialists. For a long-term effort this is rather dangerous. On the contrary, such an institute should have a clear management structure and a sound financial basis. Nevertheless, we feel that the establishment of a central clearing house has a number of disadvantages. The institute would try to impose its ideas on the outside optimization specialists (a permanent source of friction), and there is always the danger that it is too far away from a user's environment.

We would like to emphasize that the comparison of algorithms is not a separate task for a few individuals. In order to obtain an over-all picture of the available tools for optimization, one needs the contribution of many specialists, each providing a detailed comparison of methods in a restricted area. The most powerful instruments to manage this process are the journals, and these should be used to the full extent.

The proper comparison of algorithms should be a major issue in the refereeing of submitted papers on optimization. Today, it is not. In the past, referees have developed a certain intuition, experience and knowledge in order to review the style and the organization of a paper, the correctness of the mathematical argumentation, and the significance of the paper (its contribution to the existing knowledge in the related field). The organization of the computational experiments is not normally included in the review of the papers (unless there are glaring weaknesses in the reported results). This is not so surprising: even writing a computer program is often exempted

from any professional criticism. We would therefore suggest that a submitted paper be accompanied
by a listing of the relevant computer programs and
by a detailed description of the experiments that
have been carried out. We venture to say that
mere inspection of listings will supply a wealth
of information about the significance of the experiments.

It is a matter of course, that actual running of programmed algorithms will remain necessary in order to compare and to validate their
alleged qualities. A central institute for the
selection of test problems and for the collection
and validation of programs is necessary. However,
its functioning should be backed up by a common
effort of the optimization community in order to
avoid that the institute will become an isolated
unit which merely duplicates the development in
other laboratories and computing centres.

ACKNOWLEDGEMENT

I am greatly indebted to J.F. Benders and
J. de Jong (both from the University of Technology,
Eindhoven), R.J. Goffin (N.V. Philips, Eindhoven),
C. Schweigmann (AKZO, Arnhem) and U. Schendel
(Free University, Berlin) for their inspiring
comments and criticism. Moreover, it is a pleasure to acknowledge the stimulating discussions at
the Conference on Solving Large Scale Mathematical
Programming Problems (International Institute for
Applied Systems Analysis, Laxenburg, Austria) and
at the conference on Optimization Theory and Optimal Control (Oberwolfach, Germany), both in Novem-
ber 1974. Particularly the remarks by R.W. Cottle
and G.H. Golub (both temporarily at ETH, Zurich)
have been helpful and instructive. Finally, I
would like to mention the written communication
with M.J.D. Powell (AERE, Harwell) and J.A. Tomlin
(Systems Optimization Laboratory, Stanford University) about the establishment of a Working
Group on Algorithms as an extremely useful source
of ideas.

REFERENCES

Abadie, J. and Carpentier, J.(1969) "Generalization of the Wolfe reduced gradient method to the
case of nonlinear constraints", *in* "Optimization",
Ed. R. Fletcher, Academic Press, London.

Beale, E.M.L. (1968), "Mathematical Programming in
Practice", Pitman and Sons, London.

Bracken, J. and McCormick, G.P. (1968), "Selected
Applications of Nonlinear Programming", Wiley,
New York.

Colville, A.R. (1968), "A comparative study of
nonlinear programming codes", IBM New York Scientific Center, Technical Report 320-2949.

Fiacco, A.V. and McCormick, G.P. (1968) "Nonlinear
Programming: Sequential Unconstrained Minimization
Techniques", Wiley, New York.

Himmelblau, D.M. (1972) "A Uniform Evaluation of
Unconstrained Optimization Techniques", *in* "Numer-
ical Methods for Nonlinear Optimization", *Ed.* F.A.
Lootsma, Academic Press, London.

Hooke, R. and Jeeves, T.A. (1961) "Direct Search
Solution of Numerical and Statistical Problems",
J.A.C.M., **8**, 212-229.

Lasdon, L.S. (1970) "Optimization Theory for Large
Systems", McMillan, London.

Lootsma, F.A. (1972*a*) "A Survey of Methods for
Solving Constrained Minimization Problems via Unconstrained Minimization", *in* "Numerical Methods
for Nonlinear Optimization", *Ed.* F.A. Lootsma,
Academic Press, London.

Lootsma, F.A. (1972*b*) "Penalty-function Performance of Several Unconstrained Minimization Techniques", Philips Res. Repts. **27**, 358-385.

Lootsma, F.A. (1975) "Design of a Nonlinear Optimization Programme for Solving Technological Problems", *paper to be published in* "Proceedings of the
Conference on Optimization Theory and Optimal Control, Oberwolfach", *Eds.* R. Bulirsch, W. Oettli

and J. Stoer.

Nelder, J.A. and Mead, R. (1965) "A simplex method for function minimization", *Computer J.* 7, 308-313.

Rosen, J.B. and Kreuser, J.L. (1972) "A Gradient Projection Algorithm for Nonlinear Constraints", *in* "Numerical Methods for Nonlinear Optimization", *Ed.* F.A. Lootsma Academic Press, London.

Staha, R.L. (1973) "Constrained Optimization via Moving Exterior Truncations", Thesis, The University of Texas at Austin.

Staha, R.L. and Himmelblau, D.M. (1973) "Evaluation of Constrained Nonlinear Programming Techniques", Report, The University of Texas at Austin.

Tabak, D. and Kuo, B.C. (1971) "Optimal Control by Mathematical Programming", Prentice-Hall, Englewood Cliffs, New Jersey.

Optimum Design of Plate Distillation Columns

R.W.H. Sargent and K. Gaminibandara

(Imperial College, London)

SUMMARY

Distillation is the most widely used industrial process for separating a mixture into its components. The paper describes a simple model commonly used for a distillation column, and a procedure for computing its performance for given conditions. Various operational and design problems are then considered and formulated as nonlinear programming problems, and results for a few illustrative examples are presented and discussed.

Most of the design problems lead naturally to a mixed integer-nonlinear programming formulation, but it is shown that combinatorial techniques can be avoided by solving a slightly more general problem.

The problems were solved using a version of the variable-metric projection method described in an Appendix.

1. INTRODUCTION

Distillation is the most widely used industrial process for separating a fluid mixture into its components. It is based on the fact that if a vapour and liquid are brought together and allowed to come to equilibrium, the resulting vapour and liquid have different compositions. By arranging a series of such "equilibrium stages", with countercurrent flow of liquid and vapour between them, the total composition change can be enhanced and

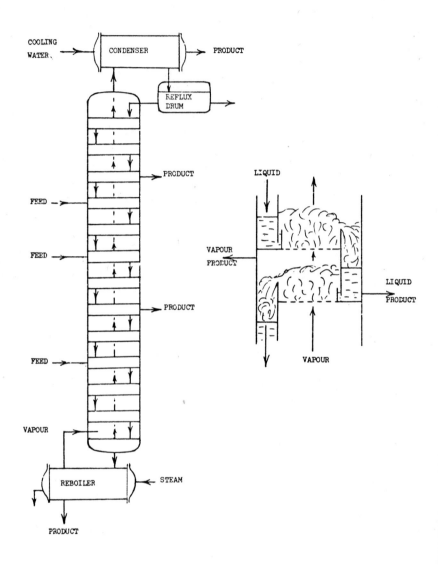

Fig. 1. Typical distillation column

the mixture thus split into two fractions, with more volatile components in one fraction and less volatile components in the other.

The stages are arranged vertically in a column so that the liquid can flow downwards by gravity from one stage to the next, and the vapour therefore flows upwards. The feed mixture enters one of the middle stages, and the countercurrent vapour and liquid streams are created by vaporizing and recycling part of the liquid arriving at the bottom, and condensing and recycling part of the vapour arriving at the top. The stages themselves often consist of perforated "plates" or "trays" across which the liquid flows, while the vapour passes upwards through the perforations and bubbles through the liquid. This typical arrangement is indicated in Fig. 1, where it is also shown that a given column may process several feed-mixtures, and fractions of intermediate composition may be withdrawn from the intermediate stages.

For the end-fractions the position of the split between the components, and the completeness of this split, depends on the number of stages, the positions of the feed streams, the amounts of recycle liquid and vapour, and the positions and relative flow-rates of the various product streams. The design problem is to choose these various quantities so that the specified separation is carried out at minimum cost.

2. THE BASIC EQUATIONS

We shall assume here a rather idealized model for a distillation column, in which the liquid and vapour phases are indeed brought to equilibrium in each stage, and the two phases leaving the stage are completely separated. This model is in fact adequate in a large number of cases, but more detailed treatments, using more realistic models may be found in Sargent and Murtagh (1969) and Sargent and Gaminibandara (1975). These further refinements do not substantially

affect the structure of the system of equations,
but merely complicate the algebraic expressions;
the optimization techniques described in this paper
can therefore be applied with equal facility to
these more realistic models. The general model
for a plate or mixing stage is shown in Fig. 2.

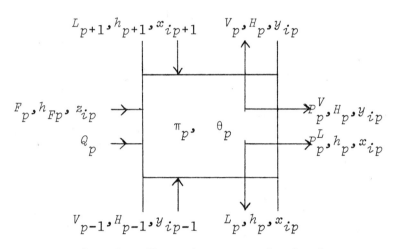

Fig. 2. Notation - typical plate

This provides for the possibility of a feed, withdrawal of a liquid or vapour product, and an indirect heat input to allow for heat losses through
the column walls or the presence of a heat-transfer
device on the plate. Normally vapour is received
from the plate below and passes to the plate above,
while the contrary is true for the liquid. However,
the model stage is sufficiently general to apply
also to the end stages, which incorporate vapori-
zation and condensation devices, and if we number
the stages from the base upwards we must write

$$V_0 = L_1 = 0; \quad V_n = L_{n+1} = 0. \qquad (1)$$

A material balance on streams entering and leaving
the stage p yields the equation

$$F_p + V_{p-1} + L_{p+1} = (V_p + P_p^V) + (L_p + P_p^L) \qquad (2)$$

270

A similar balance for a typical component i yields

$$z_{i,p}F_p + y_{i,p-1}V_{p-1} + x_{i,p+1}L_{p+1}$$
$$= y_{i,p}(V_p + P_p^v) + x_{i,p}(L_p + P_p^L) . \qquad (3)$$

An energy balance gives the equation

$$Q_p + h_{Fp}F_p + H_{p-1}V_{p-1} + h_{p+1}L_{p+1}$$
$$= H_p(V_p + P_p^V) + h_p(L_p + P_p^L) . \qquad (4)$$

Since the liquid and vapour streams leaving the stage are assumed to be in phase equilibrium we also have

$$y_{ip} = K_{ip}x_{ip} \qquad i = 1,2,\ldots c \qquad (5)$$

where K_{ip} is the equilibrium constant for the mixture, which is in general a function of pressure (π), temperature (θ) and the vapour and liquid compositions:

$$K_{ip} = K_i(\pi_p, \theta_p, \underset{\sim}{x}_p, \underset{\sim}{y}_p) . \qquad (6)$$

Similarly, the vapour and liquid enthalpies are functions of the fluid state:

$$h_p = h(\pi_p, \theta_p, \underset{\sim}{x}_p) , $$
$$H_p = H(\pi_p, \theta_p, \underset{\sim}{y}_p) . \qquad (7)$$

The pressure falls as the vapour rises through the successive stages, and to complete the model we need a means of predicting the pressure drop through each stage. For a given column the pressure drop depends on the liquid and vapour flowrates and fluid physical properties. However, it is usual to specify these pressure drops for the process design and later design the geometry of the column to achieve the specified values, so we shall assume that the pressures π_p, $p = 1,2,\ldots N$, are given.

3. DETERMINATION OF COLUMN PERFORMANCE

Distillation columns commonly separate mixtures of up to 20 components or more, and may have as many as 100 plates, so the model equations (1) to (7) form a large, sparse system of nonlinear algebraic equations. However they have a well defined structure, and the chemical engineering literature abounds with different approaches to solving them (see reviews by Sargent and Murtagh (1969), Sargent (1969) and Hutchison and Shewchuk (1974). Here we shall follow the approach described by Sargent and Gaminibandara (1975).

The basis of this approach is a procedure for determining the performance of a given column under given operating conditions. For this problem we are given the flow-rates, compositions and heat contents of the feed mixtures, the total number of stages in the column, the positions of feed introduction and of side-stream withdrawals (either liquid or vapour), the flow-rates of end and side-stream products, and the rates of heat input to each stage. In fact the column must operate continuously in a steady state, so the total material and heat leaving the column must balance the material and heat entering; we therefore leave unspecified one heat input and one product flow-rate, to be determined by the over-all material and energy balances. The problem is then to determine the internal flow-rates (L_p, V_p), temperatures (θ_p) and compositions (x_{ip}, y_{ip}) for all plates $p = 1, 2, \ldots N$, and components $i = 1, 2, \ldots c$.

The solution procedure starts with an estimate of the fluid physical properties h_p, H_p, K_{ip}, for all p and i. These initial values can in fact be quite crude - the same values for all p, computed for example from the temperature and composition of the main feed mixture

The internal flow-rates can then be computed from equations (1), (2) and (4). From (2) and (4) we obtain

$$(V_p + P_p^V)(H_p - h_{p+1}) = Q_p + F_p(h_{Fp} - h_{p+1})$$

$$+ V_{p-1}(H_{p-1} - h_{p+1}) - (L_p + P_p^L)(h_p - h_{p+1}) \quad (8)$$

and (2) can be rearranged to

$$L_{p+1} = (V_p + P_p^V) + (L_p + P_p^L) - F_p - V_{p-1} . \quad (9)$$

In view of equation (1) and remembering that Q_p, F_p, P_p^L and P_p^V are all given, equation (8) gives V_1 explicitly and (9) then gives L_2. Equations (8) and (9) can thus be used alternately to generate all the V_p and L_p. Equation (1) for the final stage is satisfied identically, provided that the Q_p, P_p^L, P_p^V were chosen to satisfy the over-all balance equations:

$$\left. \begin{array}{l} \displaystyle\sum_{p=1}^{N} (F_p - P_p^L - P_p^V) = 0 \\[4mm] \displaystyle\sum_{p=1}^{N} (Q_p + h_{Fp}F_p - h_p P_p^L - H_p P_p^V) = 0 \end{array} \right\} \quad (10)$$

With the K_{ip} and flow-rates known, the y_{ip} can be eliminated from equation (3) using equation (5) to yield a system of linear equations for the x_{ip}. This separates into independent systems for each component, which may be written

$$G^i x^i = f^i \qquad i = 1,2,3,\ldots c \quad (11)$$

where G^i is a tridiagonal matrix whose non-zero elements are

$$\left. \begin{array}{l} g_{p,p-1}^i = - K_{ip-1} V_{p-1} \\[3mm] g_{p,p}^i = (L_p + P_p^L) + (V_p + P_p^V) K_{ip} \\[3mm] g_{p,p+1}^i = - L_{p+1} \end{array} \right\} \quad (12)$$

and f^i has elements $f_p^i = z_{ip} F_p .$

273

Such a system can be solved by triangular factorization without increasing the number of non-zero elements.

Having obtained all the x_{ip} in this way, the y_{ip} are easily obtained from (5). Unfortunately, unless the K_{ip} were correctly chosen we shall not satisfy the equations

$$\sum_{i=1}^{c} x_{ip} = \sum_{i=1}^{c} y_{ip} = 1, \qquad p = 1,2,\ldots N \qquad (13)$$

as required by the definitions of the x_{ip} and y_{ip}. This can be remedied by rescaling all the x_{ip} and y_{ip} by the appropriate sums, but then of course the rescaled values no longer satisfy (5) and (11). It has therefore been found useful to introduce an inner iteration loop to improve this situation.

For many mixtures the relative volatility of any two components, defined as the ratio of their equilibrium constants, remains approximately constant over a range of compositions, so we replace the K_{ip} by $\sigma_p K_{ip}$ in (5), (11) and (12), and seek to choose the σ_p so that

$$\sum_{i=1}^{c} x_{ip} = \sum_{i=1}^{c} y_{ip} = \sum_{i=1}^{c} \sigma_p K_{ip} x_{ip}$$

whence we obtain

$$\sigma_p = \frac{\displaystyle\sum_{i=1}^{c} x_{ip}}{\displaystyle\sum_{i=1}^{c} K_{ip} x_{ip}} . \qquad (14)$$

The inner loop iteration starts with all $\sigma_p = 1$ and equations (12) and (14) are used alternately until the

$$\sum_{i=1}^{c} x_{ip}$$

are sufficiently close to unity. In effect this

solves a system of N nonlinear equations for the σ_p, and any alternative to direct substitution can be used; however it is best to avoid methods which require storage of a $N \times N$ matrix, and we have found it effective to ignore cross-coupling between the σ_p, using parabolic inverse interpolation to adjust each value separately. If the inner loop converges, it is easy to see that since (2) is satisfied, (3) and (13) are satisfied, and (5) is satisfied with the modified K_{ip}.

It now remains to determine plate temperatures (θ_p), and hence new values of the physical properties from (6) and (7). Sargent and Murtagh (1969) proposed the determination of the boiling-points of the liquids leaving each plate for this purpose, which incidentally gives new K_{ip} values. However, boiling-points are sensitive to small changes in concentration of very volatile components, and this can lead to instability of the procedure. Vapour dew-points are similarly sensitive to relatively non-volatile components, and a stable procedure which has given good results is to use a "flash equilibrium" calculation for each plate. This is described in Appendix I.

The flash calculations give θ_p and the K_{ip} for each plate so it only remains to compute new enthalpies from (7), and we are then in a position to repeat the whole calculation. This again gives a direct substitution procedure for the outer iteration loop, and again it is desirable to include some method of convergence acceleration. The best point in the loop to do this is on the V_p as they are calculated from equation (8); again we have found it effective to adjust each V_p separately, using parabolic inverse interpolation on previously calculated values.

The complete procedure can be summarized as follows.

Procedure for Performance Calculation

Given: N, F_p, h_{Fp}, z_{ip} all i; π_p; Q_p (except one); P_p^L, P_p^V (except one)

     Estimates of: h_p, H_p, K_{ip} all i

1.   Compute remaining product-rate and heat input from (10).

2.   Compute all L_p, V_p from (8) and (9), adjusting V_p by inverse interpolation.

3.   Set all $\sigma_p = 1$.

4.   Solve (12) for $i = 1, 2, \ldots c$, to give all x_{ip}.

5.   If $\left| \sum_{i=1}^{c} x_{ip} - 1 \right| > \varepsilon$ for any p, compute all σ_p from (14) and inverse interpolation, then return to step 4.

6.   Rescale $x_{ip} / \sum_{i=1}^{c} x_{ip}$ and compute

$$y_{ip} = K_{ip} x_{ip} / \sum_{i=1}^{c} K_{ip} x_{ip}.$$

7.   If the changes in V_p, x_{ip}, y_{ip} since the last iteration are all within acceptable tolerances, stop.

8.   Perform flash calculations for each plate to determine the θ_p and K_{ip}.

9.   Recompute the h_p and H_p from (7), then return to step 1.

    This procedure involves two inner loops and one outer loop of iterations. It is possible to omit the loop involving σ_p (Step 5), and to carry out only a single iteration of the flash procedure, leaving only the outer iteration loop. However, it has been found that in general convergence is improved by incorporating these inner loops, although the tolerances set for them should not be too severe.

As has been remarked already, more powerful methods could be used for adjusting the V_p in step 2 and/or the σ_p in step 5. However, in each case an extra non-sparse $N \times N$ matrix would have to be stored, and in view of the rapid convergence of the simpler methods the extra storage does not seem worthwhile.

From time to time the Newton method itself has been proposed (Hutchison and Shewchuk (1974), Stainthorp and Whitehouse (1967a), (1967b), Taylor and Edmister (1969) and Goldstein and Stanfield (1970)) for application to the whole system of equations (1) to (7) treated simultaneously, and both Stainthorp and Whitehouse (1967b) and Hutchison and Shewchuk (1974) used the known sparsity structure of the Jacobian to economize in both computation and storage. In addition Hutchison and Shewchuk propose approximations to avoid the computation of exact derivatives for the fluid physical properties (h_p, H_p, K_{ip}) with respect to temperature and concentrations. Direct comparisons with these procedures have not been made, but again each iteration would require either more storage or more computation and these extra requirements do not seem worthwhile.

4. OPTIMUM COLUMN DESIGN

4.1 *The Operational Problem*

The simplest type of optimization problem is the choice of operating conditions for a given column to achieve a specified performance at minimum cost. Assuming the column must process the given feeds, the adjustable (control) variables are the indirect heat inputs and product flow-rates (subject to over-all balance constraints) and the main items of cost are the heating and cooling requirements. It is rare for a column to have heat-transfer devices other than in the top and bottom stages, so in general only Q_1 and Q_N are available for adjustments and one of these (normally Q_N) is determined by the over-all energy balance.

Similarly one of the product rates is determined by the over-all material balance, so in the usual case the number of degrees of freedom of the system is equal to the number of product streams.

Performance specifications are usually made in terms of component concentrations or recoveries, where the recovery of a given component in a given product is the proportion of that component entering in the feeds which leaves in the given product; for example

$$\rho_{ip}^{L} = (x_{i,p} P_{p}^{L}) / \sum_{p=1}^{N} z_{ip} F_{p} . \qquad (15)$$

Thus one might require a minimum recovery or concentration of a given component in a given product, or a maximum concentration of a given impurity in a product. A product flow-rate may also be specified in some circumstances, and in some cases, especially in the oil industry, products are required to have certain physical characteristics which depend indirectly on their composition. The important point to note is that any of these specified quantities, and the cost of operating the column, can be computed once the column performance has been computed for given values of the control variables, for example by the method of the last section.

The specifications may therefore be in the form of either equality or inequality constraints, and both the constraint functions and the cost function are defined through the system equations as nonlinear functions of the control variables. We are therefore faced with a nonlinear programming problem.

It is true that the number of specifications is often equal to the number of degrees of freedom, and in most cases they are then tight enough to be satisfied as equalities, so that the problem is again reduced to solving a set of nonlinear equations.

Sargent and Murtagh (1969) treated this case, and concluded that the adjustment of the control variables to satisfy the specifications should be made an extra loop round the procedure described in the last section, rather than incorporated in the loop which adjusts the internal flow-rates.

In considering the more general nonlinear programming problem we have also treated the performance calculation as part of the procedure for determining the cost and constraint functions, rather than including all the system equations as equality constraints. The method of solution adopted was the "variable metric projection" (VMP) method of Sargent and Murtagh (1973), which is briefly described in Appendix II. This algorithm incorporates an initial phase for finding a feasible point, so it will automatically deal with the special situation of a unique feasible point, or an inconsistent specification.

It will be noted that the method requires both function and gradient values of the objective and constraint functions, and we therefore require partial derivatives of the concentrations with respect to the control variables. It is clearly not easy to obtain analytical expressions for these derivatives; for a general-purpose design program it would also be necessary to require the user to supply routines to provide derivatives of the physical property functions, equations (6) and (7), and these can also be very complicated expressions.

In such situations it is common to generate finite-difference approximations to the derivatives, using small perturbations of each variable in turn, and this provides a strong incentive for keeping the number of control variables small as in the approach adopted, rather than treating all equations and variables simultaneously.

Alternatively, we can try to find reasonable approximations to the derivatives, and this is

279

possible if we treat the fluid physical properties, h_p, H_p, K_{ip}, as constant in this approximation. Equations (8) and (9) are then linear in the remaining variables, and can be differentiated with respect to any control variable, yielding a set of recursive relations for the derivatives of the L_p and V_p with respect to this variable. From (12) for any control variable, u, we have

$$\underset{\sim}{G}^i \underset{\sim}{x}^i_u = \underset{\sim}{f}^i_u - \underset{\sim}{G}^i_u \underset{\sim}{x}^i \tag{16}$$

where suffix u denotes differentiation with respect to u. The elements of $\underset{\sim}{G}^i_u$ are simply computed from the derivatives of the L_p and V_p, and the $\underset{\sim}{x}^i_u$ can then be generated from (16), using the same factorization of $\underset{\sim}{G}^i$ as for the performance calculation itself. Thus these approximate derivatives can be computed at relatively little extra cost; we note that they are much simpler approximations than those proposed by Hutchison and Shewchuk (1974).

4.2   *The Process Design Problem*

When designing a column to achieve a specified performance at minimum cost, one must not only select the operating conditions, but also determine the total number of stages required and the optimum positions for feed introduction and side-stream product withdrawal. Since the stages are discrete elements all these extra variables are integer-valued, so in principle we now have a mixed integer-nonlinear programming problem.

For a given total number of stages there is usually a well defined optimum feed position, although the performance does become less sensitive to feed position as either the total number of stages or the internal liquid and vapour rates increase. The optimum position for a product side-stream is usually quite sharply defined, and is sensitive to the total number of stages, the operating conditions, and the positions of feeds and other side-streams. Although a single feed and side-stream can be placed by trial and error, the combinatorial problem gets rapidly out of hand as the number of feeds and side-streams increases, and it does not seem possible to devise effective bounds for use in an implicit enumeration method.

We have instead considered the more general problem of the optimum distribution of the feed-mixtures over all the stages, and correspondingly the optimum blend of side-streams from all stages for each required intermediate product. Thus we suppose that we have J feed mixtures with given values of F_j, h_{Fj}, z_{ij}, all $i, j = 1, 2, \ldots J$, and each is distributed over all the stages. We then have the relations

$$F_p = \sum_{j=1}^{J} F_{jp}, \quad h_{Fp}F_p = \sum_{j=1}^{J} h_{Fj}F_{jp}, \quad z_{ip}F_p = \sum_{j=1}^{J} z_{ij}F_{jp} \quad (17)$$

$$F_j = \sum_{p=1}^{N} F_{jp}, \quad F_{jp} \geq 0. \quad (18)$$

Equations (17) define the plate feeds in terms of the new variables F_{jp} which must satisfy the constraints in (18).

Similarly we assume that K products, P_k, $k = 1, 2, \ldots K$, are required, each defined to be either liquid or vapour and satisfying certain specifications, and each made by blending side-streams from all stages. The relations (e.g., for a liquid product) are

$$\left. P_k^L = \sum_{p=1}^{N} P_{kp}^L, \quad \bar{h}_k P_k^L = \sum_{p=1}^{N} h_p P_{kp}^L, \quad \bar{x}_{ik} P_k^L = \sum_{p=1}^{N} x_{ip} P_{kp}^L \right\} \quad (19)$$

$$\left. P_p^L = \sum_{k=1}^{K} P_{kp}^L, \quad P_{kp}^L \geq 0. \right.$$

Hence the new variables P_{kp}^L must satisfy the non-negativity constraints, and other required quantities are calculated from them using (19).

Clearly, it is not necessary to spread distributions over all stages, and to economize on computational effort provision should be made for selection of an appropriate range of plates for each feed and side-stream. However, caution is required, for we have computed optimum solutions which do not accord at all with intuitive preselections.

Varying the total number of stages in the optimization raises different problems. If stages are to be added to or removed from a given column, the appropriate location of these stages is not clear, and of course if a stage is removed its feeds and product-streams must be set to zero and redistributed. One possibility is to specify a maximum number of stages and associate a zero-one variable with each stage to indicate its actual existence; this introduces a large number of zero-one variables and many equivalent solutions, without avoiding the indirect effects on the variables F_{jp} and P_{kp}. A much better solution is to keep the adjustment of N as a separate outer loop, solving the optimization problem for each N. This has the advantage that N can be adjusted by a one-dimensional unconstrained search procedure in the outer loop simply rounding N to integer values to evaluate the solution. Effective one-dimensional procedures exist which do not require derivatives, a suitable one being that described by Sargent and Sebastian (1974). It is the difficulty of generating derivatives with respect to N, with the attendant indirect effects on the product and feed distributions and variation of the number of independent variables, which militates against treating N as an additional variable in the nonlinear programming problem.

The cost function for the design problem will in general be a sum of capital cost of the equipment and suitably discounted recurrent costs. The capital cost increases with both the number of stages and the internal liquid and vapour rates, while the operating costs depend only on the latter, mainly through the provision of heating and cooling in the top and bottom stages.

Each feed-point and side-stream withdrawal point also increases the capital cost, mainly through the extra pipework and control equipment involved. One may also wish to avoid the complexity of controlling a distribution over several feed-points or a similar blending operation, and this may be done by allocating an artificially high threshold cost to each extra non-zero feed or side-stream. Such a threshold cost introduces discontinuities into the cost function, but the size of these is usually a small percentage of the total and no serious difficulties should arise; however excessive artificial costs to suppress multiple entry or exit points should be avoided.

4.3 *Examples*

The examples given are illustrative only. They are based on the separation of real mixtures, but mixtures with few components and simply calculated physical properties have been used to economize on computation. For the same reason there is no attempt at realism for the cost functions. In all the examples the feed-mixtures are mixtures of light hydrocarbons (alkanes with three to five carbon atoms) and appropriate expressions for equilibrium constants and enthalpies were taken from Holland (1963).

Example 1.

The first example is designed to show the effect of product distribution on the optimum feed distribution. The feed is 100 lb mol./hour of a cold liquid mixture with specific enthalpy (h_F) of 4000 Btu/lb mol. and composition:

1.	C_3H_8	5	mol.%
2.	iC_4H_{10}	15	mol.%
3.	nC_4H_{10}	25	mol.%
4.	iC_5H_{12}	20	mol.%
5.	nC_5H_{12}	35	mol.%

A column of four equilibrium-stages is used, with liquid products drawn from the top (condenser) and bottom (reboiler) stages. The top product must not contain more than 7 mol.% of nC_5H_{12}, and the problem was solved for three specified values of the top product rate (P_4^L) of (a) 10 lb mol./hour (b) 15 lb mol./hour (c) 20 lb mol./hour.

The operating cost is taken as proportional to the reboiler heat input (Q_1), so the problem may be formulated as:

Minimize Q_1

Subject to $Q_1 \geqslant 0$ $\qquad\qquad\qquad$ (i)

$\qquad\qquad$ $Q_4 \leqslant 0$ $\qquad\qquad\qquad$ (ii)

$\qquad\qquad$ $x_{5,4} \leqslant 0.07$ $\qquad\qquad$ (iii)

$\qquad\qquad$ $\sum\limits_{p=1}^{4} F_p = 100$ $\qquad\quad$ (iv)

$\qquad\qquad$ $F_p \geqslant 0$ $\quad p = 1,2,3,4.$ $\quad (v)$

Thus we have five control variables Q_1, F_1, F_2, F_3, F_4, each subject to a non-negativity constraint, one linear equality constraint (iv), and two non-linear inequality constraints, (ii) and (iii), defined through the system equations.

The iterations were started with the same feasible point in each case, with uniform feed distribution and a high value of Q_1 to be sure of satisfying constraints (ii) and (iii):

$\qquad$ Initial point: $\quad Q_1 = 0.5 \times 10^7$ Btu/hour,

$\qquad\qquad\qquad$ $F_1 = F_2 = F_3 = F_4 = 25$ mol./hour.

The results are given in Table I. In all three cases Q_1 first decreases until constraint (iii) becomes active, and thereafter all variables change smoothly towards their optimum values. The solutions for the three cases show a regular variation, with increasing top product rate demanding a better separation and hence higher operating costs.

284

Table I

Results for Example 1

P_4^L	Initial Estimates	10	15	20
F_4	25	23.71	7.84	0
F_3	25	0	0	0
F_2	25	0	0	0
F_1	25	76.29	92.16	100
$Q_1 (\times 10^{-6})$	5.0	0.3375	0.3436	0.9425

$$x_{5,4} = 0.07 \text{ in all cases}$$

With no specification on bottom product purity it is not surprising to find the result of case (c), with feed in the reboiler and hence all stages used to remove nC_5H_{12} from the top product. As the separation becomes easier, it is apparently possible to use some of the cold feed as reflux in the top stage without violating the purity constraint, thus reducing the demand for condensate in the top stage and hence correspondingly the heat input at the bottom. Nevertheless, it is not easy to see why it is better to use the top stage, where the product is directly contaminated, rather than the stage below, nor why the optimum distribution is bimodal, with no feed at all to the two middle stages. Thus the optimum solution is not easily rationalized in physical terms, let alone predictable by intuitive reasoning, and the example demonstrates the need for a mathematical solution.

Example 2.

This example shows the effect of feed heat-content on the solution, for a column with 10 stages and 100 lb mol./hour of the feed mixture:

1. C_3H_8 30 mol.%

2. nC_4H_{10} 25 mol.%

3. nC_5H_{12} 45 mol.%

Three feed-conditions are examined:

(*a*)   Cold liquid (h_F = 3000 Btu/lb mol.)

(*b*)   Liquid at its boiling-point
(h_F = 5262.3 Btu/lb mol.)

(*c*)   Liquid-vapour mixture
(h_F = 8000 Btu/lb mol.)

In each case liquid products are drawn from the reboiler (stage 1) and condenser (stage 10), and a side-stream product is made up by blending liquid drawn from stages 2 to 9. Again the operating cost is taken as proportional to the reboiler heat input.

The bottom product is required to have a purity of at least 95% nC_5H_{12}, and the recovery of nC_5H_{12} in the bottom product must be at least 70%.

The mathematical formulation is:

Minimize Q_1

Subject to $Q_1 \geqslant 0$, $Q_{10} \leqslant 0$

$F_p \geqslant 0$, $P_p^L \geqslant 0$, $p = 1,2,\ldots10$

$$\sum_{p=1}^{10} F_p = 100$$

$x_{3,1} \geqslant 0.95$, $\rho_{3,1}^L \geqslant 0.70$

Thus we have 21 control variables, each subject to a non-negativity constraint, one linear equality constraint, and three nonlinear inequality constraints.

The iterations were started with equal feeds and product-rates of 10 mol./hour for all stages and the results for the three feed conditions are shown in Table IIa. It will be seen that the required reboiler heat-input decreases as the feed heat-content increases, but not in exact compensation since the condenser heat output increases. The feed is distributed over several plates, tending to be lower in the column as its heat-content increases. The minimum separation requirements are just met in each case, and there is no side-stream product. This is as expected, since there are no specifications on the top product, and indeed the whole solution is in accord with intuition.

Table IIa

Results for Example 2 - Effect of Feed Conditions

Plate Number	$h_F = 3000$		$h_F = 5262.3$		$h_F = 8000$	
	F_p	P^L_p	F_p	P^L_p	F_p	P^L_p
10	0	66.84	0	66.84	0	66.84
9	10.24	0	9.46	0	0	0
8	20.32	0	20.13	0	8.30	0
7	26.01	0	26.22	0	35.64	0
6	25.83	0	26.13	0	33.79	0
5	17.60	0	18.06	0	22.27	0
4	0	0	0	0	0	0
3	0	0	0	0	0	0
2	0	0	0	0	0	0
1	0	33.16	0	33.16	0	33.16
$Q_{10} \times 10^{-6}$ $Q_1 \times 10^{-6}$	-0.394 0.681		-0.549 0.620		-0.740 0.537	
$\rho^L_{3,1}$ $x_{3,1}$	0.70 0.95		0.70 0.95		0.70 0.95	

287

Table IIb

Results for Example 2 - Compositions

for ^hF = 5262.3

Plate Number	F_p	P_p^L	x_1	x_2	x_3
10	0	66.84	0.4489	0.3497	0.2013
9	9.46	0	0.1778	0.3423	0.4799
8	20.13	0	0.1052	0.2816	0.6132
7	26.22	0	0.0804	0.2411	0.6785
6	26.13	0	0.0627	0.2143	0.7230
5	18.06	0	0.0406	0.1887	0.7707
4	0	0	0.0162	0.1539	0.8299
3	0	0	0.0061	0.1154	0.8786
2	0	0	0.0021	0.0796	0.9183
1	0	33.16	0.0007	0.0493	0.9500

Example 3.

The conditions are exactly as for Example 2, except for the product specifications. In this example the top product is required to have a purity of at least 95% C_3H_8, with a recovery of C_3H_8 in the top product of at least 70%. The formulation is:

Minimize $\quad Q_1$

Subject to $\quad Q_1 \geqslant 0, \quad Q_{10} \leqslant 0$

$$F_p \geqslant 0, \quad P_p^L \geqslant 0, \quad p = 1,2,\ldots 10$$

$$\sum_{p=1}^{10} F_p = 100$$

$$x_{1,10} \geqslant 0.95, \quad \rho_{1,10}^L \geqslant 0.70 .$$

The starting conditions were as in Example 2, and the results are given in Tables IIIa and b. Again the minimum separation requirements are just met. With a specification on the top product instead of the bottom product, as in Example 2, the feed is displaced towards the bottom of the column, and there is even some feed direct to the reboiler. Again the reboiler heat input decreases, and the condenser heat output increases, as the feed heat-content increases.

Table IIIa

Results for Example 3 - Effect of Feed Conditions

Plate Number	$h_F = 3000$		$h_F = 5262.3$		$h_F = 8000$	
	F_p	P_p^L	F_p	P_p^L	F_p	P_p^L
10	0	22.11	0	22.11	0	22.11
9	0	0	0	0	0	0
8	0	0	0	0	0	0
7	0	0	0	0	0	0
6	0	0	0	0	0	0
5	15.69	0	10.10	0	5.70	0
4	25.79	0	23.39	0	18.74	0
3	26.29	0	28.70	0	26.75	0
2	19.09	0	24.65	0	23.93	0
1	13.14	77.89	13.16	77.89	24.88	77.89
$Q_{10} \times 10^{-6}$	-0.0274		-0.0947		-0.2161	
$Q_1 \times 10^{-6}$	0.3387		0.1773		0.0253	
$\rho_{1,10}^L$	0.70		0.70		0.70	
$x_{1,10}$	0.95		0.95		0.95	

Table IIIb

Results for Example 3 - Compositions

for $^h F$ = 5262.3

Plate Number	F_p	P_p^L	x_1	x_2	x_3
10	0	22.11	0.9500	0.0498	0.0002
9	0	0	0.8641	0.1345	0.0014
8	0	0	0.7154	0.2771	0.0075
7	0	0	0.5264	0.4400	0.0336
6	0	0	0.3710	0.5229	0.1061
5	10.10	0	0.2605	0.4898	0.2497
4	23.39	0	0.2014	0.4163	0.3823
3	28.70	0	0.1707	0.3637	0.4656
2	24.65	0	0.1483	0.3339	0.5178
1	13.16	77.89	0.1139	0.3071	0.5790

It is of interest to note that, whereas in Example 2 the nC_4H_{10} concentration (x_2) increases steadily with plate number, in Example 3 there is a maximum in x_2 at plate number 6.

For this example the calculations were repeated for two products only, but with each obtained by blending streams from all ten stages. The result was identical, confirming the intuitive conclusion based on thermodynamic reasoning that the optimum should correspond to products drawn only from the end stages.

Example 4.

Again the conditions are the same as in Example 2, except for the product specifications. Here we require the specifications of Example 2 and 3 to be satisfied simultaneously, giving the formulation:

Minimize Q_1

Subject to $Q_1 \geqslant 0, \quad Q_{10} \leqslant 0$

$$F_p \geqslant 0, \quad P_p^L \geqslant 0, \quad p = 1,2,\ldots 10$$

$$\sum_{p=1}^{10} F_p = 100$$

$$x_{1,10} \geqslant 0.95, \quad \rho_{1,10}^L \geqslant 0.70$$

$$x_{3,1} \geqslant 0.95, \quad \rho_{3,1}^L \geqslant 0.70 \;.$$

The starting conditions were as before, and the results are given in Tables IVa and b. All the minimum separation requirements are just met, and since the nC_4H_{10} is virtually excluded from both end products it must leave in the side-stream product. This side-stream is almost all drawn from Stage 6 where the nC_4H_{10} concentration reaches a maximum and the feed is also limited to a single introduction point, without the need for artificial threshold costs to achieve this.

A better separation of the feed is achieved in this example, reflected in the much higher heating and cooling requirements, and the trends with feed heat-content are similar to those in Examples 2 and 3.

Example 5.

This last example treats the feed of Example 1 in a seven-stage column, except that the feed now enters at its boiling-point (h_F = 7700 Btu/lb mol.). All products are again drawn off as liquids. The bottom product must not contain more than 5.5% of nC_4H_{10}, while the top product must not contain more than 5.5% of iC_5H_{12} with a minimum flow-rate of 10 lb mol./hour.

Table IVa

Results for Example 4 - Effect of Feed Conditions

Plate Number	$h_F = 3000$		$h_F = 5262.3$		$h_F = 8000$	
	F_p	P_p^L	F_p	P_p^L	F_p	P_p^L
10	0	22.11	0	22.11	0	22.11
9	0	0	0	0	0	0
8	0	0	0	0	0	0
7	0	1.41	0	0.70	0	1.64
6	0	43.32	0	44.03	0	43.09
5	0	0	0	0	0	0
4	100	0	100	0	100	0
3	0	0	0	0	0	0
2	0	0	0	0	0	0
1	0	33.16	0	33.16	0	33.16
$Q_{10} \times 10^{-6}$	-2.9537		-3.1648		-3.2930	
$Q_1 \times 10^{-6}$	3.2990		3.2329		3.1383	
$\rho_{1,10}^L$	0.70		0.70		0.70	
$x_{1,10}$	0.95		0.95		0.95	
$\rho_{1,10}^L$	0.70		0.70		0.70	
$x_{3,1}$	0.95		0.95		0.95	

Table IVb

Results for Example 4 - Compositions

for h_F = 5262.3

Plate Number	F_p	P_p^L	x_1	x_2	x_3
10	0	22.11	0.9500	0.0495	0.0005
9	0	0	0.8651	0.1312	0.0037
8	0	0	0.6730	0.3004	0.0266
7	0	0.70	0.4073	0.4749	0.1178
6	0	44.03	0.1895	0.4998	0.3107
5	0	0	0.0783	0.3808	0.5409
4	100	0	0.0407	0.2512	0.7081
3	0	0	0.0116	0.1574	0.8310
2	0	0	0.0031	0.0907	0.9062
1	0	33.16	0.0008	0.0492	0.9500

Thus the split required is essentially between nC_4H_{10} and iC_5H_{12}, but the additional specification on flow-rate would seem to require an extra degree of freedom, so we allow a side-stream blended from liquids drawn from stages 2 to 6. Again the operating cost is taken as proportional to the reboiler heat input, so the mathematical formulation is:

Minimize Q_1

Subject to $Q_1 \geqslant 0$, $Q_7 \leqslant 0$

$\qquad x_{3,1} \leqslant 0.055 \qquad x_{4,7} \leqslant 0.055$

$\qquad F_p \geqslant 0 \quad , \quad P_p^L \geqslant 0, \quad p = 1,2,\ldots 7$

$\qquad \sum_{p=1}^{7} F_p = 100 \quad , \quad \sum_{p=1}^{6} P_p^L \leqslant 90$

293

The results are given in Table Va. The product specifications are all met, and as expected there is a side-stream (essentially drawn from stages 4 and 5); the feed is distributed over three stages.

However the result could scarcely be what was intended, for although the bottom product satisfies its purity specification its product-rate is zero! The total side-stream is not far from the feed composition, so clearly it is used as a device for minimizing the amount of feed which must be separated to provide the specified amount of top product.

This demonstrates the need for care in specifications when side-streams are possible. The two purity specifications alone would be quite satisfactory if only end products were allowed, although in fact they would not be attainable in this case with only seven stages, even with $Q_1 = \infty$. With a side-stream allowed, these two purity specifications alone are satisfied with Q_1 close to zero and no end products at all. To obtain a sensible result it is necessary to add specifications on either desired recoveries or minimum product-rates; such a solution is given in Table Vb, for minimum end-product rates of 30 lb mol./hour each.

<u>Table Va</u>

Results for Example 5 for $P_7^L \geqslant 10$

Plate Number	F_p	P_p^L	x_1	x_2	x_3	x_4	x_5
7	0	10.00	25.0	30.4	33.0	5.5	6.1
6	0	0.10	10.6	25.6	35.3	12.1	16.4
5	12.84	26.81	4.3	17.7	29.5	18.7	29.8
4	42.62	60.31	2.1	11.6	22.1	22.8	41.4
3	44.54	2.80	1.5	8.3	16.9	24.3	49.0
2	0	0	0.4	4.0	10.0	25.7	59.9
1	0	0	0.1	1.8	5.5	25.2	67.4

$Q_1 = 24867$ Btu/hour $Q_7 = -51409$ Btu/hour

Table Vb

Results for Example 5 for $P_1^L \geqslant 30$, $P_7^L \geqslant 30$

Plate Number	F_p	P_p^L	x_1	x_2	x_3	x_4	x_5
7	0	30.00	15.27	32.86	41.49	5.50	4.88
6	0	0	6.31	26.80	42.83	11.61	12.45
5	0	5.00	2.33	18.38	36.23	18.93	24.13
4	100	35.00	0.85	10.96	25.99	24.92	37.26
3	0	0	0.25	5.82	16.79	28.55	48.60
2	0	0	0.07	2.83	9.90	29.63	57.57
1	0	30.00	0.02	1.30	5.50	28.92	64.26

$$Q_1 = 7.359 \times 10^6 \text{ Btu/hour}$$

$$Q_7 = -7.361 \times 10^6 \text{ Btu/hour}$$

5. MULTIPLE COLUMN SYSTEMS

Since a single column can make a substantial-ly complete split into only two fractions, we need $(n - 1)$ columns to split an n-component mixture into essentially pure individual components. Clearly, the successive splits can be made in a large number of different ways, and the number is further increased by the possibility of processing "side-streams" from intermediate stages of some of the columns. Thus the problem of selecting the best configuration of columns for separation of given mixtures into several well-defined fractions is a large combinatorial problem.

Some possibilities for the separation of a ternary mixture into its components are shown in Fig. 3. These are far from exhaustive, but there is a logical pattern of development. The three basic cases are shown in Figs. 3a, b and c; at first sight case (c) looks uneconomic since it requires three columns rather than two, but it may in fact require fewer stages over-all than either (a) or (b).

The condensate at the top of Column 1 in Fig. 3a is a mixture of $(A + B)$ and clearly similar liquid mixtures are descending in Column 2. We can therefore eliminate the condenser at the top of Column 1, and draw the required liquid reflux from the middle of Column 2 instead, thus arriving at the configuration in Fig. 3d. The upper section of Column 2 is now in a sense common to both columns; it receives vapour from Column 1 and the lower section of Column 2 and distributes liquid to each of them. Thus it can be transferred to Column 1, to give the functionally equivalent system of Fig. 3g.

A similar argument can be used to develop Figs. 3e and h from Fig. 3b, and Figs. 3f and i from Fig. 3c. In the latter case, both the reboiler and condenser of Column 1 can be eliminated by drawing liquid from Column 2a and vapour from Column 2b. Also the reboiler in Column 2a generates vapour from the descending liquid which is essentially pure B, and hence the same as the vapour condensed in the condenser of Column 2b to produce liquid reflux; thus both reboiler and condenser can be eliminated and the two columns joined into a single column, as in Fig. 3f. The transfer of end-sections then produces Fig. 3i.

An apparently different solution is depicted in Fig. 4a. Here a side-stream is taken from the lower section of Column 1, low enough to contain satisfactorily small quantities of the most volatile component A; this mixture of $(B + C)$ is then separated in Column 2. However the bottom sections of both columns are performing essentially the same function, and can be combined into a single section, producing either (4a) or (4b); these are identical to Figs. 3h and e, respectively.

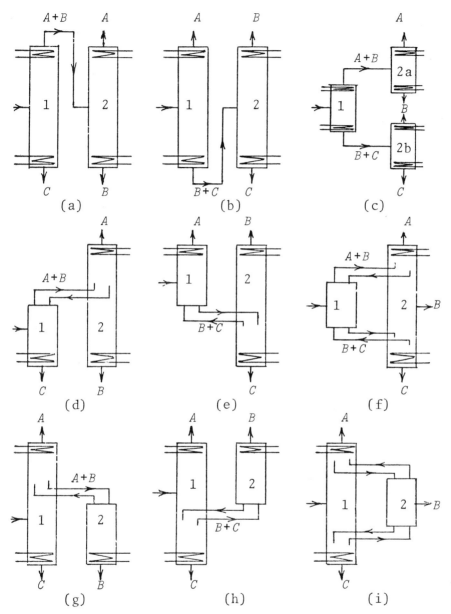

Fig. 3. Separation of a ternary mixture into its components

297

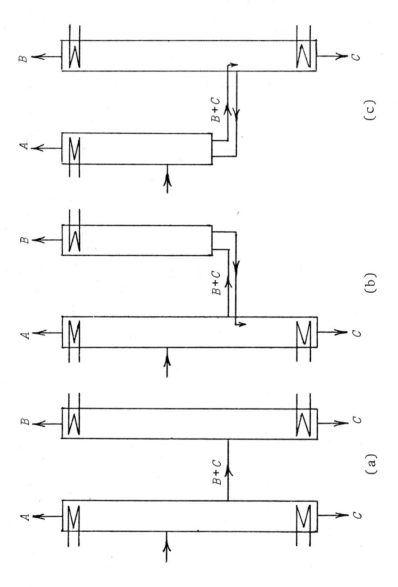

Fig. 4. Equivalent configurations

From these considerations it can be seen that all possible functionally distinct configurations are contained in the system shown in Fig. 5a. Any particular configuration can be obtained from this general system by deletion of column-sections, heat-transfer devices, or interconnecting streams. Since these deletions correspond to setting variables to zero, the optimum configuration can in principle be found by optimum design of this general system, without having to generate all possible configurations and select from them by combinatorial techniques.

The general system is extended to four components in Fig. 5b, from which the further extension to n components is obvious. For an n-component system there will be $(n - 1)$ "levels" and a maximum of $\frac{1}{2}n(n - 1)$ columns.

In the figure only one external feed is shown, but it is clear that external feeds may enter any of the columns provided that they fit the composition range for the column in question; in principle therefore one should consider the optimum distribution of all feeds over all stages in the system. However, to attempt an optimum distribution over all stages for all feeds and interconnecting streams would clearly result in a nonlinear program with an enormous number of variables and prohibitive computational effort. Indeed this leads to a network of equilibrium-stages with feeds and heat-inputs distributed to all of them, and the liquid and vapour streams of each distributed over all the others; products would be blends of streams from all stages. For J feeds, K products and a total of N stages, this yields $N(2N + J + K-1)$ control variables, and the system would be completely dense.

Clearly we must be content with a less general solution, and if we assume a unique introduction or withdrawal point for each feed, product, or pair of interconnecting streams, we can take as control variables the flow-rates of these streams, the number of stages in each column-section, and

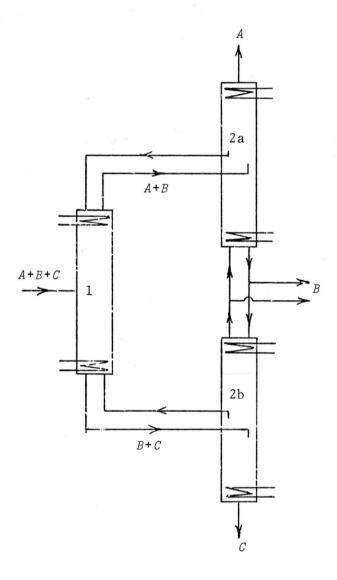

Fig. 5a. General system for ternary mixtures

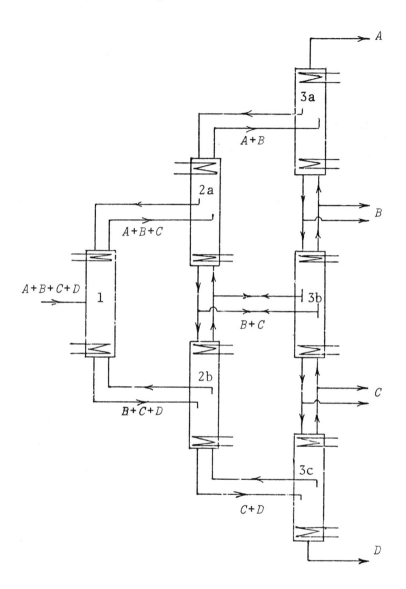

Fig. 5b.   General system for four components

the indirect heat transfer rates at the top and bottom of each column. With this plan, liquid and vapour rates can be calculated by recursion separately for each column, and of course plate temperatures can be determined individually, as in the single-column procedure of Section 3. However, the interconnecting streams make the compositions interdependent and these must be determined simultaneously for the whole system; fortunately the linear system corresponding to equation (11) in Section 3 has a well-defined sparsity structure which can be exploited.

Our studies of multiple-column systems are still at an exploratory stage, and we hope to report on them in a later paper. However, it is clear that several approaches are possible, and the problem emphasizes the need for development of powerful techniques to deal with large-scale non-linear programming problems.

REFERENCES

Goldstein, R.P. and Stanfield, R.B. (1970) "Flexible Method for the Solution of Distillation Design Problems Using the Newton-Raphson Technique", *I. & E.C. Process Design and Development*, **9**, 78.

Holland, C.F. (1963) "Multicomponent Distillation", 494, Prentice Hall, Englewood Cliffs.

Hutchison, H.P. and Shewchuk, C.F. (1974) "A Computational Method for Multiple Distillation Towers", *Trans. Instn. Chem. Engrs.*, **52**, 325.

Sargent, R.W.H. (1969) "Column Design and Performance - Report by the Rapporteur", *in* "Distillation - I. Chem. E, Symposium Series No. 32", 5:3.

Sargent, R.W.H. (1974) "Reduced Gradient and Projection Methods", *in* "Numerical Methods for Constrained Optimization", 149, *eds*. Gill, P.E. and Murray, W. Academic Press, London.

Sargent, R.W.H. and Gaminibandara, K. "A Model for Multicomponent Plate Efficiency and the Computation of Column Performance", *to be published*.

Sargent, R.W.H. and Murtagh, B.A. (1969) "The Design of Plate Distillation Columns for Multicomponent Mixtures", *Trans. Instn. Chem. Engrs.*, 47, 85.

Sargent, R.W.H. and Murtagh, B.A. (1973) "Projection Methods for Nonlinear Programming", *Math. Prog.*, 4, 245.

Sargent, R.W.H. and Sebastian, D.J. (1974) "The Minimization of Single-Variable Functions", presented at meeting on "Optimization Problems in Engineering and Economics", Naples.

Stainthorp, F.P. and Whitehouse, P.A. (1967a) "General computer programs for multi-stage counter current separation problems, Part I: Formulation of the problem and method of solution", *in* "Efficient Computer Methods for the Practising Chemical Engineer", Symposium Series No. 23, 181, Institution of Chemical Engineers, London.

Stainthorp, F.P., Whitehouse, P.A. and Kandela, H. (1967b) "General computer programs for multi-stage counter current separation problems, Part II: Description of the computer program", *Ibid*, 189.

Taylor, D.L. and Edmister, W.C. (1969) "General Solution for Multi-component Distillation Processes" *in* "Distillation", I. Chem. E. Symposium Series No. 32, 5, 23.

APPENDIX I

FLASH EQUILIBRIUM

We consider the situation depicted in Fig. 6, where a given feed mixture is brought to equilibrium and separates into a liquid and vapour phase.

By total material balance on the system we have

$$F = L + V \qquad (I.1)$$

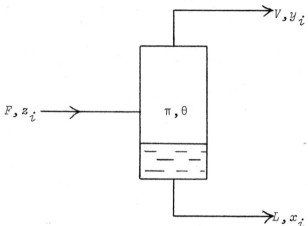

Fig. 6. Flash equilibrium

A similar balance for component i yields

$$z_i F = x_i L + y_i V .$$ (I.2)

The equilibrium between the two phases is expressed by

$$y_i = K_i x_i \quad i = 1,2,\ldots c$$ (I.3)

where in general

$$K_i = K_i(\pi,\theta,\underset{\sim}{x},\underset{\sim}{y}) .$$ (I.4)

We also have by definition of the mole-fractions:

$$\sum_{i=1}^{c} x_i = \sum_{i=1}^{c} y_i = 1 .$$ (I.5)

We are interested in the case where the feed (F, z_i for all i), the pressure (π), and the vapour rate (V) are specified, and we wish to find the temperature (θ) and the phase compositions (x_i, y_i for all i).

We first consider the case of ideal solutions, for which (I.4) simplifies to

$$K_i = K_i(\pi,\theta) .$$ (I.4a)

Since π is given, this means that all the K_i are functions of the single unknown θ.

From (I.2) and (I.3) we have

$$x_i = \frac{Fz_i}{L + VK_i}; \quad y_i = \frac{Fz_i K_i}{L + VK_i} \quad \bullet \quad (I.6)$$

Then from (I.5) and (I.6) we have

$$S_x = \sum_{i=1}^{c} x_i = \sum_{i=1}^{c} \frac{Fz_i}{L + VK_i} = 1 \quad (I.7)$$

$$S_y = \sum_{i=1}^{c} y_i = \sum_{i=1}^{c} \frac{Fz_i K_i}{L + VK_i} = 1 \quad (I.8)$$

$$S = S_y - S_x = \sum_{i=1}^{c} \frac{Fz_i (K_i - 1)}{L + VK_i} = 0 \quad \bullet \quad (I.9)$$

From (I.1) and (I.2) and the assumption that the given z_i sum to unity, it is easy to see that if any of the equations in (I.5) is satisfied then all three of them are. Thus if any one of (I.7), (I.8) and (I.9) is satisfied, then so are the others. Each contains only the K_i as unknowns and so together with (I.4a) suffices to determine θ. We must therefore consider which is the best choice.

We note that as $V \to 0$ the solution θ approaches the boiling point of the feed mixture at pressure π, while conversely θ approaches the dew-point as $V \to F$ and hence $L \to 0$; we may be interested in both of these limits.

$$\frac{dS_x}{d\theta} = -\sum_{i=1}^{c} \frac{FVz_i}{(L + VK_i)^2} \cdot \frac{dK_i}{d\theta}; \quad \frac{dS_y}{d\theta} = \sum_{i=1}^{c} \frac{FLz_i}{(L + VK_i)^2} \cdot \frac{dK_i}{d\theta}$$

$$(I.10)$$

$$\frac{dS}{d\theta} = \sum_{i=1}^{c} \frac{F^2 z_i}{(L + VK_i)^2} \cdot \frac{dK_i}{d\theta} \quad \bullet$$

Thus, $dS_x/d\theta \to 0$ as $V \to 0$, making (I.7) ill-conditioned, and similarly (I.8) becomes ill-conditioned as $L \to 0$. On the other hand $dS/d\theta$ remains non-zero over the whole range of V, so we use (I.9) to determine θ.

For ideal solutions all the K_i increase monotonically with θ and it follows from (I.10) that S also increases monotonically with θ. From (I.9) we see that S is never infinite, so equation (I.9) determines a unique value of θ.

If we are given simple analytical formulae for the K_i, it will be easy to compute $dS/d\theta$ from (I.10) and hence the Newton-Raphson rule could be used to solve (I.9). However parabolic inverse interpolation (or even the secant rule) converges so fast that it is not worth the bother of coding derivatives.

Once θ and the K_i are known, the x_i and y_i are easily evaluated from (I.6).

For non-ideal solutions the problem does not decompose in this way. However, unless the non-ideality is severe the K_i remain relatively insensitive to compositions, and it is more efficient to retain the above procedure, and add an outer iteration loop to correct the K_i for composition effects. Simple successive substitution is usually adequate, but a convergence acceleration procedure can be applied to the recomputed K_i. Our own algorithm incorporates parabolic inverse interpolation on each K_i, treated independently.

APPLICATION TO THE DISTILLATION PROBLEM

To determine the stage temperatures in the performance calculation of the distillation column (Section 3) we determine the total feed to each stage p, as indicated by the left-hand sides of equations (2) and (3) in Section 2, then perform a flash calculation as above for a specified vapour-rate of $(V_p + P_p^V)$. This yields new θ_p, and modi-

306

fied x_{ip}, y_{ip} which sum correctly to unity. In many non-ideal cases quantities computed to determine the K_{ip} are also used to compute enthalpies, and clearly these quantities should be retained for this purpose. It is therefore logical to use the x_{ip} and y_{ip} determined by the flash calculations to compute enthalpies, although it must be remembered that these do not satisfy the column material balances unless over-all convergence to high accuracy has been achieved.

APPENDIX II

THE VARIABLE-METRIC PROJECTION METHOD

A general discussion of projection methods will be found in Sargent (1974). The version of the variable-metric projection algorithm used in this work is a development of that described by Sargent and Murtagh (1973); a brief description is given here, together with a statement of the algorithm as implemented.

The problem is to find the minimum of the function $f^0(x)$ of the n-vector x, subject to the constraints

$$a \leqslant f(x) \leqslant b \qquad (II.1)$$

where a, b and f are m-vectors, with elements a^j, b^j, $f^j(x)$, respectively. The algorithm generates a sequence $\{x_k\}$, $k = 0,1,2\ldots$ of estimates of the solution, each of which satisfies the constraints to within a given tolerance:

$$a^j - \varepsilon \leqslant f^j(x) \leqslant b^j + \varepsilon \qquad j = 1,2,\ldots m \quad (II.2)$$

At each iteration k, local quadratic approximations to the functions are made:

$$f^j(x_k + p_k) \simeq f_k^j + p_k^T g_k^j + \tfrac{1}{2} p_k^T H_k^j p_k \qquad j = 0,1,2,\ldots m \quad (II.3)$$

where suffix k denotes evaluation at x_k. To make these approximations reasonable, we add an additional constraint on the size of the step:

$$\tfrac{1}{2}p_k^T D_k p_k \leqslant \delta^2 \qquad\qquad (II.4)$$

where δ is small and D_k is a given positive-definite matrix.

Using the approximations (II.3) the Kuhn-Tucker conditions for a minimum of $f^0(x)$ at $x_{k+1} = x_k + p_k$, subject to the constraints (II.2) and (II.4) are

$$(g_k^0 + H_k^0 p_k) + \sum_{j=1}^{m} \lambda_k^j (g_k^j + H_k^j p_k) + \mu_k D_k p_k = 0, \quad \mu_k \geqslant 0$$

and for $j = 1,2,\ldots,:$

$$a^j - \varepsilon \leqslant f_{k+1}^j \leqslant b^j + \varepsilon \qquad\qquad (II.5)$$

$$\lambda_k^j (f_{k+1}^j - a^j + \varepsilon) \geqslant 0, \qquad \lambda_k^j (f_{k+1}^j - b^j - \varepsilon) \geqslant 0$$

where μ_k and the λ_k^j are Kuhn-Tucker multipliers.

Rather than evaluate the Hessian matrices H_k^j at each iteration, we build up an estimate, S_k, of the inverse of the matrix

$$(H_k^0 + \sum_{j=1}^{m} \lambda_k^j H_k^j)$$

using the rank-one secant formula:

$$S_{k+1} = S_k + z_k z_k^T / c_k$$

where

$$z_k = p_k - S_k q_k, \qquad c_k = z_k^T q_k, \qquad (II.6)$$

$$p_k = x_{k+1} - x_k, \qquad q_k = (g_{k+1}^0 - g_k^0) + (G_{k+1} - G_k)\lambda_k$$

λ_k is the m-vector with elements λ_k^j, and G_k is the $n \times m$ matrix with columns g_k^j, $j = 1,2,\ldots m$.

We start with $S_0 = 1$ and if necessary c_k is modified to ensure that successive S_k remain bounded and strictly positive-definite. It is then convenient to take $D_k = S_k^{-1}$ and define $\alpha_k = 1/(1 + \mu_k)$. Then $0 < \alpha_k \leq 1$, and for $\alpha_k = 1$ the step-size is not limited by (II.4). Making these substitutions in (II.5) we obtain

$$p_k = - \alpha_k S_k (g_k^0 + G_k \lambda_k)$$

and for $j = 1, 2, \ldots,$

$$a^j - \varepsilon \leq f_{k+1}^j \leq b^j + \varepsilon \qquad (\text{II}.7)$$

$$\lambda_k^j (f_{k+1}^j - a^j + \varepsilon) \geq 0, \qquad \lambda_k^j (f_{k+1}^j - b^j - \varepsilon) \geq 0$$

To solve this system of inequalities, we use a simplified Newton procedure, generating a sequence of approximations $\lambda_{k,L}$, $p_{k,L}$, $L = 1, 2, \ldots$ according to the scheme:

$$p_{k,L} = - \alpha_k S_k (g_k^0 + G_k \lambda_{k,L})$$

$$\phi_{k,L} = f(x_k + p_{k,L-1}) + G_k^T (p_{k,L} - p_{k,L-1})$$
$$a^j \leq \phi_{k,L}^j \leq b^j \qquad (\text{II}.8)$$

$$\lambda_{k,L}^j (\phi_{k,L}^j - a^j + \varepsilon) \geq 0, \qquad \lambda_{k,L}^j (\phi_{k,L}^j - b^j - \varepsilon) \geq 0$$

with $p_{k,0} \equiv 0$. To facilitate the solution of (II.8), α_k is chosen small enough to ensure that the set of active constraints (those with non-zero $\lambda_{k,L}^j$) does not change as α_k is increased from zero up to the chosen value.

The active set is established recursively by testing each constraint in turn, adding and dropping constraints from the active set as required to satisfy the conditions in (II.8); α_k is chosen so that the step $p_{k,L}$ satisfies the bounds on the $\phi_{k,L}^j$ for the inactive constraints.

This is a finite process which produces a solution of (II.8) for an appropriate α_k, and in

general repeated iterations of (II.8) will converge
to a solution of (II.7). This must happen for sufficiently small α_k, so if a solution of (II.7) is
not attained in a given number of iterations (typically L_{max} = 8), α_k is reduced (typically by a factor of θ = 0.3) and the iterations restarted. To
ensure convergence of the major sequence $\{x_k\}$ to a
Kuhn-Tucker point, α_k must similarly be reduced if
the convergent step p_k does not satisfy the "stability condition":

$$f^0(x_k) - f^0(x_{k+1}) \geqslant \delta \cdot p_k^T g_k^0 \qquad (II.9)$$

for some given δ in the range $0 < \delta < 1$.

The formulae for updating the $\lambda_{k,L}^j$ on changing the active set are obtained from (II.8) by
eliminating $p_{k,L}$:

$$\phi_{k,L} = f_{k,L-1} - G_k^T p_{k,L-1} - \alpha_k G_k^T S_k (g_k^0 + G_k \lambda_{k,L})$$

or

$$(G_k^T S_k G_k) \lambda_{k,L} = (f_{k,L-1} - \phi_{k,L}$$
$$\qquad (II.10)$$
$$- G_k^T p_{k,L-1})/\alpha_k - G_k^T S_k g_k^0 = \psi_{k,L-1} .$$

We may extract from (II.10) the subset of equations
corresponding to the active constraints (with
$\lambda_{k,L}^j \neq 0$) and write these in the form:

$$U_k^A \lambda_{k,L}^A = \psi_{k,L-1}^A \quad or \quad \lambda_{k,L}^A = R_k^A \psi_{k,L-1}^A ,$$

where $\qquad\qquad\qquad\qquad\qquad\qquad\qquad (II.11)$

$$U_k^A = (G_k^A)^T S_k G_k^A \quad and \quad R_k^A = (U_k^A)^{-1}$$

and of course the $\phi_{k,L}^j$ used to compute $\psi_{k,L-1}^A$ are
the bounding values (a^j or b^j) for the active constraints.

If constraint j is added to the active set,
U_k^A and R_k^A both have an extra row and column, and

$\lambda^A_{k,L}$ has an extra element, giving new values:

$$\begin{bmatrix} U^A_k & u_j \\ u^T_j & u_{jj} \end{bmatrix}, \begin{bmatrix} R^A_k + r_j r^T_j / r_{jj} & r_j \\ r^T_j & r_j \end{bmatrix}, \begin{bmatrix} \lambda^A_k - \lambda^j_k R^A_k u_j \\ \lambda^j_{k,L} \end{bmatrix}$$

where $u_j = (G^A_k)^T S_k g^j_k$, $\quad u_{jj} = (g^j_k)^T S_k g^j_k \qquad (II.12)$

$$r_j = -r_{jj} R^A_k u_j, \quad r_{jj} = (u_{jj} - u^T_j R^A_k u_j)^{-1}$$

$$\lambda^j_{k,L} = r_{jj}(\psi^j_{k,L-1} - u^T_j \lambda^A_k).$$

Using these formulae we can also obtain an estimate of the objective function decrease obtained when constraint j is dropped from the active set:

$$\beta_j = -(g^0_k)^T \Delta p = \{\alpha_k \lambda^j_{k,L} - (f_{k,L-1}$$
$$(II.13)$$
$$- \phi_{k,L} - G^T_k p_{k,L-1})^T (R^A_k)_j\} \lambda^j_{k,L} / r_{jj}$$

where $(R^A_k)_j$ is the jth column of R^A_k.

The details of the procedure are as follows:

Given: S_k, det $|S_k|$, $Tr(S_k)$, x_k, f^0_k, g^0_k, f_k, G_k

1. Set $\bar{\alpha} = 1$

2. Set $L = 0$, $x_{k+1} = x_k$, $f' = f_k$

3. Set $g^c = g^0_k$, $\lambda = 0$

4. Compute direction: $d = -S_k g^c$, set $\alpha = \bar{\alpha}$

5. Set up active set. Set $j = 1$

   (a) If j is already active, go to step $(5f)$

   (b) Feasible step: If $a^j - \varepsilon \leqslant f'^j + \alpha d^T g^j_k \leqslant b^j + \varepsilon$, go to step $(5f)$

311

(c) Interior point: If $a^j + \varepsilon \leqslant f'^j \leqslant b^j - \varepsilon$, set $\alpha = (\phi^j - f'^j)/d^T g_k^j$ then go to step ($5f$).

(d) Add constraint j to the active set:

(i) Compute u_{jj}, u_j, Ru_j, $u'_{jj} = u_{jj} - u_j^T Ru_j$

(ii) Linear dependence: If $u'_{jj} < \delta . u_{jj}$ go to step 10.

(iii) Update R, λ, g^c: $r_{jj} = 1/u'_{jj}$,

$$r_j = -r_{jj}.Ru_j, \quad R = R - Ru_j.r_j^T$$

$$\Delta\lambda^j = \lambda^j = r_{jj}(\psi_{k,L-1}^j - u_j^T\lambda), \Delta\lambda = -\lambda^j Ru_j$$

$$\lambda = \lambda + \Delta\lambda, \quad g^c = g^c + G_k^A \Delta\lambda$$

(e) Test for dropping a constraint:

(i) Find $\beta'_i = \max \beta_i$ over all constraints i in the active set not satisfying the conditions on $\lambda_{k,L}^i$ in (II.8)

(ii) If $\beta'_i \leqslant \varepsilon$ return to step (4).

(iii) Drop constraint i' from the active set and update R, λ, g^c:

$$R = R - r_{i'}.r_{i'}^T/r_{i'i'}, \quad \Delta\lambda = -\lambda^{i'} r_{i'}/r_{i'i'}$$

$$\lambda = \lambda + \Delta\lambda \quad , \quad g^c = g^c + G_k^A \Delta\lambda$$

(iv) Return to step ($5ei$)

(f) If $j < m$, set $j = j+1$ and return to step ($5a$)

6. Compute the new point $x_{k+1} = x_k + \alpha d$ and the change $\|\Delta x_{k+1}\|$

7. Nonlinear constraint correction:

(a) If $\|\Delta x_{k+1}\| < \varepsilon$, or if there are no non-linear constraints, go to step (8).

(b) If $L = L_{max}$, go to step (10).

(c) Compute $f_{k+1} = f(x_{k+1})$, $f' = f_{k+1} - \alpha G^T d$, set $L = L+1$, then return to step (3).

8. Compute $f_{k+1}^0 = f^0(x_{k+1})$. If max ($\|d\|$, $\|g^c\|$, $|d^T g_k^0|$) $< \varepsilon$, STOP - a solution has been found.

9. Stability test: If $f_k^0 - f_{k+1}^0 \geqslant \delta \cdot \alpha |d^T g_k^0|$, go to step 11.

10. Step reduction:

(a) If $\theta \alpha \|d\| \geqslant \varepsilon$, set $\bar{\alpha} = \theta \alpha$ and return to step 2.

(b) If $S_k \neq I$, set $S_k = I$ and return to step 1.

(c) Otherwise STOP (errors).

11. Compute new gradients: $g_{k+1}^0 = g^0(x_{k+1})$, $G_{k+1} = G(x_{k+1})$

12. Update S_k to S_{k+1}:

(a) Compute q_k, z_k, $z_k^T z_k$, $z_k^T g^c$, $c_k = z_k^T q_k$

(b) Small correction: If $z_k^T z_k < \varepsilon_c |c_k \cdot Tr(S_k)|$, go to step (13).

(c) Small denominator: If $\varepsilon_c z_k^T z_k > |c_k \cdot T_r(S_k)|$ set $c_k = \varepsilon_c z_k^T z_k / Tr(S_k)$ and go to step (12e)

(d) Positive-definiteness: If $-z_k^T g^c / c_k < \delta$, set $c_k = z_k^T z_k / Tr(S_k)$

313

(e) Update S_k, det $|S_k|$, $Tr(S_k)$

(f) Determine eigenvalue bounds for S_{k+1}

(g) Scale S_{k+1} or set $S_{k+1} = I$ as necessary.

13. Set $k = k + 1$ and return to step 1.

The Development of Computer Optimization Procedures for Use in Aero Engine Design

A.H.O. Brown

(Rolls Royce (1971) Ltd.)

SUMMARY

The design process is seen to be one of optimization, using functions developed empirically by subgroups of the over-all design team. The generalization of this process in a form suitable for the application of nonlinear constrained optimization techniques is examined. The recursive equality quadratic programme algorithm due to Biggs was selected and adapted for the purposes of the investigation.

Application to the project design of a bypass engine replacement for the Olympus 593 in the Concorde supersonic aeroplane is described and the problems encountered are detailed. These problems follow in the main from the nature of the functions involved.

Finally, the experience gained is summarised as a series of recommendations regarding the development and use of optimisation software suitable for application to large scale engineering and industrial problems.

1. INTRODUCTION

The purpose of the work reported in this paper was to develop the software for a constrained optimization facility for general use by technologists and engineers in the Aero Engine Design Organisation at the Bristol Engine Division of

315

Rolls Royce (1971) Limited.

The work can be seen to involve the following stages.

1. Search and identification of suitable routines.

2. Implementation of them on our computers; and validation against the usual examples.

3. Development by application to suitable engineering problems.

4. Education of the prospective users in the application of the software.

The author has been chiefly concerned with stages 3 and 4 and this note is therefore directed primarily at prospective users of constrained optimization software.

The note suggests that software is already available which can be applied, with some care and understanding, to major engineering problems, and illustrates this by describing the application of the chosen software to a large aero-engine optimization problem.

It goes on to discuss with illustrations the ways that the user can increase the speed of the solution and the help that can be afforded by the software supplier.

It is hoped that there may be some interest to the supplier as well as the prospective user in the topics referred to in the example and discussion.

2. THE CHOSEN ROUTINE

After initial interest in SUMT routines our attention was directed to the Biggs algorithm described in Lootsma (1972) and the software was obtained from the Numerical Optimization Centre, The Hatfield Polytechnic, together with their routine for supplying, by numerical differencing, approximations to the first derivatives of the objective functions and constraints. This was transcribed for our computers and modified in one important particular.

If the user routine recognizes that some part of $F(x)$, $c(x)$ or $g(x)$ (which are the objective function, the equality constraint residuals and the inequality constraint residuals, respectively) is not calculable, then it can hand control back immediately to the optimization routine which halves the step from the last successful point and tries again, repeating this process as required.

In our version the routine is called by the codeword NOLIN.

3. ORGANIZATION OF A TYPICAL INDUSTRIAL PROBLEM

Fig. 1 illustrates the organization of a large problem to a prospective user audience.

The core of the organization, common to the simplest possible problem, is the circuit shown following from array x_i (i from 1 to N) to the object function (shown as OPT:F), the equality constraint residuals and the inequality constraint residuals, and then back to NOLIN which will alter at least one of the x_i and cause the cycle to be repeated until the problem is converged.

The diagram shows how in the large problem the x_i to be used are selected at run time from an array of all likely independent variables, the remainder forming part of the usually large set of input data. Typically this may consist of the order of 10 000 items, many of which may be arranged as tabular functions of 1, 2 or 3 input variables, the function values to be obtained by linear interpolation.

317

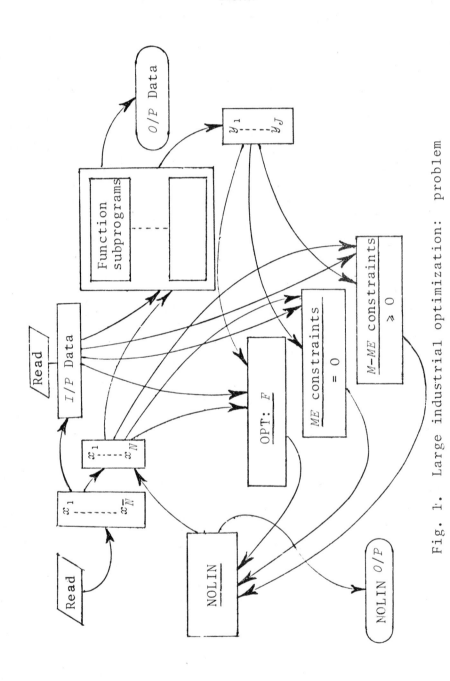

Fig. 1. Large industrial optimization: problem

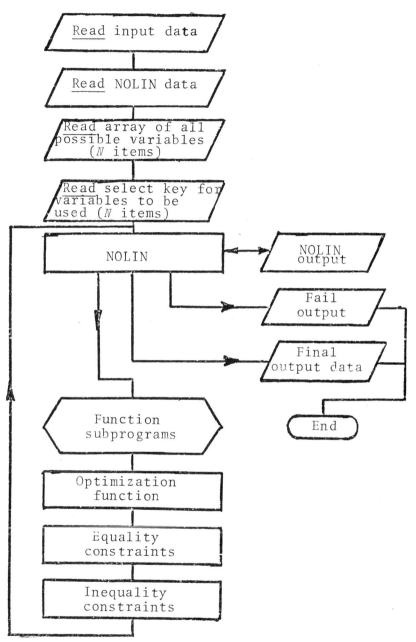

Fig. 2.

This type of problem is characterized by a set of subprograms. These subprograms may well be large in their own right. They will probably use tables of tabular functions (introducing disconti- nuities into the calculation of first derivatives) and they will probably contain iterative procedures involving solutions to given tolerances. They may also be the responsibility of different individuals or departments drawn from essentially different disciplines.

Their output will be of two kinds, the first of technical interest when the problem has converged, but of no immediate interest to the optimization procedure. The second consists of an array of dependent variables, y_j (j from 1 to J).

The objective function and the constraint residuals, therefore, draw their data from the input data direct, the independent variable array x and the dependent variable array y.

Fig. 2 is a block flow diagram of such a program.

4. APPLICATIONS

4.1 The problem

The problem taken to illustrate the use of NOLIN is as follows.

Optimize a replacement engine for Concorde to meet proposed new noise requirements. Take the existing Olympus 593 components but add a front fan as a radial extension to the existing low pres- sure compressor. Other minor internal alterations may also be considered.

It must be emphasized that the above is only one possible strategy and the results of the investigation, while technically interesting, have no significance regarding the political future of Concorde, nor the actual performance of any genuine

Olympus developments. In particular, you will notice that the by-pass air is *not* mixed with the exhaust gas.

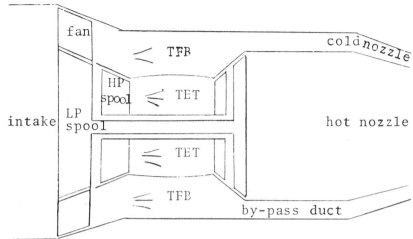

Fig. 3. Twin-spool unmixed front fan with fan burning

Fig. 3 is a schematic drawing of the proposed engine.

The fan, which is shown fixed to the low pressure compressor, does not "supercharge" that compressor.

The Olympus uses exhaust burning (reheat) to boost its thrust during take-off and transonic climb. With a fan engine it is known to be more efficient to replace the reheat by burning in the fan duct.

An unmixed ducted fan was expected to produce on balance an inferior aircraft performance to the existing engine, but it is a most effective method for reducing noise, and the problem was therefore to find the fan which would satisfy the noise limits and cause the least loss of payload for a typical Atlantic flight. The optimum required is therefore the maximum payload.

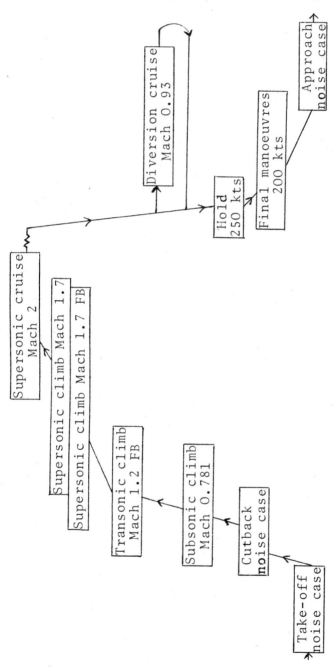

Fig. 4. Flight plan

Fig. 4 shows the flight path. The change in payload from a given datum is based on a finite difference formula which allows for change in thrust and fuel flow at the eight specified flight cases shown in the figure. The noise limits are applied at three cases also shown.

The extra weight of the fan engine is a direct loss to the aircraft payload and the bigger engines also cause more drag and hence increase the thrust requirement at each flight condition.

4.2 The variables

Table I is a table of the variables in the problem.

Table I
Olympus 593 Fan: Variables

1. Fan design mass flow	10. TET mach 0.781
2. Fan design enthalpy rise	11. TET take-off
3. TFB Mach 1.7	12. LP rpm Mach 0.93
4. TFB Mach 1.2	13. LP rpm 250 kts
5. TFB take-off	14. LP rpm 200 kts
6. TET Mach 2.0	15. LP rpm cutback
7. TET Mach 1.7 FB	16. LP rpm approach
8. TET Mach 1.7	17. HP turbine stator area
9. TET Mach 1.2	18. LP turbine stator area

Variables 1 and 2 define the diameter of the fan and the work it does and hence its size and weight.

Variables 3 to 5 are the combustion temperatures in the fan duct for those cases requiring thrust boost.

Variables 6 to 11 are the temperatures after combustion and just prior to entry to the turbines for the flight cases usually needing high temperatures.

Variables 12 to 16 are the revolutions per minute of the low pressure compressor and turbine for the flight cases which must be matched to a given thrust. Turbine entry temperatures are not usually high in these cases.

When the fan is added the balance between flow, pressure rise and rpm will be altered for the low pressure compressor and this in turn will affect the relationships for the high pressure compressor. We say that the engine will rematch. Redesigning the throat cross-sectional areas of the respective turbine stators is a possible means of readjusting this balance and the inclusion of these areas as variables allows them to readjust in an optimum manner.

This point may be better appreciated by reference to Fig. 5, which also affords an example of the extent of tabular data required by the program.

The top figure shows the classical way of plotting compressor performance as a relationship between rpm, pressure ratio and mass flow. There would be a similar plot replacing the pressure ratio by temperature ratio and the efficiency of the compressor would then be implied by the two plots.

For computer work it is customary to use a linear interpolation routine to operate on three tabulations. In each case the input variables are rpm and an entirely arbitrary parameter, β, and the dependent variables are pressure ratio, temperature ratio and mass flow.

An important line on the top diagram is the surge line above which the compressor is unusable and this line is given a constant β value. Other

constant β lines are drawn in, lower values being
assigned below the surge line, and higher values
above.

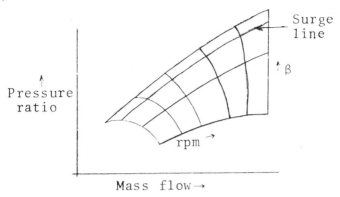

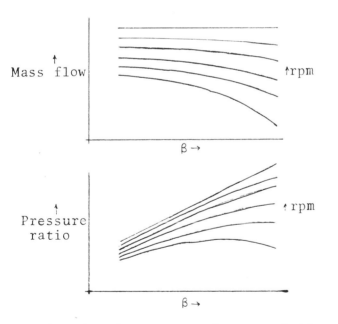

Fig. 5. Compressor characteristic

The lower two plots show two of the three with rpm and β as independent variables and pressure ratio, temperature ratio and mass flow, respectively, as the dependent variable. The object in drawing the β lines is to get smooth plots that interpolate in a smooth manner with a minimum of tabular values. Nonetheless upwards of 650 points will be required to define a compressor this way.

4.3 The constraints

Table II gives the constraints.

Table II
Olympus 593 Fan: Inequality Constraints

1. Minimum thrust Mach 2.0	10. TET Mach 2.0
2. Minimum thrust Mach 1.7 FB	11. TET Mach 1.7 FB
3. Minimum thrust Mach 1.7	12. TET Mach 1.7
4. Minimum thrust Mach 1.2	13. TET Mach 1.2
5. Minimum thrust Mach 0.781	14. TET Mach 0.781
6. Minimum thrust take-off	15-25. Maximum βLP for each flight condition
7. Maximum noise take-off	26. Maximum $N/\sqrt{\theta}$ at Mach 2.0
8. Maximum noise cutback	27. Maximum ΔT_{fan} take-off
9. Maximum noise approach	28. Minimum ΔT_{fan} take-off
	29. Maximum by-pass ratio

Only inequality constraints were used in this problem although similar problems have been run successfully using equality constraints as well.

Constraints 1 to 6 ensure that the engine develops sufficient thrust at the given flight conditions.

Constraints 7 to 9 ensure that the noise is acceptable.

The turbine entry temperatures (constraints 10 to 14) were, for stress reasons, kept at 100° K below the temperature at take-off.

Constraints 15 to 25 ensured that the low pressure compressor did not surge.

At supersonic cruise the engine flow must be matched to a supersonic intake and hence a constant flow, and constraint 26 ensures that the rpm do not exceed a safe limit at these conditions.

Constraints 27 and 28 restrict the solution to a given number of rows of blades in the fan. As the number altered the weight of the engines, and hence the payload, it was necessary to optimize successively for 1, 2 and 3 fan rows and select the over-all optimum.

Constraint 29 limited the by-pass flow ratio to a maximum reasonable value.

4.4 The results

Table III summarizes the results obtained.

It will be seen that at the solution the problem with optimum turbine stator areas was distinctly better than the case with unaltered turbine stator areas and that it had 11 active constraints, and six degrees of freedom.

There was one "loose variable", defined as a variable which is not involved in the objective

327

function and is only involved in constraints which are not active at the solution. Such variables may have been altered during the program run, but their value is, in fact, indeterminate.

Table III

Results of Olympus 593 Fan Problem

60% by-pass fan with 3 stages. Take-off TET=1430° K

    Δ payload, with Olympus 593 areas = -4380 lb
    Δ payload, with optimum areas   = -2960 lb

Active constraints:

    Minimum thrust at Mach 2.0, Mach 1.7 FB, Mach 1.2, Mach 0.781, take-off.

    Maximum noise at take-off, cutback.

    TET, 100° K below take-off at Mach 2.0, Mach 1.7 FB, Mach 1.7, Mach 1.2.

Loose variable: LP rpm at approach.

Degrees of freedom for problem with optimum areas=

    18 variables - 1 loose variable - 11 active constraints = 6.

---

5. DISCUSSION

The discussion takes the form of a commentary on the development and running of the problem already presented, making more general points as they arise.

5.1 Program structure

Fig. 6 shows the structure of the problem and is laid out in the same way as Fig. 1, showing the eleven subprograms, one for each flight case identified in Fig. 4.

As already remarked, there are no equality constraints. In this instance each subprogram

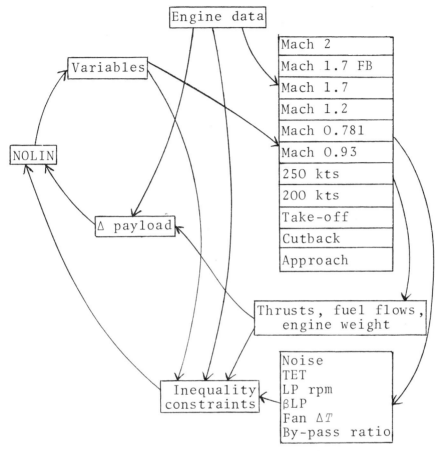

Fig. 6. Elements of Olympus 593 fan problem

consisted of a common engine matching routine, but
supplied with different data, the data including
parameters which determined the mode of operation.

Fig. 7 shows the structure of the engine
matching routine. A subroutine for the solution
of a set of nonlinear equations adjusts the three
variables, βLP, βHP and HP rpm. These and other
engine data feed the engine routine shown in Fig.
8 and this outputs three residuals which have zero
value when the engine is matched. They are the

329

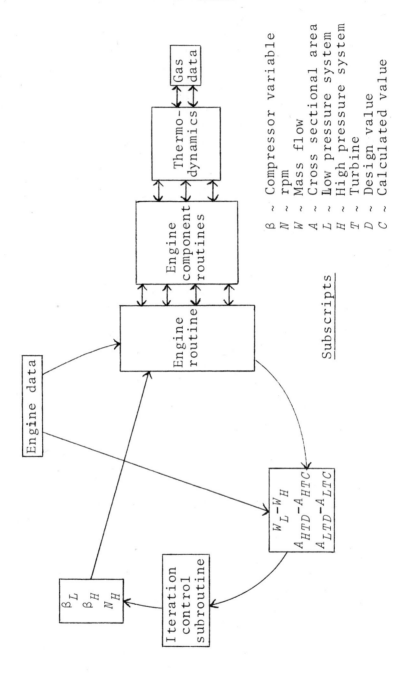

Fig. 7. Elements of engine matching

β ~ Compressor variable
N ~ rpm
W ~ Mass flow
A ~ Cross sectional area
L ~ Low pressure system
H ~ High pressure system
T ~ Turbine
D ~ Design value
C ~ Calculated value

Subscripts

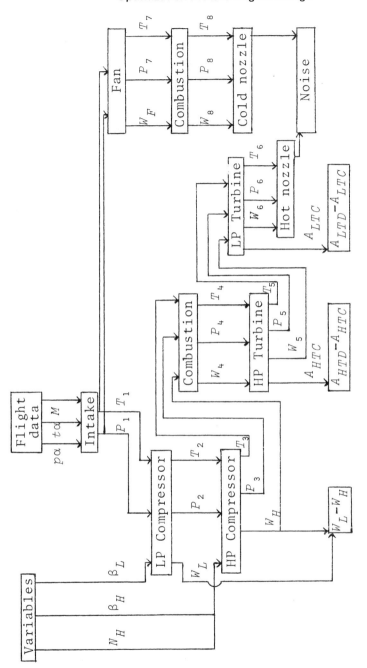

Fig. 8. Engine routine

difference between low and high pressure compressor flows, and the differences between the required and calculated values of turbine stator throat areas. These residuals are in turn input into the solution subroutine which uses a developed Newton-Raphson technique.

Not all the subprograms use the same variables or residual functions and the number of variables and residuals also varies. Each of the residuals will be matched to a selected tolerance which can be input or programmed.

The engine routine as shown in Fig. 7 draws on three tiers of subroutines each supplied as a package. The first is a package of engine component routines and these are shown as the boxes labelled "Compressor", "Turbine", etc. in Fig. 8. They draw on a further package of basic thermodynamic routines and a still more basic set which provides gas property data. At all levels there are iterative routines and therefore a tier of tolerances to be set.

At the start of development one run of the function subprograms took 20 minutes on our KDF9, admittedly by present standards a slow machine, but as we expected of the order of 1000 such function evaluations before convergence, it was clear that every method possible would be required to speed up the program as a whole.

Clearly, the general conclusion is that whatever can be done with the functions, large engineering problems require the most sophisticated optimization software they can obtain. However, we confine ourselves to the improvements that can be affected by the user or by user orientated changes in the software package.

5.2 Simplifying the user program

It can be misleading to simplify the model in a way that alters its main structure or the

balance of its partial derivatives (*i.e.*, changing
the number of variables, the object function, the
number of inequality constraints or in our case
(see Fig. 7) the engine routine). As an example,
an early version of the routine limited the take-
off TET to a maximum permissible value by an inequality constraint and assumed a fixed turbine cooling
flow. The solution ran to the maximum TET allowed.
When the cooling flow was programmed as a function
of the TET, the optimum TET was found to be well
below the previous limit.

However at the level of the basic thermodynamics it was possible to simplify without altering
the nature of the problem, or the balance of the
partial derivatives. We found that by using a
modified form of constant specific heat theory for
the elementary thermodynamic routines, but true
averaged specific heat for each process, we could
double the speed. As this only entailed a loss in
accuracy in the function calculation of the order
of 0.1% and no loss of partial derivative balance,
it was accepted as the basic thermodynamics for the
problem, but as a general point, it may be possible
to speed the whole process by adopting a series of
subprogram packages and using the optimum obtained
with the simplest pack as the starting point for
the next.

5.3 *The machine and compiler*

It can be extremely important to check on
the idiosyncracies of the machine and compiler
software being used. In our case we found that
logic was very slow in relation to arithmetic. We
took two actions. The first was to remove the
generality of many of our subroutines. Where an
input parameter could alter the logic of the subroutine, we removed the logic and used separate
subroutines.

Secondly we added a further compiler pass,
producing an intermediate source listing which was
very simple, but much larger than the original.

These two actions produced an over-all increase in subprogram speed of 4 to 1.

5.4 Which variable

The optimization routine alters all the variables at once on a pattern move, one variable only at the start of a run to get the partial derivatives, and two at once where one is being reset and the next one perturbed.

The user will know which variables from the possible variable array affect any given subprogram and if the optimization routine provides information about the changes which have been made, it can be arranged for only those parts affected in the user routines to be rerun.

The user will need to arrange for the dependent variable array acquired at the last pattern move to be stored in order to replenish the current array on each partial derivative move.

NOLIN does not provide the required information but by keeping a copy of the last independent variable values and comparing it with the current one it proved possible to use this technique to reduce the running time for a complete NOLIN iteration by a factor of 3.

The general principle implied here must be worthwhile with almost any optimization routine on large problems of the category being discussed.

5.5 Bisection

Fig. 9 illustrates the three regions which usually exist in constrained optimization.

The first is the feasible region where no constraints are violated. The solution is obviously inside or on the edge of this region.

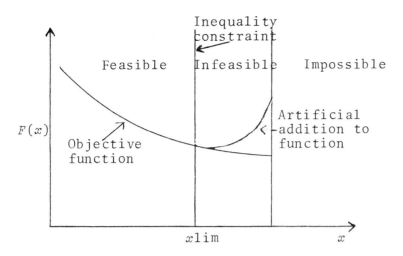

Addition = $K \times [\,(x - x\mathrm{lim})^2\,]_{\;x > x\mathrm{lim}}$

Fig. 9.   Function modification in the infeasible region

The second is the region where at least one constraint is violated but where the user routine will produce values for all its required outputs.

The third or impossible region is where the user routine cannot produce values for all its required outputs.

In paragraph 2 it was explained that if an independent variable array is supplied by NOLIN in this impossible region, the user routine can trigger NOLIN to bisect back towards the last successful array.

This feature is very useful in the early stages of a problem when the relaxation weighting coefficient is limiting the effects of the constraints on the penalty function.

It is important that the user set traps as early as possible in his programs to catch impossible input values so that no unnecessary calculation is done before the run is identified to the optimization routine as in the impossible region.

If bisection occurs in the later stages of a problem it can cause the problem to chatter along very close to the boundary between the impossible and infeasible regions and finally stop altogether. Clearly this must be avoided.

One simple way which is sometimes possible is to use some form of extrapolation to extend the infeasible region into the impossible one. This may be necessary if the impossible boundary is very close to the infeasible boundary.

A preferred method is to add a penalty to the function in the infeasible region (see Fig. 9). This method has been used with success on both βLP and βHP in our problem, the addition being of the form

$$K [(\beta - \beta_{surge})^2] \quad \beta \gtrless \beta_{surge}$$

where K is a suitable constant.

5.6 Scaling

In NOLIN

$$P(\underset{\sim}{x}_k, r_k) = F_k + \frac{1}{r_k} \cdot \underset{\sim}{g}_k^T \cdot \underset{\sim}{g}_k$$

where P is the penalty function,

$\underset{\sim}{x}_k$ is the kth set of independent variables

r_k is the kth relaxation weighting coefficient or penalty parameter

F_k is the kth optimization function or object function

$\underset{\sim}{g}_k$ is the kth vector of "active" constraint residuals.

The vector of so called "active" constraint residuals includes those of the constraints likely to be violated at the solution in addition to

336

those violated on the kth iteration.

The main scaling between the object function and constraint residuals is therefore affected by r_k.

There is a hierarchy of strategies for deciding r_k but for the preferred methods,

$$r_k = f(\alpha, \underset{\sim}{\lambda}_k)$$

where α is a coefficient fixed by the user

and λ_k is the vector of approximate Lagrange multipliers of the active constraints.

$$\lambda_k = (W_k B_k^{-1} W_k^T) W_k B_k^{-1} \nabla F_k$$

where W_k is the Jacobian of the active constraints and B_k is an approximation to the Hessian of the objective function.

5.7 Penalty coefficient

Simply, the smaller the value of α the faster and more unstable the working becomes.

Without experience of a particular problem a value of 0.5 is recommended. With a well known problem it might be safe to reduce it to 0.1.

5.8 The objective function

Scaling between the constraints and the objective or optimization function is adequately affected by r but in view of the interrelationship with convergency tolerances it is probably best to scale F to the order of 1

i.e., let $$F_s = F/F_a$$

where F_s is the scaled value

and F_a is some anticipated final value.

The sign of F_s is, of course, decided by whether a maximum or minimum is required.

5.9 *The variables and the constraint residuals*

As an initial strategy NOLIN copies and modifies the user's array of variables, such that

$$x_N = x_u / x_{ui}$$

where x_N is a NOLIN variable

x_u is a user variable

and x_{ui} is the initial input value of the user variable.

Scaling of the constraints is left to the user. Our initial strategy (and often the only one required) was to scale a constraint by dividing by the limiting or required value of the principle variable involved

e.g., if $\beta_s - \beta \geqslant 0$ where β_s is the surge value then we would scale the constraint by using

$$\frac{\beta_s - \beta}{\beta_s} \geqslant 0.$$

The Lagrange multipliers are printed out every time they are successfully calculated, and they provide evidence of trouble. We have experienced occasions both when individual values in the array have been very different (say factors of 10^6 to 10^{12}) and when successive values of the same item in the array vary by the same order. On one occasion progress ceased altogether and the program went into an infinite loop.

These troubles are not simply related to the values of the Jacobian, but all have so far been cured by the following procedure.

Scale all terms in the Jacobian to the same order by scaling the constraint residuals relative to each other, and scaling the input variables relative to each other by substituting

$$\underset{\sim u}{x} + \underset{\sim}{b} \quad \text{for} \quad \underset{\sim u}{x}$$

where $\underset{\sim u}{x}$ is the user array of variables

and $\underset{\sim}{b}$ is an array of arbitrary scaling factors.

It would seem to us that scaling is important enough to warrant serious consideration by the software supplier. Until there is a "respectable" theory for scaling he should not only provide the best "rule of thumb" method available, but should also provide "override" facilities so that the user can take charge of the scaling when necessary.

5.10 Tolerances

NOLIN only uses one tolerance and it is used in the following two ways.

1. To decide when the problem has converged.

2. To decide the step lengths for the calculation of partial derivatives.

We feel that the use of a single parameter for both rôles is insufficient and understand it is not the practice in the later versions of REQUP available from the NOC.

From the point of view of convergence, we feel that the user should be able to supply separate convergence parameters for each independent variable, each constraint residual, and the optimization function itself. These parameters should apply to the function as calculated by the user routines and should be independent of any scaling undertaken by the optimization routine.

With regard to the step lengths for partial calculation, there is an opportunity to make further time savings.

All the tolerances used for iterative processes in the user subprograms could be scaled to a single parameter, and it is clear that time could be saved if this parameter were kept as coarse as

possible consistent with the stability of the optimization routine. The value of the parameter
should therefore be provided by the optimization
routine, and obviously the routine should allow
much larger steps for partial derivative calculation during the early iterations.

6. CONCLUSIONS

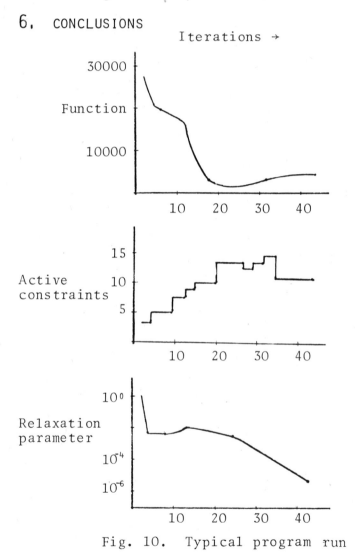

Fig. 10. Typical program run

Fig. 10 shows for the current version of the
problem we have described the way in which the
function converged. It also shows how the number
of active constraints and the relaxation parameter
varied during a run. Between 40 and 50 iterations
were required and the run time was about 4 to 5 hours
on the KDF9. We expect this to reduce to the order
of 5 to 10 minutes on a CYBER machine, and it demonstrates the ability of existing optimization software to handle large engineering problems.

Finally, Table IV summarizes our recommendations to the user of routines in this class, and
Table V summarizes our recommendations to the software supplier.

Table IV - Recommendations to User
1. Simplify user program
(a) do not alter main structure or partial derivative balance,
(b) possibly supply several versions of routine, using optimum obtained with simplest pack as start for the next.
2. Can user routine be improved by suiting style of program to the given computer? Can the compiler be improved?
3. Check variables
(a) which variables have been changed? Only recalculate affected subprograms.
(b) can the calculation be performed? Check as early as possible.
4. Add penalty functions as necessary in infeasible region to avoid "impossible region".
5. Scaling and tolerances
(a) divide optimization function by some guessed final value,
(b) divide all constraints by limiting or required value of principle variable involved,
(c) see "recommendations to supplier". Make up for any deficiencies in routine available.

Table V
Recommendations to Optimization Routine Supplier

1. Checking variables. Supply check list to user routine of variables modified.
 	
 2. Scaling (routines of the type under review). Monitor need for scaling by checking Lagrange multipliers.
 	
 If scaling required: scale independent variables and constraint residuals to bring Jacobian terms to the same order.
 	
 3. Tolerances.
    (a) Provide user with the ability to supply comprehensive data for checking convergence,
    (b) keep step length for partial derivative calculation as large as possible throughout program run,
    (c) update a tolerance parameter for use in the user's subprograms. Keep this as coarse as possible while maintaining stability.

REFERENCE

Lootsma, F. *ed.* (1972) "Numerical Methods for Nonlinear Optimization", Academic Press, Chapter 29.

Optimization of Aircraft Designs at the Initial Project Stage

B.A.M. Piggott

(Mathematics Department, RAE, Farnborough)

SUMMARY

The paper describes part of the work done at the Royal Aircraft Establishment, Farnborough on the use of numerical optimization techniques to assist in preliminary project studies of new aircraft.

The first part of the paper describes the type of problem which was tackled and how it was brought into a form suitable for solution by numerical optimization techniques.

The second part of the paper gives some details of the constrained optimization techniques which are being used. The main method used is the sequential unconstrained minimization technique combined with a variable metric method, gradients being evaluated numerically. A projection-restoration method has also been used.

Finally, the paper discusses the resulting computer programs and how their use was extended as the work developed.

1. THE AIRCRAFT DESIGN PROBLEM

The specification to which an aircraft is designed arises from an analysis of the circumstances in which it is expected to operate. For a civil transport aircraft, these will include the distances to be flown, the characteristics of the airfields to be visited, safety and noise regulations which must be satisfied and, of course, economic factors. From these considerations, we derive the design specification in terms of cruise speeds, airfield performance, operating costs, etc.

For our example, we have chosen the following simplified specification. The aircraft must be able to carry a given number of passengers over the specified range at a given cruise speed and altitude. It must be able to take off within a given runway length and climb away at a gradient sufficient to meet the civil airworthiness requirement relating to engine failure during take-off. A further airworthiness requirement specifies that the landing approach speed must exceed the stalling speed by an adequate safety margin. Subject to all these, we seek to minimize the cost to an airline of purchasing and operating the aircraft.

In general, there will be a limited number of choices for the basic layout of the aircraft, that is the fuselage/wing/tailplane configuration and the number and position of the engines; several competing layouts may be chosen for preliminary study. To meet the specification outlined above, we choose a conventional swept-wing design with two rear-mounted engines (see Fig. 1) and focus particular attention on the high-lift devices on the wings. The high-lift devices improve aerodynamic performance at the cost of increased structural weight; only by examining the implications of these two effects for the whole design can we arrive at an optimum configuration.

Within the basic layout, there are a multitude of specific designs, determined by the values of the *design variables*. Thus, in our example, five variables specify the geometry of the wings, four specify the size of the high-lift devices, four more the operation of these devices at take off and landing, one the size of the engines, and two the aerodynamic design of the wings - a total of 16 variables. Other values needed to complete

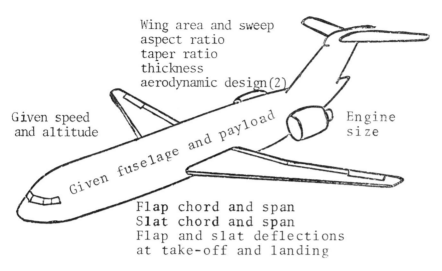

Wing area and sweep
aspect ratio
taper ratio
thickness
aerodynamic design(2)

Given speed
and altitude

Given fuselage and payload

Engine
size

Flap chord and span
Slat chord and span
Flap and slat deflections
at take-off and landing

Fig. 1. Variables to be optimized

the design are determined by simple relationships,
for example the fuselage size is calculated from
the specified payload and the tailplane size from
stability considerations.

The next step is to establish the relationships between the design variables and the specification. Considering for example the range
requirement, we must be able to calculate, from
the values of the design variables, the aerodynamic
characteristics of the aircraft in the specified
cruise condition and to combine these with the
engine performance data to give the amount of fuel
needed. The space available for the main fuel
tanks must be adequate to contain this fuel; hence
we arrive at the constraint

$$V_T(x) \geqslant V_R(x)$$

where V_T denotes the fuel tank volume available in
the wings, V_R the fuel volume needed to meet the
range requirement, and x the vector of design
variables. Four more constraints are derived in a
similar manner:

345

$$m_E \geqslant m_1(x)$$
$$m_E \geqslant m_2(x)$$
$$m_E \geqslant m_3(x)$$
$$S \geqslant S_L(x)$$

where m_E denotes the mass of each engine, m_1 the engine mass needed to meet the cruise requirement, m_2 and m_3 the engine masses needed to meet the two take off requirements, and S_L the wing area needed to meet the landing requirement. All these constraints involve nonlinear functions of the design variables. The cost requirement leads us to minimize a composite cost function called the direct operating cost, which is also a nonlinear function of the design variables. Thus we have to minimize a nonlinear function of 16 variables subject to five nonlinear constraints; in addition, most of the design variables have fixed upper and lower bounds.

The design calculations used in formulating this optimization problem are necessarily of a simplified nature, but it is essential that they characterize the interactions between design changes in the different components of the aircraft, since it is the importance of these interactions which makes over-all design optimization a valuable exercise. At the level of detailed design and development, the complexity of the design procedures for each section of the aircraft renders over-all optimization impracticable; however, where detailed work on one aspect of the design leads to significant changes in its characteristics, the results may be fed back into the simpler over-all optimization model in order to assess the implications for other aspects of the design.

2. USE OF A PENALTY FUNCTION METHOD

We have shown how an aircraft design problem can be formulated as a constrained optimization problem:

minimize $f(x)$

subject to $c(x) \geqslant 0$ and $x_L \leqslant x \leqslant x_U$

where x is a vector of n design variables, each bounded above and below, $f(x)$ is the cost function and $c(x)$ is a vector of m constraint functions.

The main method used to solve this problem has been the sequential unconstrained minimization technique (SUMT), together with a variable metric method for minimization subject only to fixed bounds on the variables. At the kth stage in the solution, the variable metric method is used to minimize

$$P(x,r_k) = f(x)/f_s + r_k \sum_{i=1}^{m} B_k(c_i(x)/c_{si})$$

subject to $x_L \leqslant x \leqslant x_U$

where r_k is a decreasing sequence of positive numbers, f_s and c_{si} are scales for the various functions, and $B_k(t)$ is a penalty function given by

$$B_k(t) = \begin{cases} (2r_k^2 - t)/r_k^4 & \text{if } t < r_k^2 \\ 1/t & \text{if } t \geqslant r_k^2. \end{cases}$$

This penalty function is a minor modification of the reciprocal barrier function; it is defined for all values of t and thus we avoid the need for a preliminary stage to find a feasible starting point. At the end of each stage after the first, the latest two stage solutions are combined to give the extrapolated solution

$$x = [x^{(k)} + h.r_k^{\frac{1}{2}}/(r_{k-1}^{\frac{1}{2}} - r_k^{\frac{1}{2}}).(x^{(k)} - x^{(k-1)})]_b$$

where $x^{(k)}$ is the kth stage solution, obtained by minimizing $P(x,r_k)$, $h(0 \leqslant h \leqslant 1)$ is a factor introduced to ensure that the extrapolated solution satisfies the constraints and gives $f(x) < f(x^{(k)})$, and $x = [y]_b$ denotes

$$x_i = \begin{cases} x_{Li} & \text{if } y_i < x_{Li} \\ y_i & \text{if } x_{Li} \leqslant y_i \leqslant x_{Ui} \\ x_{Ui} & \text{if } y_i > x_{Ui} . \end{cases}$$

The derivatives $\partial P/\partial x_i$ required by the variable metric method are computed using finite difference formulae. After some experiment, we have settled on the use of a central difference formula with a fixed step length; thus $2n$ function evaluations are required for each gradient evaluation. The derivatives of the cost and constraint functions are calculated and combined to give the derivatives of $P(x, r_k)$:

$$\frac{\partial P}{\partial x_i} = \frac{1}{f_s} \frac{\partial f}{\partial x_i} + r_k \sum_{j=1}^{m} \frac{1}{c_{sj}} B_k'(c_j/c_{sj}) \partial c_j/\partial x_i .$$

The variables and the cost and constraint functions are all scaled to be of order unity, the necessary scaling factors being supplied by the user of the program. This is not entirely satisfactory and occasionally leads to difficulties, especially for new users not familiar with the technique, but no better procedure appears to be available at present.

3. USE OF A PROJECTION-RESTORATION METHOD

An alternative method of solving the constrained optimization problem is as follows. Let $x^{(k)}$ be the value obtained for x at the end of the kth iteration and suppose that $x = x^{(k)}$ satisfies the constraints, in which we now include the bounds on the variables. In particular, suppose that $a(x^{(k)}) = 0$, where a is the vector of active constraint functions. The next iteration consists of a step from $x^{(k)}$ to $x^{(k+1)}$ given by

$$x^{(k+1)} = x^{(k)} + d^{(k+1)} + r^{(k+1)}$$

where $d^{(k+1)}$ is a projected descent step and $r^{(k+1)}$ is a constraint restoration step. The descent step is

$$d^{(k+1)} = -h_{k+1}(f_x - a_x p)$$

where

$$p = (a_x^T a_x)^{-1} a_x^T f_x ,$$

h_k is a positive step factor, and the gradients f_x and a_x are calculated at $x = x^{(k)}$. The restoration step is

$$r^{(k+1)} = -e a_x (a_x^T a_x)^{-1} a_x$$

where e is a step factor chosen to ensure that $x^{(k+1)}$ satisfies the constraints, and a and a_x are calculated at $x = x^{(k)} + d^{(k+1)}$.

The step factor h_k is carried from iteration to iteration, being increased when the process is going well and decreased when necessary to ensure that $x = x^{(k)} + d^{(k+1)}$ satisfies the inactive constraints and that $f(x^{(k+1)}) < f(x^{(k)})$. New constraints are introduced into the active set when they are encountered and, at each iteration, an attempt is made to drop from the active set the constraint corresponding to the largest negative component of p.

This is a first order method and, after the first few iterations, convergence naturally tends to become slow. However, the method has been found useful for sensitivity analysis, that is for solving problems which differ only slightly from previously solved ones, so that excellent initial estimates for the variables are available. The corresponding second order method was not found to give a significant improvement in the rate of convergence; the reason for this disappointing result is not known.

4. SOME PRACTICAL DETAILS

The aircraft design optimization programs in use at RAE are divided into two physically separate parts, the "designer" and the "optimizer". The "designer" part of the program subdivides into three sections. The design set-up section prepares the relevant data and carries out any calculations which are not affected by changing values of the design variables; this section is executed once only, before optimization commences. The design evaluation section, which is executed many times during the optimization process, calculates the values of the cost and constraint functions for any acceptable values of the design variables. The output section prints details of the final optimized design and of any other intermediate or neighbouring designs which may be required. The "optimizer" part of the program may be an implementation of either of the methods discussed above - the same "designer" may be used with either "optimizer". The projection-restoration "optimizer" incorporates a sensitivity analysis procedure which enables the design to be re-optimized with one design parameter fixed at a sub-optimal value or with one item of data changed.

In solving a typical example of the aircraft design problem, using the penalty function method, the design evaluation section of the program was called about 3000 times, execution time on the ICL 1907 computer being a little under 5 minutes. The projection-restoration method required an average of 700 design evaluations for each additional case solved in the course of a sensitivity analysis.

5. PROBLEM DEVELOPMENT

The aircraft design problem described above was an early attempt to put this type of problem into a form suitable for the application of numerical optimization techniques. It was highly simplified - deliberately so, in order to enable experience in using the methods to be gained with as few complications as possible. As the work has progressed, several modifications have been introduced in order to produce a more realistic and useful model.

The design variables - As confidence in the performance of the program has increased, there has
been a tendency to treat as design variables quantities which at first were fed in as data. Cruise
speed and altitude are examples of such quantities.

The cost function - A range of possible cost criteria for transport aircraft has been developed,
including purchase price, various forms of the
direct operating cost function, and "passenger
cost" which makes allowance for the value of time
saved by increasing speed. Any aircraft design is
a compromise and no single criterion can lead to
the "right" aircraft, but our approach provides
information on the relative costs of various possible compromises and the ability to calculate this
information quickly is valuable.

The constraints - Various constraints have been
added to the original set, either because their
omission led to impractical ·designs or because new
requirements have been introduced. For example,
it has been found necessary in some cases to impose
a constraint on the angle of the fuselage to the
horizontal at take off and landing.

The design relationships - Early versions of the
program simply took over the design relationships
already used in initial design project work.
Although these gave good estimates of the characteristics of particular designs, it was found that
they often failed to represent accurately the
effects of changing design variables singly and
this led to an inaccurate picture of the important
interactions between different components in the
design. Thus it has been necessary to devote considerable effort to the development of improved
design relationships which are, incidentally, a
valuable byproduct of the optimization exercise.
As a result, the design evaluation section of the
program has increased from less than 100 to over
300 Fortran statements.

Discussion Session on Constrained Optimization

R. Fletcher: If I may I would like to make one comment which relates not only to Dr. Piggott's talk but also to others in which the use of SUMT techniques is reported. It is my experience that although SUMT was adequate and appropriate at the time when the author started to solve his problem I would strongly recommend that a different penalty function should now be considered as the first choice by research workers setting about a nonlinearly constrained problem. This is the ideal penalty function described in Murray's review paper, which was developed by Powell, Hestenes and Rockafeller independently. The method has many advantages and virtually no disadvantages. It has global convergence at an ultimately superlinear rate, and the computational effort per minimization falls off rapidly. The conditioning is good with no difficult problems to solve relating to singularities or to the function being undefined outside the feasible regions. No initial feasible point need be supplied and the function is defined for all values of the parameters. Finally and of prime importance the method is very easy to program by using an established unconstrained minimization technique entirely without modification. The method can be implemented whether or not derivatives are available.

A.H.O. Brown: It might be of interest to record that Rolls Royce used one of Dr. Piggott's problems namely that described in RAF TR 71074* as a test example for the NOC's program OPRQP. Some difficulties were encountered at first but eventually the problem was solved quite quickly. Would Dr. Biggs care to add to this comment?

---

*TR71074 April 1971 "Application of Numerical Optimisation Techniques", B.A.M. Piggott and B.E. Taylor.

M.C. Biggs: This problem was discussed with A.
Clarke of Rolls Royce and it was found to be made
easier by being reformulated by a rescaling of the
variables. The solution time using OPRQP on the
Dec System 10 computer was between 1 and 2 minutes
depending on the precise reformulation used.

L.C.W. Dixon: In Mr. Brown's talk he noted that
infeasible areas exist where OPRQP had to be modified as the function values could not be calculated there.   Following Dr. Fletcher's comment it
seems only fair to point that at such points the
ideal penalty function can also not be calculated.
Also Dr. Biggs has proved the convergence and
ultimate superlinear rate of convergence for OPRQP
on a similar limited subclass of functions as the
proof for the global convergence of the ideal
penalty function, namely that for which the ideal
penalty function is convex or has only 1 stationary point.   Otherwise the ideal penalty function
can have many local minima, some of which may not
even be feasible.

M.J.D. Powell: Mr. Brown you mentioned that whenever an optimization subroutine changes only a few
variables then in your type of problem the user
can save computing effort in his function subroutine by taking advantage of this fact.   Now most
optimization subroutines, apart from that which
changes one variable at a time, are not written
to allow the user to take advantage of this, and
very few writers of algorithms have taken account
of the fact that the user might find it beneficial
if only a few variables were changed at a time.
This could be very important as it could open a
new field of research.   How much importance do
you put on this?

A.H.O. Brown: I would suggest that it is a matter
for the software supplier to investigate the
possibilities, as the importance must depend very
much on the advantages that he can take from the
fact in a particular problem.   Applying OPRQP to
our problem, the algorithm regularly made success-

ive changes to individual variables to estimate numerical differences and we could make enormous savings in computing costs by just recalculating those subroutines in the function evaluation that were effected by the variable being changed.

W. Murray: If you are only changing a variable to obtain the partial derivatives, then you do not need to difference those expressions where it is known the partial derivative is zero, as these could be preset. What we really need is a routine that enables the user to either calculate numerical estimates or enter analytic expressions when these are known, *i.e.,* if

$$\frac{\partial F}{\partial x_5} = \frac{\partial E_1}{\partial x_5} + \frac{\partial E_2}{\partial x_5} + \frac{\partial E_3}{\partial x_5} \quad \text{and}$$

$$\frac{\partial E_2}{\partial x_5} = \frac{\partial E_3}{\partial x_5} = 0$$

because E_2 and E_3 are independent of this variable then one only needs to difference E_1. This is a special case of the more general situation where $\frac{\partial F}{\partial x_5}$ may consist of a number of expressions (not necessarily a sum) the derivatives of some of which are known analytically. What we would like to be able to do is mix differencing and analytic derivatives at will.

Anon: Obviously OPRQP was developed at Hatfield on one machine and transferred to another machine at Rolls Royce. Could you comment on the difficulties experienced in this move.

A.H.O. Brown: The OPRQP algorithm was supplied in Fortran and because of the state of our gas dynamics software at that time we found it convenient to rewrite it in our own language - hence NOLIN. We see an increasing rôle for Fortran in the future both because it is a means of coordination between machines, and because Fortran compilers are becom-

354

ing more efficient. I feel that suppliers of software must take account of the fact that the solution of optimization problems is largely a thing we do ourselves and we expect to do it on our own machine in the most efficient way possible.

Anon: I had in mind the difficulties of repeatability of results when a routine is transferring from one machine to another because of change in word length, etc.

A.H.O. Brown: We have to be able to guarantee that anything we design will work within the standards laid down in the specification. We have to guarantee a certain degree of repeatability and accuracy when say our performance calculations are repeated on other machines. We therefore check that any deck we send out conforms to our standards of transferability. So far as transfer to our machines is concerned, both the KDF9 and the CYBER have very long word lengths and give very little trouble of the sort you were suggesting.

D. Goldfarb: From Laurence Dixon's remark it would seem that in your problem it was not a matter of sparsity in the derivatives that caused the saving when only a few variables were changed at a time but rather the fact that only part of the objective calculation had to be repeated. Am I correct?

A.H.O. Brown: Yes. Each subroutine only involves a few of the variables.

Anon: Professor Sargent could you comment on the number of variables involved in your problems.

R. Sargent: It depends on which level of optimization you are referring to. In the paper we consider a column with 10 plates separating 3 components. For optimum feed and sidestream distribution this gives 21 "control" variables for the optimization, with 2 linear constraints and 3 non-linear constraints (apart from non-negativity of

the variables), but each column performance calculation for given values of these control variables
involves 141 "state" variables. So far we have
deliberately used simple problems to test the
method and to learn to interpret the results, and
we are very much at the exploratory stage for
multiple- column systems. A typical industrial
column might involve 40 plates and 5 components,
usually only two end products, and a single feed
perhaps spread over 10 plates. This gives 12
control variables but 802 state variables, so
the optimization problem is small compared with
the calculation of the performance for given
control variables.

M.C. Biggs: Both the objective functions described
by Arthur Brown and Professor Sargent involve the
solution of a set of nonlinear equations, say
$\phi(x,y) = 0$, to obtain a dependent variable y.
Thus the function routine can be very expensive.
If the y variables were treated as independent
variables and equations $\phi(x,y)=0$ handled as equality constraints then their iterative solution is
avoided. In this approach the nonlinear equations
would only be solved once, as the iterative process
would solve $\phi(x,y) = 0$ simultaneously with optimizing the function. In this way the expense of each
function evaluation would be greatly reduced.

A.H.O. Brown: That is a very interesting observation and might well speed up the solution of many
problems. In other cases it may not be advantageous. Definitely it would not be feasible on the
problem I described. We have 2 or 3 dependent
variables in each subroutine. If we treated all
these as independent we would have had over 50
variables. In one particularly difficult engine
we did find it advantageous to treat 1 dependent
variable in each subroutine as independent. In
this problem there was only a very restricted area
with any solutions and the optimization algorithm
could hardly come into play at all. By treating
these variables as independent we gave the optimization algorithm a much wider field of play and
it converged much more quickly.

356

R. Sargent: The question as to whether one should solve internal loops as quickly as possible or open them and solve a large loop is very problem dependent.

J. Bennett: It might be a good idea to initiate a research programme to see whether some general rules can be formulated.

R. Sargent: In my own problem we did some experiments and found that closing some loops as quickly as possible and treating the performance problem as part of the objective function calculation, all speeded up the rate of convergence.

Anon: You were presumably rather unsure about the values of the efficiencies of the plates in your distillation tower. I would be interested to hear your comments on how stable your optimal solution is with respect to changes in these efficiencies. If the solution is entirely dependent on these efficiencies your calculations would not be particularly meaningful.

Secondly, the possibility of both controlling the process and using the input/output information to determine the internal parameters (which are varying) is a very difficult problem. The efficiency of the plates must be a parameter of this type as it will change with age. Are you considering using the actual performance information to evaluate the parameter values and thereby change the optimal operating conditions?

R. Sargent: I will try to be brief. First we are concerned with design studies in our paper and in our test problems have used simple models. More generally, we assume that appropriate correlations will be available for the thermodynamic and performance data. The sensitivity of the optimum solution to variations in these data is an interesting question and one we have not explored; however the minima tend to be rather flat so we should not expect great sensitivity. When we come to the on-line situation, and the use of measurements to estimate parameters in order to compute optimum operating conditions, the problem is more difficult, and I refer you to Peter Young's review paper later in this conference. The quick answer is that you do not have time to do the type of calculation I was talking about on-line, and you have to be content with a much cruder model.

M.J.D. Powell: Dr. Murray in your talk you described a method for treating nonlinear constraints, in which the Lagrangian was minimized subject to linear approximations of the constraints being satisfied. In this method what criteria are used to determine when to update the linear approximation to the constraints?

W. Murray: I am describing other people's algorithms which I do not advocate but I believe many such algorithms are analogous with the penalty function approach but instead of a sequence of unconstrained problems being solved, a sequence of linearly constrained problems is solved. This is not always true however and is probably unreasonably inefficient.

A.H.O. Brown: I believe in OPRQP the linearization is changed at each iteration.

W. Murray: That is often true when a quadratic approximation of the Lagrangian is also made but not in the more general approaches in which a complete nonlinear function is minimized.

Anon: I wish to question whether the runtime of an optimization algorithm is a relevant measure of its efficiency. We usually find that in engineering problems about 95% of the time is taken up in first formulating the model and then interpreting the results and that compared with this 6 hours runtime is almost irrelevant.

A.H.O. Brown: Whilst I agree with the first half of your comment, it all depends on the length of runtime, your situation and how frequently your optimization problems occur. Six hours runtime when you are fighting everyone else and only getting 3 minute slices is a lot! Once you can get your 6 hours down to 5 minutes it becomes easy.

What we have noticed over the years is that once our people have got their runtime down to the time that gets them $\frac{1}{2}$ day turn-round then they do not bother to speed the computation up any more.

When we were running reasonably small numbers of performance calculations at a time we were achieving this turn-round. We thought they were very fast, we knew they were accurate, but when we combined them together in an optimization exercise we suddenly realized that it was necessary and possible to make them much more efficient.

M.C. Biggs: Arthur Brown commented in his paper that the Lagrange multiplier estimates λ_k can sometimes, in practice, take on implausible values, either being extremely large or else showing excessive variations. This sort of behaviour is almost certainly due to poor conditioning of the matrix $(W_k \, B_k^{-1} W_k^{\,T})$. This in turn is probably caused by the rows of W_k (*i.e.* the normals of the active constraints) being approximately linearly dependent. If the constraint normals really are linearly dependent then it may be possible for some redundant constraint to be dropped from the problem. It is more common, however, for the ill-conditioning to be a symptom of loss of accuracy which may be due to poor scaling among the elements of W_k and can often be remedied in ways like those discussed by the author. Obviously the calculation of λ_k in OPRQP/NOLIN contains checks upon the singularity of $W_k \, B_k^{-1} W_k^{\,T}$: but such checks, involving arbitrary "threshold" numbers, may not trap every "suspicious" value of the Lagrange Multipliers.
    It is worth remarking that later versions of OPRQP compute λ_k from

$$(\frac{r_k}{2} \, I + W_k \, B_k^{-1} \, W_k^{\,T}) \, \lambda_k = W_k \, B_k^{-1} \, \nabla F_k - g_k$$

rather than

$$W_k \, B_k^{-1} \, W_k^{\,T} \, \lambda_k = W_k \, B_k^{-1} \, \wedge F_k .$$

From practical experience the calculation of λ_k is then less susceptible to such errors. In these later versions, B_k is also an approximation of the Hessian of the Lagrangian rather than the objective function.

[Added in proof]

A.H.O. Brown: I would like to draw attention to the relationships between the step lengths used in numerical differencing techniques and the toleran- ces set in iterative processes in the function subprograms referred to in my paper. We have discovered that a failure to recognize these relationships can cause much longer runtimes and can, in fact, result in a total failure to converge.

Let P ($\underset{\sim}{x}$,r) be the penalty function, where $\underset{\sim}{x}$ is the vector of independent variables and r is the relaxation weighting coefficient

Let the vector $(\frac{\partial P}{\partial \underset{\sim}{x}})$ be approximated using step lengths $\Delta \underset{\sim}{x}$.

Consider the iterative procedure in one of the function subroutines. In our cases, at least, this involved modifying a set of n variables $\underset{\sim}{z}$ until a set of n parameters $\underset{\sim}{y}$ were matched to required values within a set of n tolerances $\partial \underset{\sim}{y}$.

A particular variable may appear in both $\underset{\sim}{x}$ and $\underset{\sim}{y}$.

Let such a variable be x_i in $\underset{\sim}{x}$ and y_j in $\underset{\sim}{y}$.

Consider the calculation of $\frac{\partial P}{\partial x_i}$.

We can write P as,

$P(\phi (x[\text{ excluding } x_i]),\ \theta (x_i),\ \psi (x_i),r)$

where θ is a direct function, not involving the iterative procedure, and ψ is the function involving the iterative procedure.

Then

$$\frac{\partial P}{\partial x_i} = (\frac{\partial P}{\partial x_i})_\theta + (\frac{\partial P}{\partial x_i})_\psi = (\frac{\partial P}{\partial x_i})_\theta + (\frac{\partial P}{\partial y_j}) . \quad (1)$$

There will be random errors in the calculation of P at x_i and $(x_i + \Delta x_i)$ (assuming forward differences) due to all variables used in iterative procedures throughout the function subprograms. However, as explained in the paper, if results from other subprograms are stored, it is only necessary on the second calculation to obtain the contribution of the subprogram containing y_j.

Let the random errors in y (within the tolerances ∂y) be $E(y)$.

Let $E(y)$ give rise to errors in z of $E(z)$ and to an error in P of $E(P)$.

Then $E(P) = E(z)^T . \frac{\partial P}{\partial z}$

and $E(z) = [\frac{\partial z}{\partial y}].E(y)$

therefore $E(P) = E(y)^T . \frac{\partial P}{\partial y}$ (2)

where $\frac{\partial P}{\partial y} = [\frac{\partial z}{\partial y}]^T . \frac{\partial P}{\partial z}$.

The greatest magnitudes that the terms of $E(y)$ can have are those of δy, and as these can be arbitrarily positive or negative, the maximum error in P from equation (2) is given by

$$E_{max}(P) = \delta y^T \; ABS\left\{\frac{\partial P}{\partial y}\right\} \quad (3)$$

where $ABS\left\{\frac{\partial P}{\partial y}\right\}$ is taken to mean the vector of absolute values.

If ΔP_i is the true difference in P due to Δx_i, bearing in mind that $E_{max}(P)$ can occur both

361

at x_i, and $(x_i + \Delta x_i)$ in opposite senses, then it is essential that

$$2\, E_{\max}(P) << ABS(\Delta P_i) \; .$$

If q is some safely large number, incorporating the factor 2, this can be written

$$q \cdot E_{\max}(P) = ABS(\Delta P_i) = \Delta x_i \cdot ABS\left(\frac{\partial P}{\partial x_i}\right) \; .$$

Therefore substituting from equation (3),

$$\frac{\Delta x_i}{q} \cdot ABS\left(\frac{\partial P}{\partial x_i}\right) = \delta y^T \cdot ABS\left(\frac{\partial P}{\partial y}\right) \; .$$

A reasonable strategy for determining suitable values for δy would be to divide the permissible error equally among the n terms.

Then the general term of the tolerance vector, δy_k say, would be given by

$$\delta y_k = \frac{\Delta x_i}{n \cdot q} \cdot \frac{ABS\left(\frac{\partial P}{\partial x_i}\right)}{ABS\left(\frac{\partial P}{\partial y_k}\right)} \; . \tag{4}$$

Substituting in this expression from equation (1), this gives

$$\delta y_k = \frac{\Delta x_i}{n \cdot q} \cdot \frac{ABS\left(\left(\frac{\partial P}{\partial x_i}\right)_\theta + \left(\frac{\partial P}{\partial y_j}\right)\right)}{ABS\left(\frac{\partial P}{\partial y_k}\right)} \; . \tag{5}$$

362

In the failure case which drew our attention to this subject,

$(\frac{\partial P}{\partial x_i})_\theta$ was zero, and $(\frac{\partial P}{\partial y_j})$ was very small in relation to the other $\frac{\partial P}{\partial y}$ terms. We were using a single value of δy for all n terms of the tolerance vector, and the consequent errors in the calculation of ΔP, were sufficient completely to obscure the true value of $\frac{\partial P}{\partial x_i}$. On using an initially calculated vector δy based on the above analysis the failure was overcome and non-failure cases were solved much more economically.

It is hoped that this note provides sufficient information to enable the user to recognize the problem and to cope with simple examples of it as they arise.

For the software supplier or the user who wishes to take it more seriously, it can only be an introduction to a subject that obviously warrants further research. He will wish to provide for more general equivalence between x and y and to monitor and adjust δy throughout a program run. One interesting suggestion that has come to light is that Δx_i in equation (5) might be replaced by $Max(\Delta x_i, \delta \bar{x}_i)$ where $\delta \bar{x}_i$ is the step taken in the ith direction on the previous move.

Further work on this whole subject is already under discussion.

Our acknowledgements are due to Ian Dawson whose tape recording of the sessions form the basis of the above.

363

An Approach to the Optimal Scheduling of an Electric Power System

M.C. Biggs

*(Numerical Optimization Centre,
The Hatfield Polytechnic)*

SUMMARY

In this paper a nonlinear programming approach to a scheduling problem is described. The particular application for which computational experience is given is for an electrical power system and concerns the operation of generators to minimize the cost of meeting a prescribed demand for electric power over a given time period. It is expected, however, that some aspects of the approach described here will be relevant to other scheduling problems. Essentially we consider a system which has to provide some specified "output" over a given period. The problem is to allocate the various resources in the system in some optimal way, given bounds on the amount of each resource that can be used at any time. Limitations are also placed on the *flexibility* of the system since, in practice, a schedule must make allowance for the time taken (and sometimes the cost incurred) when the system is changed from one state to another.

1. INTRODUCTION

In this paper we consider the problem of determining the operation of generators in an electric power system so as to minimize the cost of meeting some prescribed demand for power over a given period of time. More specifically, suppose

364

that at times $t^{(0)}$, $t^{(1)}$,...$t^{(T)}$ in the period under consideration it is known that the demands on the system are $D^{(0)}$, $D^{(1)}$,...$D^{(T)}$. The problem is to determine appropriate outputs from the generators at each of these times so as to minimize the cost of satisfying the demands.

In a very simplified representation of the system one might attempt to solve the problem by performing a separate optimization at each time $t^{(k)}$. This approach, however, overlooks the fact that the generators are not infinitely flexible: that is, it may not in practice be possible, in the time available, to increase generator output from some low level which appears optimal at time $t^{(k)}$ to a high level which may be needed at some later time $t^{(k+1)}$. In particular the time to "run up" a generator from rest is not negligible and generally a cost is being incurred during this time without any power being injected into the system. (Similar remarks apply to the "running down" of a generator.) No optimization procedure that is concerned *only* with events at time $t^{(k)}$ can make provision for start up or shut down costs because the decision to switch in or switch out a unit must be taken on the basis of the future load on the system.

In the next section we express the generator scheduling problem in a form suitable for solution by nonlinear programming techniques, bearing in mind the points mentioned above.

As this is a preliminary study the electric power system representation is kept very simple. It is assumed that the intervals $t^{(k)}$, $t^{(k+1)}$ are sufficiently short that we may regard the system as experiencing step changes in demand at each time point and as operating in a constant fashion between times. We make no attempt to represent an electrical network between the generators and the demand. Hence there are no line losses or transmission restrictions to contend with. (Such features could however be included as constraints in

365

an extended formulation.) The only variables we shall use to describe the power system are those pertaining to the generators; and all engineering and safety limitations will be interpreted as constraints on the operation of the generators.

As a concluding remark in this section we observe that some of the features of this scheduling problem may be common to resource allocation in other sorts of system. The essential points about the problem are the varying demand for an "output" or "product" and the existence of sources of this product which can operate within certain limits and which require a finite time (and possibly incur a cost) if they are to be changed from one state to another. There would seem to be analogies with the problem of scheduling machines on a production line or with some transportation problems.

2. A FORMULATION OF THE POWER SCHEDULING PROBLEM AS A NONLINEAR PROGRAMMING PROBLEM

Suppose that there are N generators in the system and that the operating level of the ith generator at time $t^{(k)}$ is denoted by $P_i^{(k)}$. Let the initial generator setting $P_i(0)$ be given. Then for our purposes the power system over the period in question is represented by NT variables $P_i^{(k)}$ ($i = 1...N$, $k = 1...T$). Suppose also that the cost of operating the ith generator during the interval $[t^{(k)}, t^{(k+1)}]$ is given by a function $C_i(P_i^{(k)})$.

Clearly the operating levels of each generator must be bounded and we first assume that the bounds are of the form

$$P_i^{max} \geqslant P_i^{(k)} \geqslant 0 \quad (k = 1...T) . \quad (1)$$

Furthermore, in order to reflect the fact that in practice the generator operating levels can only be changed by a limited amount in the interval $[t^{(k)}, t^{(k+1)}]$ the further constraints

$$u_i^{max} \geqslant P_i^{(k)} - P_i^{(k-1)} \geqslant u_i^{min} \qquad (k = 1 \ldots T) \quad (2)$$

are imposed.

Finally, since the power generated must satisfy the demand,

$$\sum_{i=1}^{N} P_i^{(k)} \geqslant D^{(k)} \qquad (k = 1 \ldots T) . \qquad (3)$$

Thus the problem of minimizing the operating cost over time $[t^{(0)}, t^{(T)}]$ can be expressed as

$$\text{Min} \sum_{k=1}^{T} \sum_{i=1}^{N} C_i(P_i^{(k)})$$

$$\text{s.t.} \quad P_i^{max} \geqslant P_i^{(k)} \geqslant 0 \qquad \begin{array}{l} i = 1 \ldots N \\ k = 1 \ldots T \end{array}$$

$$u_i^{max} \geqslant P_i^{(k)} - P_i^{(k-1)} \geqslant u_i^{min} \qquad \begin{array}{l} i = 1 \ldots N \\ k = 1 \ldots T \end{array} \quad (4)$$

$$\sum_{i=1}^{N} (P_i^{(k)}) \geqslant D^{(k)} \qquad k = 1 \ldots T.$$

(Note that some initial values $P_i^{(0)}$ $(i = 1 \ldots N)$ are assumed given).

It is obvious that (4) is a linearly constrained problem, and as such may be expected to be reasonably easy to solve. In a subsequent section we shall discuss some schedules obtained for a small example by solving (4). First however we consider ways in which the constraint expressions (1) to (3) may be modified to give a more realistic system representation.

In (1) we have placed no lower limit on the power that can be supplied by a generator. For stability reasons, however, one would normally insist that the power being injected into the system must exceed some non-zero limit. To replace (1) by

$$P_i^{\max} \geqslant P_i^{(k)} \geqslant P_i^{\min}$$

is however not a satisfactory alternative because this excludes the possibility of a generator being switched off altogether. What we require is a condition which implies that any generator whose operating level is less than $P_i^{\min}$ must be disconnected from the system. Suppose therefore that we replace (3) by

$$\sum_{i=1}^{N} \phi(P_i^{(k)}) \geqslant D^{(k)} \qquad k = 1 \dots T$$

$$\text{where} \quad \phi(P_i^{(k)}) = P_i^{(k)} \quad \text{if} \quad P_i^{(k)} \geqslant P_i^{\min} \qquad (5)$$

$$= 0 \qquad \text{if} \quad P_i^{(k)} < P_i^{\min} .$$

A constraint like (5) defines, in a sense, three generator "modes". If $P_i^{(k)} = 0$ the generator is switched off (and can be regarded as having no cost). If $P_i^{(k)} \geqslant P_i^{\min}$ the generator is injecting power into the system at a cost $C_i(P_i^{(k)})$. If $0 < P_i^{(k)} < P_i^{\min}$ then the generator is incurring a cost but providing no power. This last mode can be regarded as representing a start up or shut down phase.

If the bounds in (2) are such that

$$u_i^{\max} < P_i^{\min} \quad \text{and} \quad u_i^{\min} > -P_i^{\min}$$

then no generator can, in a single time step, make the transition between a zero cost condition and a state of injecting power into the system. We can regard the time spent in the intermediate phase as giving an estimate of start up and shut down costs, as well as reflecting the need to make decisions concerning the starting of a generator *before* the output from that generator is required.

The constraint (5) requires some further consideration because, as it stands, it is discon-

tinuous and hence is liable to present difficulties to a constrained minimization algorithm. Let
$\Theta_i(P_i)$ be a continuous function, defined for
$P_i \leqslant P_i^{\min}$, and having the following properties

$$\Theta(P_i^{\min}) = P_i^{\min}$$

$$\frac{d}{dP_i}\Theta(P_i^{\min}) = 1 .$$

(6)

Now let (5) be replaced by the continuous constraint

$$\sum_{i=1}^{N} \phi_i(P_i^{(k)}) \geqslant D^{(k)}$$

where $\phi_i(P_i^{(k)}) = P_i^{(k)}$ $P_i^{(k)} \geqslant P_i^{\min}$

$\phi_i(P_i^{(k)}) = \Theta_i(P_i^{(k)})$ $P_i^{(k)} < P_i^{\min}$.

(7)

Fig. 1 shows the general form that is required for
$\phi_i(P_i)$.

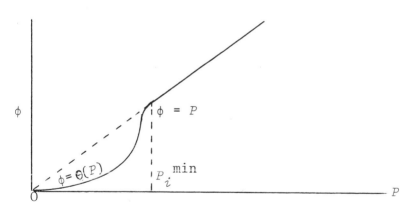

Fig. 1.   A continuous approximation to the constraint function in (5)

One of many suitable expressions giving the
required general form for $\Theta_i(P_i)$ is

369

$$\Theta_i(P_i) = aP_i + bP_i^{n-1} + cP_i^{n+1} \tag{8}$$

for $n \gg 1$. (The linear term must be present in (8) because the constraint (7) needs a non-zero derivative when $P_i = 0$, otherwise the system will be completely insensitive to any generators that are switched off.)

Hence an alternative formulation of the power scheduling problem is

$$\text{Min} \sum_{i=1}^{N} \sum_{k=1}^{T} C_i(P_i^{(k)})$$

s.t. $P_i^{\max} \geqslant P_i^{(k)} \geqslant 0 \qquad\qquad i = 1 \ldots N$

$u_i^{\max} \geqslant P_i^{(k)} - P_i^{(k-1)} \geqslant u_i^{\min} \qquad \begin{array}{l} k = 1 \ldots T \\ i = 1 \ldots N \\ k = 1 \ldots T \end{array} \tag{9}$

$\sum_{i=1}^{N} \phi_i(P_i^{(k)}) \geqslant D^{(k)} \qquad\qquad k = 1 \ldots T$

where ϕ_i is given by (7) and (8).

It should be noted that a schedule produced by (9) will contain values of $P_i^{(k)}$ lying between 0 and $P_i^{\min}$. Because of the form of the constraints however, such values should tend to lie close to zero or close to $P_i^{\min}$ and the optimum should not rest heavily upon generators operating continuously at unrealistic power settings. Computational experience using (9) is given in the next section.

3. COMPUTATIONAL RESULTS

In order to illustrate some of the points made in the preceding section we consider a simple scheduling problem, involving a system consisting of three generators, whose characteristics are summarized in Table I.

Table I

Generator cost data and performance limits

Generator	P_i^{max}	Cost Function
1	90	$2.3P_1 + 0.0001P_1^2$
2	120	$1.7P_2 + 0.0001P_2^2$
3	60	$2.2P_3 + 0.00015P_3^2$

The limits on the changes in $P_i^{(k)}$ appearing in (2), are assumed to be as follows:

$$u_1^{max} = 6. \qquad u_2^{max} = 7. \qquad u_3^{max} = 6.$$

$$u_1^{min} = -7. \qquad u_2^{min} = -7. \qquad u_3^{min} = -7.$$

We suppose that the period under consideration consists of 5 equal time segments, during each of which the demand D is assumed to remain constant. A scheduling problem for the system is completely specified by a value for the demand at the start of each time segment and values for the output of each of the generators prior to the start of the first time segment. Table II gives this information for the three cases that will be considered.

Table II

Details of problem considered

Problem number	$P_1^{(0)}$	$P_2^{(0)}$	$P_3^{(0)}$	$D^{(1)}$	$D^{(2)}$	$D^{(3)}$	$D^{(4)}$	$D^{(5)}$
1	30	30	30	100	110	115	130	150
2	30	30	30	80	70	55	40	30
3	15	50	10	60	50	70	85	100

It will be observed that in problem 1 the demand is monotonically increasing; in problem 2 it is monotonically decreasing and in problem 3 there is a decrease followed by an increase.

We shall seek solutions to these problems using both formulations (4) and (9). The constrained minimization method used to solve these problems is the recursive quadratic programming approach (CPRQP) is due to Biggs (1972, 1974).

We first consider the schedules produced using (4). Note that the cost functions given in Table I imply that (4) becomes a straight-forward quadratic programming problem in 15 variables and with 65 constraints and, as might be expected, OPRQP solves such a problem rapidly. Calculations were done in single precision Fortran on a DEC System computer. Starting from initial solution approximations $P_i^{(k)} = P_i^{(0)}$ $(i = 1...N, k = 1...T)$ the three problems were each solved in about 30 seconds of computer time. The resulting schedules are given in Tables III to V.

Table III

Schedule produced by (4) for problem 1

Segment	D	P_1	P_2	P_3
$t^{(1)}, t^{(2)}$	100	27	37	36
$t^{(2)}, t^{(3)}$	110	24	44	42
$t^{(3)}, t^{(4)}$	115	17	51	47
$t^{(4)}, t^{(5)}$	130	20	58	53
$t^{(5)}, t^{(6)}$	150	26	65	59

Total cost = 1220.4

Table IV

Schedule produced by (4) for problem 2

Segment	D	P_1	P_2	P_3
$t^{(1)}$, $t^{(2)}$	80	23	34	23
$t^{(2)}$, $t^{(3)}$	70	16	38	16
$t^{(3)}$, $t^{(4)}$	55	9	37	9
$t^{(4)}$, $t^{(5)}$	40	2	36	2
$t^{(5)}$, $t^{(6)}$	30	0	30	0

Total cost = 523.3

Table V

Schedule produced by (4) for problem 3

Segment	D	P_1	P_2	P_3
$t^{(1)}$, $t^{(2)}$	60	8	49	3
$t^{(2)}$, $t^{(3)}$	50	1	56	0
$t^{(3)}$, $t^{(4)}$	70	1	63	6
$t^{(4)}$, $t^{(5)}$	85	3	70	12
$t^{(5)}$, $t^{(6)}$	100	5	77	18

Total cost = 664.8

It can be seen that the optimum solutions obtained for problems 2 and 3 make use of the assumption that the generators can be operated satisfactorily at any level between zero and P_i^{max}. We may therefore expect these schedules to be altered (and the operating cost to increase) when a non-zero lower limit P_i^{min} is imposed on the power that can be supplied by each generator and (9) is used to seek an optimum operating policy.

Assume now that $P_1^{min} = P_2^{min} = P_3^{min} = 10$, and consider the function $\Theta(P)$ appearing in the constraints of (9). Let Θ be defined as the polynomial

$$\Theta(P) = 0.01P + 0.000297P^5 - 0.00000198P^7. \quad (10)$$

To illustrate the extent to which this function approximates the discontinuous constraint (5) we may tabulate $\Theta(P)$ for values $1 \leqslant P \leqslant 10$.

Table VI

Values of $\Theta(P)$ given by (10)

P	$\Theta(P)$
1	0.01
2	0.03
3	0.1
4	0.31
5	0.82
6	1.8
7	3.4
8	5.7
9	8.1
10	10.0

It should be noted that the schedule produced by (9) will still contain generator output lying between zero and P_i^{min}. The nonlinear form of the supply/demand constraints will however tend to force such generator settings to be either close to zero or close to P_i^{min}, thus *approximating* the situation where a generator can be considered either as operating at some minimum level or as being disconnected from the system.

The highly nonlinear constraints make (9) a more difficult problem than (4). We seek to obtain solutions more rapidly by using starting approximations based on the schedules calculated using (4). Consider first problem 2. (The solution for problem 1 will be the same using both (4) and (9) since no generator outputs less than P_i^{min} are involved.) Starting from the values of $P_i^{(k)}$ given in Table IV, OPRQP solves (9) in 25 seconds of computer time and obtains the schedule shown in Table VII.

Table VII

Schedule produced by (9) - with constraint function (10) - for problem 2

Segment	D	P_1	P_2	P_3
$t^{(1)}, t^{(2)}$	80	23	34	23
$t^{(2)}, t^{(3)}$	70	16	38	16
$t^{(3)}, t^{(4)}$	55	9	38.7	9
$t^{(4)}, t^{(5)}$	40	2	39.9	2
$t^{(5)}, t^{(6)}$	30	0	32.9	0

Total cost = 537.9

375

Comparison between Tables IV and VII shows that the same over-all strategy (shutting down generators 1 and 3 as rapidly as possible) is given by both methods. The output of generator 2 has to be increased in Table VII to compensate for the reduced contribution of the other two towards meeting the demand.

While Table VII appears to offer the *best* schedule for problem 2 that can be obtained using formulation (9), it was also discovered during numerical tests that there is at least one other local solution. This was found when the starting guesses for solving (9) with OPRQP were taken to be the generation levels of Table IV with the modification that values below P_i^{min} were reset to P_i^{min}. REQP then required 90 seconds of computer time to obtain the solution shown in Table VIII.

Table VIII

Alternative schedule produced by (9) for problem 2

Segment	D	P_1	P_2	P_3
$t^{(1)}, t^{(2)}$	80	23	34	23
$t^{(2)}, t^{(3)}$	70	16	38	16
$t^{(3)}, t^{(4)}$	55	10	34.2	10.9
$t^{(4)}, t^{(5)}$	40	7	27.1	9.6
$t^{(5)}, t^{(6)}$	30	0	20.1	9.9

Total cost = 542.9

In Table VIII only the least economical generator is shut down, and the other generator 3 is operated at about its minimum permitted level.

We also found multiple solutions when problem 3 was attempted using the formulation (9). When started from the generation levels given by (4) UPRQP solved (9) in 27 seconds and gave the schedule shown as Table IX.

Table IX

Schedule produced by (9) - with constraint function (10) - for problem 3

Segment	D	P_1	P_2	P_3
$t^{(1)}, t^{(2)}$	60	8	54.2	3
$t^{(2)}, t^{(3)}$	50	1	56.8	2.2
$t^{(3)}, t^{(4)}$	70	0	63.8	8.2
$t^{(4)}, t^{(5)}$	85	0.1	70.8	14.2
$t^{(5)}, t^{(6)}$	100	6.1	77.8	20.2

Total cost = 692.4

The schedule of Table IX differs only slightly from that of Table V, in that generator 3 is somewhat more, and generator 1 somewhat less, heavily used. A rather more extensively modified schedule is obtained when OPRQP is applied to problem (9) starting from the initial estimate given by Table V with low values of P_i either rounded down to zero or up to P_i^{min}. This solution (requiring 30 seconds of computing) is given in Table X.

This is a more economical schedule than the previous one. The most efficient generator bears more of the load and the least efficient one is shut down completely. The only drawback is that quite a large amount of unwanted power is being generated during the second time interval. A

377

Table X

Alternative schedule produced by (9) for problem 3

Segment	D	P_1	P_2	P_3
$t^{(1)},t^{(2)}$	60	8	54.2	3
$t^{(2)},t^{(3)}$	50	1	61.2	0
$t^{(3)},t^{(4)}$	70	0	68.2	6
$t^{(4)},t^{(5)}$	85	0	75.2	11.8
$t^{(5)},t^{(6)}$	100	0	82.2	17.8

Total cost = 687.8

third schedule which is also a local solution of (9) - and which is the best solution found in several trials - was found when the starting approximation given to OPRQP had all generation levels greater than or equal to P_i^{min}. The operating policy is completely different (see Table XI) but the cost is in fact only slightly reduced.

Table XI

Second alternative schedule produced by (9) for problem 3

Segment	D	P_1	P_2	P_3
$t^{(1)},t^{(2)}$	60	8.9	43	9.4
$t^{(2)},t^{(3)}$	50	1.9	49.9	2.8
$t^{(3)},t^{(4)}$	70	7.9	56.8	8.8
$t^{(4)},t^{(5)}$	85	10.0	63.9	11.1
$t^{(5)},t^{(6)}$	100	11.9	70.9	17.1

Total cost 687.5

4. CONCLUSIONS

It is not our purpose to draw practical conclusions about power scheduling from the small and
simplified example of the previous section. What
we have shown is that the problem formulations of
section 2 can actually be applied to numerical
examples. The simpler form (4) is likely to be
quite easy to solve by nonlinear programming techniques. The more complicated problem (9), while
yielding solutions that are somewhat more realistic, has a number of drawbacks. It is obviously
more difficult to solve. Indeed it may be remarked
that OPRQP does well to cope with such highly nonlinear constraints as those given by (10). Moreover it has been found that the non-convex form of
the supply/demand constraints in (9) permits the
existence of multiple local minima. At this stage
we can say nothing very constructive about this
question of multiple minima except that the use of
a number of different starting points should, in
practice, enable a good schedule (if not the best)
to be calculated.

A final word should perhaps be said about
extensions of this work. It still seems possible
that a nonlinear programming approach like that
presented in section 2 can be used to solve larger
and more realistic scheduling problems. In particular, if we consider the question of obtaining
preliminary estimates of optimum operating policies
it may well be quite acceptable to neglect many of
the system constraints (in the way that we have
neglected voltage restrictions etc. in this paper).
One of the most important features of a schedule
produced by (4) or (9) may be the fact that it
provides an estimate of which generators should be
operating. The approaches described in this paper
allow this choice of generators to be made taking
account of the whole period under consideration.
Once it is known which generators are to provide
the power, the optimum output levels can be
revised, if desired, using a more sophisticated
problem formulation involving more of the system
constraints.

ACKNOWLEDGEMENT

Some of the work described in this paper was carried out as part of the author's research for the degree of PhD at the University of London, under the supervision of Dr. H.M. Liddell and Dr. M.A. Laughton.

REFERENCES

Biggs, M.C. (1972) "Constrained Minimisation using Recursive Equality Quadratic Programming", *in* "Numerical Methods for Nonlinear Optimization", *Ed.* F.A. Lootsma, Academic Press.

Biggs, M.C. (1974) "Constrained Minimisation using Recursive Quadratic Programming: Some alternative subproblem formulations", The Hatfield Polytechnic, Numerical Optimization Centre, Technical Report 51.

A Dynamic Programming Algorithm for Optimizing System Performance

A.W. Clarke

*(Central Electricity Generating Board
Berkeley Nuclear Laboratories)*

SUMMARY

The problem of optimizing the operating conditions of certain systems as a function of time
may be formulated in terms of maximizing a nonlinear cost function subject to a few nonlinear constraints. It was found that a Dynamic Programming
(DP) Algorithm was a particularly efficient way of
solving this problem compared with an alternative
nonlinear programming method tried. In addition,
the DP method has the big advantage that it may be
generalized to take account explicitly of uncertainties which are inevitably present in the constraint equations, provided their probability distributions may be quantified as a function of time.

A purpose built DP algorithm has been written
to evaluate the expected costs of following particular strategies of system operation. It was
found that additional constraints which did not
fit neatly into a DP framework had to be imposed
and the modified problem was solved iteratively by
linking the algorithm to a Monte Carlo simulation
program. The same technique was also used to
assess the economic risk of following particular
strategies and to assess the cost effectiveness of
particular monitoring schemes aimed at reducing
the uncertainties.

As an example, results are presented which

show how the technique may be used to help a Nuclear Reactor Operator decide how he may use the measurements made at intervals on his plant to plan his future operating strategy.

1. INTRODUCTION

On any large industrial plant there will always be certain components for which the cost of replacement is comparable with the cost of closing down the plant and building a new one; *i.e.*, these components are effectively irreplaceable. This is particularly true for nuclear reactors where the high radiation fields within the concrete shielding make access difficult. In this paper, we will be considering a particular problem associated with Magnox reactors. Analogous problems exist for all reactor systems. A typical Magnox reactor is shown schematically in Fig. 1 and consists basically of an array of graphite bricks (known as the moderator) arranged in the shape of a right circular cylinder contained within a pressure vessel made either of steel or concrete. Natural uranium fuel elements are inserted down channels in the graphite bricks and the heat generated by fission is extracted by passing pressurized carbon dioxide up the fuel channels and thence to a conventional boiler. The presence of CO_2 implies that component oxidation cannot entirely be eliminated but by adjusting the composition (*i.e.*, the impurity concentration), temperature and pressure of the coolant gas, the rate of oxidation can be controlled.

In the problem under consideration, two irreplaceable components are important, namely the graphite core and the steel support structure above the core. Graphite oxidation (which manifests itself as an integrated weight loss, y) increases with the reactor pressure, p, decreases with water concentration, w, and to first order is independent of the temperature T. Steel oxidation, however, increases with pressure, temperature and water concentration and thus there is an immediate conflict in that a good coolant for the steel will

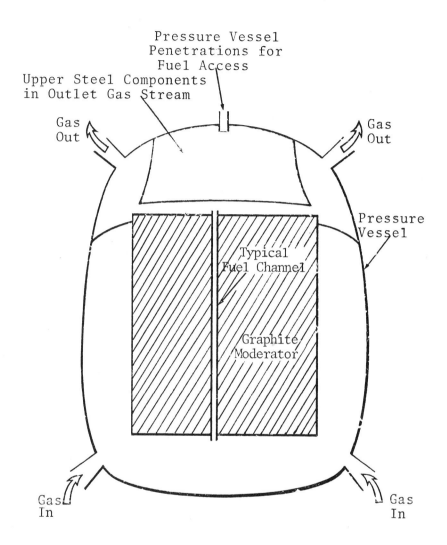

Fig. 1 Schematic Diagram of a Magnox Reactor

be a bad one for the graphite moderator and *vice
versa.*

Models to represent the dependence of the
steel and graphite oxidation rates as functions of
the coolant conditions have been developed and
estimates of the limiting amounts of steel oxidation, L, and graphite weight loss, M, have been
made. Consequently it is possible to formulate
the problem of choosing an acceptable set of temperatures, pressures and water concentrations as
an optimization problem. The objective or cost
function may be taken as a suitably weighted integral of the electricity generated during the reactor's life time. Electrical output depends upon
the coolant pressure and temperature. Two possible methods of solution of this essentially
deterministic problem are discussed in Section 2.
The first part of Section 2 considers how the
problem may be solved by a separable programming
method and the second shows that the problem may
be reformulated as a dynamic programming problem.

The usefulness of solving the deterministic
problem is limited because in practice there are
large uncertainties in the models representing the
constraint equations and in the values of the constraints L and M. In the presence of uncertainties,
it is appropriate to take the cost function as the
expected value of the electricity generated.
Since, the problem is nonlinear, the effect of the
uncertainties may be not only to change the expected value of the cost function but also to influence the initial operating strategy. The initial
operating strategy may also be influenced by measurements made at future times on the reactor even
though the outcome of the measurement is not known
at the time of fixing the operating conditions.
This is simply because a wrong strategy early in
life (*i.e.*, one which oxidises components at a
faster rate than the optimum) may be detected if
measurements are made and a compensating strategy
adopted to bring the oxidation rates down. It is
shown in Section 3 that the effect of uncertainties

and of measurements may be accommodated within the dynamic programming framework.

In practice, the cost function adopted here does not describe the whole picture and there are additional constraints on the system which are difficult to interpret in purely economic terms. As an example of these, the operator may require an assurance that a proposed operating scheme will yield only a small probability that the station would have to be shut down after a given number of years, in the event that the constraints L and M turn out to be more restrictive than their current best estimate. Such a constraint cannot easily be accommodated within a dynamic programming framework and Section 4 discusses an approximate simulation method which may be used in conjunction with a dynamic programming algorithm.

2. THE DETERMINISTIC PROBLEM

At the design stage, an assessment is made of the maximum period for which it will be economic to generate electricity from the reactor. It is assumed that at the end of this period, which is known as the design life time and is typically 30 years, the reactor will be shut down whether or not any component has reached its limiting oxidation value. It is convenient to split the design reactor life time into 15 equal periods and to assume that conditions remain constant within a given period. This is quite a good assumption since at a large power plant information about the oxidation rates can only be collected and processed on a timescale of about 2 years.

(a) *A Formulation Suitable for Separable Programming Solution*

In this section, it is assumed that the equations describing the constraints and the limiting values of the constraints are well defined deterministic quantities. Then the problem may be written as:

$$\text{maximize} \sum_{i=1}^{15} a_i \; K(\underset{\sim}{x}_i) \qquad (1)$$

$$\text{subject to} \sum_{i=1}^{15} a_i \; G(\underset{\sim}{x}_i) \leqslant L \qquad (2)$$

$$y_{15} \leqslant M \qquad (3)$$

with y_{15} defined by the series of equations

$$a_i H_1(\underset{\sim}{x}_1) = H_2(y_i) - H_2(y_{i-1}) \qquad i = 1,15 \quad (4)$$

$$\text{and} \quad y_0 = 0. \qquad (5)$$

The variables $x_i \equiv \begin{Bmatrix} P_i \\ w_i \\ T_i \end{Bmatrix}$, y_i describe the operating

state during period i and are bounded above and
below. The variables a_i are allowed to take values
0, 1 only and are there to allow the station to be
shut down before the end of the design life time.
We must have:

$$a_{i-1} \geqslant a_i \qquad i = 2,15$$

In the above equations, (1) describes the
cost function; $K(\underset{\sim}{x}_i)$ is the value of electrical
output attributable to period i. Equation (2)
describes the steel oxidation constraint. The
function $G(\underset{\sim}{x})$ is of the form $b \; w^\alpha p^\beta \exp(-k/T)$. In
this expression, the water concentration, pressure
and temperature are the primary operational controls and b, d, β, k are constants.

Equations (3) to (5) describe the graphite
constraint and in this case H_2 is of the form:

$$\log(1 + \delta y) - \varepsilon y$$

with y equal to the percentage weight loss of graphite and δ, ε are constants. H_1 is proportional
to $p^{5/3}$ and to a tabulated smooth function of w.

This set of equations was solved by representing the nonlinear functions as a series of linear segments and using the separable programming facility of the MPSX-MIP program package. This particular method was chosen because there was a need to produce results on an urgent time scale and a large amount of experience in the use of this package was available within the CEGB. Because of the large number of additional variables that this technique introduces, what had appeared to be a relatively simple problem became very time consuming to solve and this was the main reason why an alternative method of solution was sought. The structure of the equations suggested that a dynamic programming approach might be fruitful.

(b) *A Dynamic Programming Formulation*

To get the equations in the right form for a dynamic programming approach, equations (1) to (5) are rewritten in the form:

$$\max_{\underset{\sim}{x}_1}\left[a_1 K(\underset{\sim}{x}_1) \ + \ \max_{\underset{\sim}{x}_2}\left[a_2 K(\underset{\sim}{x}_2) \ + \ .. \ + \ \max_{\underset{\sim}{x}_{15}} a_{15} K(\underset{\sim}{x}_{15})\right] ..\right] \tag{6}$$

$$\text{subject to} \quad a_{15} G(\underset{\sim}{x}_{15}) \leqslant L - \sum_{i=1}^{14} a_i G(\underset{\sim}{x}_i) \equiv L_{15} \tag{7}$$

$$y_{15} \leqslant M \tag{8}$$

$$a_i H_1(\underset{\sim}{x}_i) = H_2(y_i) - H_2(y_{i-1}) \quad i = 1,15 \tag{9}$$

$$y_0 = 0 . \tag{10}$$

Then if it is assumed that L_{15} and y_{14} are known (as would be the case in practice when that stage was reached) the 15th stage problem may now be solved. This is done over the possible range of values of L_{15} and y_{14} to give, symbolically,

$$\max_{\underset{\sim}{x}_{15}} a_{15} K(\underset{\sim}{x}_{15}) = F(L_{15}, y_{14}) . \tag{11}$$

Then writing $F(L_{15}, y_{14})$ as $F(L_{14}) - a_{14} G(\underset{\sim}{x}_{14})$, $H_2^{-1}(H_2(y_{13}) - a_{14} H_1(\underset{\sim}{x}_{14}))$ the 14th stage problem may be solved for all possible values of L_{14}, y_{13}. Proceeding in this way, the 1st stage problem may eventually be solved as a function of $L_1 \equiv L$ and y_0. We are only interested in $y_0 = 0$ but information on the range of values of L is useful.

The benefit of the approach in solving this problem is that the optimization problem at any stage is relatively simple although of course a large number of such computations have to be carried out. A very simple algorithm (similar to the simplex method) was used to find the optima of these smaller problems. Because, from physical considerations, a good idea of the behaviour of the equations was already available, it was usually possible to compute such a good starting guess that convergence was rapidly achieved. In general a solution was obtained at about one tenth of the cost of the MPSX-MIP method.

As stated in the introduction, the solution of the deterministic problem is of only limited value and it is necessary to consider how uncertainties in the constraint equations may be accommodated and also to consider how a series of measurements might influence the early strategy. The formulation in 2(a) is not at all suitable for dealing with these factors but the DP formulation in 2(b) may readily be modified.

3. THE TREATMENT OF UNCERTAINTIES IN THE DYNAMIC PROGRAMMING FORMULATION

In order to simplify the analysis, it will be assumed that all of the uncertainty in the modelling of the constraint may be translated into uncertainties in the limits L and M and that the information gained at shut downs may be used to reassess their values and associated uncertainties.

Once uncertainties are allowed to enter

explicitly into the equations, there are several cost functions that might be adopted. In this problem, it is supposed that at any stage the operators will wish to maximize the value of the electricity generated at the next stage plus the expected value $E\{\ \}$ of the electricity generated over all remaining stages.

Thus, mathematically we may write the problem as

$$\max_{\underset{\sim}{x}_1}\left[a_1 K(\underset{\sim}{x}_1)+E\max_{\underset{\sim}{x}_2}\left[a_2 K(\underset{\sim}{x}_2)+\ldots+E\{\max_{\underset{\sim}{x}_{15}} a_{15} K(\underset{\sim}{x}_{15})\}\ldots\right]\right]$$

$$(12)$$

$$\text{subject to } a_{15}G(\underset{\sim}{x}_{15}) \leqslant L_{15} \qquad (13)$$

$$y_{15} \leqslant M_{15} \qquad (14)$$

$$a_i H_1(\underset{\sim}{x}_i) = H_2(y_i) - H_2(y_{i-1}) \qquad (15)$$

$$y_0 = 0 \qquad (16)$$

$$L_i = L_{i-1} - a_{i-1}G(\underset{\sim}{x}_{i-1}) + \xi_i \qquad i = 2,15 \qquad (17)$$

$$M_i = M_{i-1} + \eta_i \qquad i = 2,15 \qquad (18)$$

In these equations, ξ_i, η_i are random variables whose distribution is known and the expectations in the cost functions are taken over the distribution functions of the appropriate random variables at each stage. L_i and M_i are the estimates of the steel and graphite limits at the beginning of the ith stage. This set of equations (12) to (18) may be solved in just the same way as (6) to (10) except that now results have to be tabulated progressively as functions of L_i, y_{i-1} and M_i and it is also necessary to find the expectation of a function of two variables. The expectation was found by direct simulation, sampling simultaneously from the distributions of ξ_i and η_i. Typically a sample of 500 points was required to give reasonable statistical accuracy. The additional complexity of this formulation increased the computation time required but the cost per case was still comparable with that achieved with the original linear programming formulation.

By allowing different values of ξ_i, n_i for each i it is possible to estimate the value of improved inspection and monitoring techniques.

4. THE TREATMENT OF PROBABLISTIC CONSTRAINTS

It has been stated that the cost function adopted in the previous section is not always sufficient in itself to yield all of the information required by a reactor operator. In this section constraints of the form

$$\text{prob}\ (a_{i+1} = 0) \leqslant \gamma \qquad (19)$$

for a given value of γ will be considered.

It has not yet been possible for us to incorporate constraints of this type into a dynamic programming framework and the following *ad hoc* procedure has been adopted.

4.1 Firstly equations (12) to (18) are solved without considering the additional constraint (19) to give an optimal ith stage strategy for any given values of L_i, M_i, and y_{i-1}.

The optimum first stage strategy is selected by applying the results of the DP algorithm to the best estimate values of L_1, M_1 and $y_0(\equiv 0)$. This also enables y_1 to be calculated. L_2 and M_2 are then evaluated by sampling from the distributions of ξ_1 and n_1. This enables an optimal second stage strategy to be determined (dependent only on the values of ξ_1, n_1 chosen). This procedure is repeated for each stage until the reactor is shut down. By repeating the procedure a large number of times and choosing different values of ξ_i, n_i for each run, the expected value of the output and the probability of having to shut down after a given number of years may be evaluated. If constraint (19) is not satisfied the following action is taken.

4.2   The procedure adopted in 4.1 is repeated but instead of taking the optimum strategy at each stage, the temperature is reduced by $x°C$ (to reduce the steel oxidation rate) and water concentration is increased by z parts per million (to reduce the graphite oxidation rate). Again the probability of having to shut down after a given number of years is evaluated.

4.3   The method outlined in 4.2 is repeated for a suitable range of values of x and z and the optimum strategy is chosen as that set of conditions that yields the maximum expected value of electricity generated subject to the constraint (19) being satisfied.

Although the procedures described above do not necessarily lead to a true optimum solution and are not mathematically rigorous, any errors in strategy at one time step will tend to be compensated at the next time step and intuitively it is felt that the method will produce solutions not far removed from the optimum.

5.   RESULTS

Application of the above algorithms gives rise to a large number of results and a selection of these are given in the following figures. Fig. 2 gives a comparison of the solutions of the MPSX method and the dynamic programming method for a range of values of the constraint L and for a particular value of the constraint M. There are minor differences in the results obtained arising from the different approximations employed in the two routes but in general the agreement is good confirming that no errors are present. Fig. 3 shows the effect on initial operating strategy of a 15% constant-in-time uncertainty in both the steel and graphite limits (assuming that ξ_i, η_i are normally distributed and that 15% is a 1σ value) for a range of values of L_1 and M_1. It may be seen that the effect on the expected value of the cost function is of order of 5% which repre-

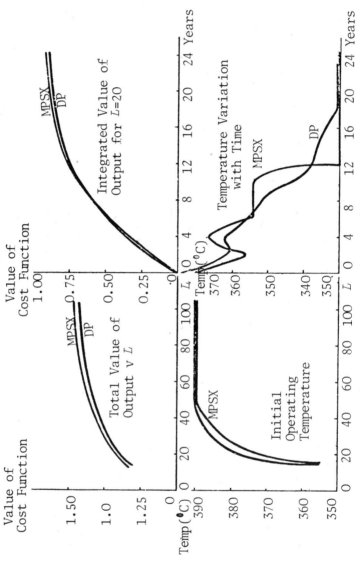

Fig. 2 A Comparison between MPSX and DP (M=20% throughout)

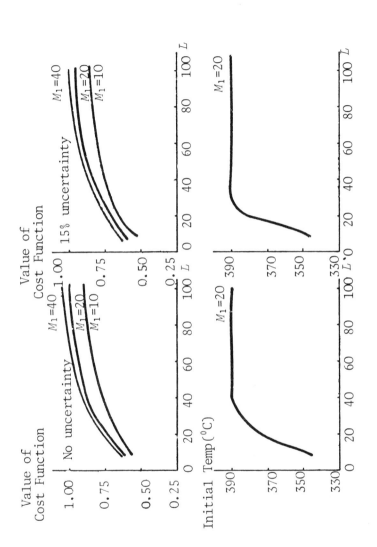

Fig. 3   Effect on Initial Operating Strategy of a 15% Uncertainty

sents a very considerable sum of money but that
the effect on the initial operating strategy is
small in this case. One reason for the small differences here is that over much of the range of L_1,
the optimum strategy is to run at as high a temperature as is allowed by the other plant constraints.

Figs. 4 and 5 demonstrate the effect of
including each of the probability constraints
$p(a_{10} = 0) \leqslant \gamma$ and $p(a_{13} = 0) \leqslant \gamma$ for two possible
monitoring schemes and a range of values of γ.
The first of these assumes that the uncertainties
on both the steel and graphite limits stay constant
at 15% and the second assumes that the uncertainties can be progressively reduced from 15% at the
start to zero after 30 years. It may be seen that
for both of these constraints, the effect of the
improved monitoring scheme becomes quite dramatic
when γ is reduced. Eventually, the situation is
reached where the constraints cannot be satisfied
at all with one scheme but can be with the other.

6. CONCLUSIONS

A complex problem of investigating coolant
conditions which will yield improved performance
throughout the life of a typical Magnox Reactor
has been formulated as a nonlinear optimization
problem.

1. If the equations describing the constraints and their limiting values are well
defined, it is shown that a dynamic programming
algorithm is an efficient method of solution.

2. When the constraint equations and their
limiting values are uncertain, the dynamic programming approach may still be used with very little
reduction in efficiency. A conventional optimization technique could not deal with this problem at
all.

3. The dynamic programming method cannot

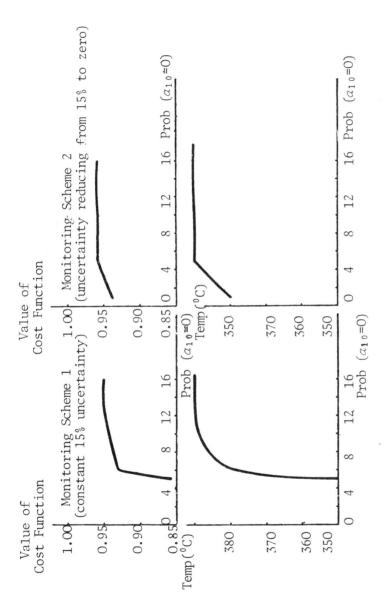

Fig. 4 Variation in Initial Strategy as a Function of Prob(α_{10}=0)

395

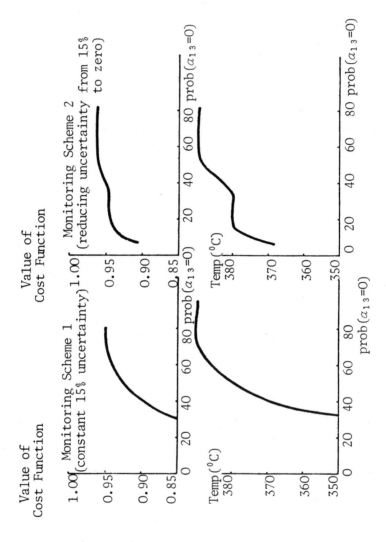

Fig. 5 Variation in Initial Strategy as a Function of prob$(a_{13}=0)$

treat probability constraints explicitly. However, feasible near optimum solutions have been obtained by coupling the dynamic programming algorithm to a direct simulation technique and iterating.

ACKNOWLEDGEMENT

This paper is published by permission of the United Kingdom Central Electricity Generating Board.

Reflections on the Global Optimization Problem

L.C.W. Dixon, J. Gomulka and S.E. Hersom

*(The Numerical Optimization Centre,
The Hatfield Polytechnic)*

SUMMARY

In this paper we review the global minimization problem and describe some of the methods that
have been proposed for locating the global minimum
of a function numerically. A major section of
this paper is devoted to the behaviour of techniques written for local minimization when applied
to functions with more than one minimum. Other sections are devoted to the principles behind probali-
stic and space covering techniques and to methods
based on descent from local minima and the trajectory approach proposed by Branin (1970). Methods
for solving this problem based on cutting plane or
branch and bound principles are not discussed.

1. INTRODUCTION

Many practical engineering applications can
be formulated as optimization problems in which
the objective function is not convex and possesses
many local minima. In this paper we consider a
few of the methods that have been suggested for
solving this problem and attempt to suggest a
framework for future investigations.

The problem under consideration can be stated
as determine that $\underset{\sim}{x}^*$ which minimizes $f(x)$

$$\text{subject to } x \in S \subset E^n.$$

It will be assumed that $\underset{\sim}{x}^*$ exists and is in the
interior of S. The problem under consideration is

that termed the "essentially unconstrained" problem by Hartman (1972). Whenever it is possible
from engineering considerations that a solution on
the boundary of S could be desirable then the constraints defining S must be handled more rigorously
and we have an essentially different problem. A
review of the latter problem was given by McCormick
(1972).

For convenience we shall assume that the
function to be minimized is thrice differentiable,
so our problem is

$$\min_{x \in S} \{f(x)\}, \ f \in C^3.$$

We shall be considering numerical algorithms therefore we should not hope to determine x^* or $f(x^*)$
exactly, but must be content to define a finite
region which we hope to locate. In local minimization three such regions are used as alternatives

(i) CRITERION x: $x \in A_x$ where
  $A_x = \{x: \| x - x^* \| < \varepsilon_x\}$

(ii) CRITERION g: $x \in A_g$ where
  $A_g = \{x: \| g(x) \| < \varepsilon_g\}$

(iii) CRITERION F: $x \in A_F$ where
  $A_F = \{x: f(x) - f(x^*) < \varepsilon_F\}$

where ε_x, ε_g, ε_F are small preselected parameters,
and g is the gradient vector of f. The second
criterion is clearly inappropriate for global
optimization since it is satisfied in the neighbourhood of every stationary point. The first can
also be unsatisfactory if there are some scattered
minima with function values sufficiently close to
$f(x^*)$. If that is the case then A_F would typically
consist of several small regions, each containing
one "nearly global" minimum. In most cases we
would probably be interested in locating one point
inside each region.

Each of the different concepts introduced in
this paper will be illustrated on the same function,
the 6 hump camel back function:

$$f(x) = 4x_1^2 - 2.1x_1^4 + \tfrac{1}{3}x_1^6 + x_1 x_2 - 4x_2^2 + 4x_2^4.$$

The region S considered in this paper for this function is defined by

$$S = \{x: -3 \leqslant x_1 \leqslant +3, -1.5 \leqslant x_2 \leqslant 1.5\}.$$

The contours of this function are shown in Fig. 1: where the various local minima, maxima and saddle points are indicated.

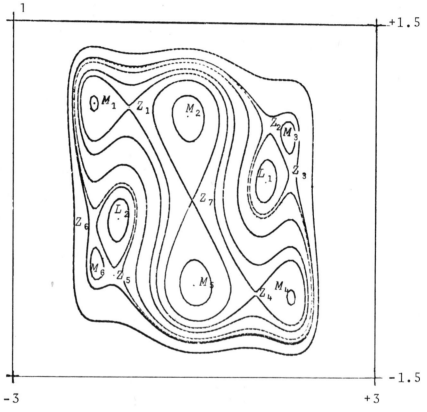

Fig. 1. The six hump camel back function: contour plot

For any multiminimal function we shall refer to the sets of such points as M, L and Z, so on the 6 hump camel back function M contains 6 points, L contains 2 points and Z contains 7 points.

The region in the neighbourhood of the two saddle points Z_5 and Z_6 is complex as $0.0001 < |f(Z_5) - f(Z_6)| < 0.0002$. This region was used as a test of the accuracy of a contour plotting routine available at the NOC,(Hersom private communication (1975)). For interest an enlarged set of contours of this region is shown in Fig. 2 including one that splits the function value at these two saddle points.

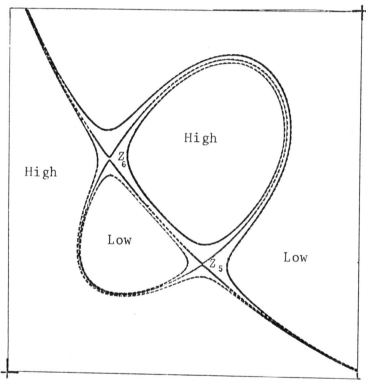

Fig. 2.   Enlarged view of saddle points Z_5, Z_6.
Note $|f(Z_6) - f(Z_5)| < 0.0002$

2. THE BEHAVIOUR OF LOCAL MINIMIZATION ALGORITHMS

Probably the most frequently used approach to the problem of locating a global minimum is to select a series of starting points $x^{(0)}$ at random

and to use a local minimization algorithm from
each starting $x^{(0)}$.

The Multistart Algorithm: MS

STEP 1: Select $x^{(0)}$ at random.

STEP 2: Start a local minimization algorithm
from $x^{(0)}$ with tolerance ε.

STEP 3: The user decides if he is satisfied
that this is the global minimum and
if so stops.

STEP 4: Return to STEP 1.

We should note that STEP 3 is difficult and
will require further discussion.

As this is the most frequently used method,
we shall consider the behaviour of local minimization algorithms in some detail. For some functions
and most algorithms, starting points exist from
which the search will leave the set S. For any
algorithm let us define this as the set U. If the
search remains in S then many algorithms exist
which theoretically converge to a point in A_g,
i.e., in the ε_g neighbourhood of a local minimum
or saddle point and we shall restrict our consideration to such algorithms. It is important to
remember that other algorithms do exist that do
not possess this property (*i.e.*, the Nelder and
Mead Simplex Algorithm (1965), and unsafeguarded
Variable Metric algorithms, Dixon (1974)). We
shall define the sets S_{Mj} and S_{Zj} for a particular
algorithm as consisting of those points $x^{(0)}$ for
which that local minimization algorithm terminates
in an ε_g neighbourbood of M_j or Z_j, respectively.
The sets S_{Mj} and S_{Zj} may well depend on the values
given to parameters internal to the routine,
but for a constant algorithm the sets U, S_{Mj} and
S_{Zj} will cover S. From the literature the impression could be gained that these sets S_{Mj}, S_{Zj} are
each connected, for most algorithms however this
is not the case. S_{Mj}, S_{Zj} will not be connected

sets and for some algorithms they may even contain isolated points.

To be able to state anything useful about the behaviour of algorithm MS we must specify the algorithm used at step 2.

Let us first consider the behaviour of the local minimization algorithm which follows the trajectory

$$\dot{x} = \frac{-g(x)}{1 + \|g(x)\|}$$

g being the gradient of the function and the differentiation being with respect to an arbitrary monotonically increasing parameter. We will term this the gradient trajectory algorithm (GTA). For this algorithm the sets S_{M_j} are connected and the sets S_{Z_j} are of zero measure and form the boundaries of the sets S_{M_j}.

Definition:

The *Region of attraction* R_j of a local minimum M_j is its convergence set S_{M_j} for the gradient trajectory algorithm.

The regions of attraction of the local minima of the six hump camel back function are shown in Fig. 3.

The GT Algorithm is however well known to be an inefficient method for finding local minima. Most efficient local minimization routines are of the descent type (Wolfe (1970), Dixon (1974)) in which

$$x^{(k+1)} = x^{(k)} - \alpha_k p^{(k)}$$

where $p^{(k)}$ is chosen to satisfy

$$I \qquad p^{(k)^T} g^{(k)} > \varepsilon_1 \|p^{(k)}\| \|g^{(k)}\|$$

and α_k is chosen to satisfy

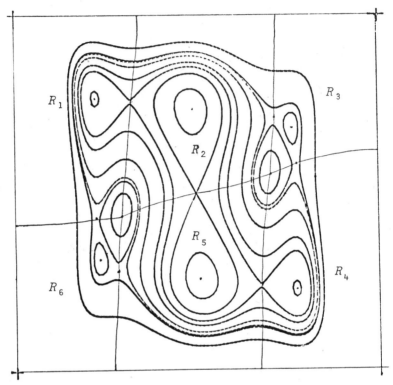

Fig. 3. Regions of attraction

$$II \quad f(x^{(k)}) - f(x^{(k+1)}) \geqslant \varepsilon_2 \alpha_k \, \|p^{(k)}\| \, \|g^{(k)}\|$$

$$III \qquad g^{(k+1)^T} p^{(k)} < \varepsilon_3 g^{(k)^T} p^{(k)}.$$

For algorithms which satisfy these three conditions convergence to a local minimum or saddle point is proven. However, to be able to determine which local minimum they will locate, we must restrict the linear search still further by insisting that α_k is also chosen to satisfy

$$IV \quad \alpha_k = \min_{\alpha > 0} \{\alpha : g(x^{(k)} - \alpha p^{(k)})^T p^{(k)} = \mu g(x^{(k)})^T p^{(k)}\}$$

for the appropriate value of μ. This implies that no local unidirectional maxima are jumped during

404

the linear search and is identical to the condition
introduced by Stoer (1974) when determining the
rate of superlinear convergence of variable metric
algorithms. Unfortunately as far as we are aware
no implemented descent method satisfies this criterion. Even with this criterion the sets S_{M_j} for
descent algorithms can contain disconnected
regions. It will be useful at this point to introduce a few additional definitions.

Definitions

A *level set* $L(v)$ can be defined by
$$L(v) = \{x : f(x) < v\}.$$

A *component* of $L_c(v)$ is a connected subset
of $L(v)$.

A *contour* $c(v)$ is the boundary of a component
of $L(v)$.

In terms of these definitions we may state
the following theorems.

THEOREM 1

If a descent algorithm, which satisfies conditions I to IV, is started from a point $x^{(0)}$ that
is inside a closed contour completely contained in
R_j then the algorithm will converge to M_j.

THEOREM 2, Trecanni (1971, 1975)

Let v_c be the maximum value of v with an
associated closed contour completely contained in
R_j, then $c(v_c)$ contains a saddle point Z_j which is
itself on the boundary of R_j.

Definition

The *basin* B_j of M_j can be defined by
$$B_j = \{x : f(x) < f(Z_j), \ x \in R_j\}.$$

405

Corollary

If a descent algorithm, which satisfies conditions I to IV is started from a point $x^{(0)}$ in B_j then it will converge to M_j.

The basins of the 6 hump camel back function are shown in Fig. 4.

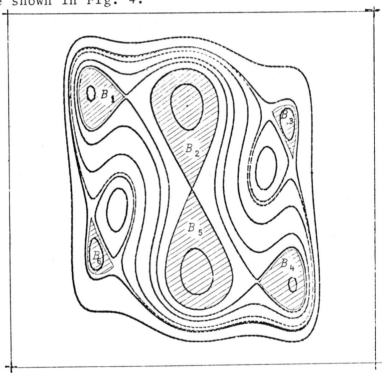

Fig. 4. Basins

Whilst discussing the behaviour of descent algorithms it is perhaps advisable to stress that when they are started from a point in R_j but not in B_j they do not necessarily converge to M_j. If we wish to consider their behaviour further we can introduce the idea of a combined basin.

THEOREM 3

If the closed contour $c(v)$ only contains a finite number of minima M_j and saddle points Z_j, and if it is also contained in the regions of attraction of these minima, then a descent algorithm started from a point within it will converge to one of the minima or saddle points contained therein.

Definition

If v_c is the largest function value such that $c(v_c)$ contains a set of minima M_j and saddle points Z_j and is contained in their regions of attraction, then we may define their combined basin, as the component of the level set whose boundary is $c(v_c)$.

Corollary 1

The boundary of a combined basin $c(v_c)$ contains at least one saddle point.

Corollary 2

Any descent algorithm started at a point in the combined basin will converge to a stationary point contained therein.

On Fig. 5, M_2 M_5 and Z_7 define one combined basin, and Z_1 and Z_4 are saddle points on its boundary. Similarly, M_1, M_2, M_4, M_5, Z_1, Z_4, Z_7 define a larger combined basin.

The above discussion has ignored the region near the boundaries of S. When the set U for the GTA approach is non-zero and when the contours $c(v)$ intersect S some of the above results probably require modification.

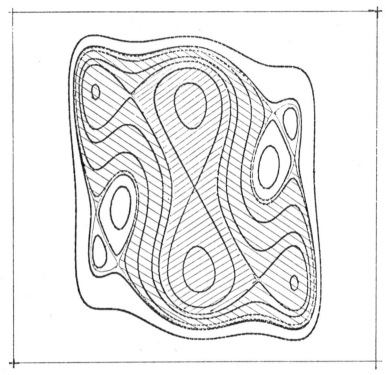

Fig. 5.   Combined basins

3. SOME PROBABILISTIC RESULTS

Many techniques proposed for global optimization contain a random element. These techniques
rely for convergence essentially on the fact that,
if a sufficient number of points are distributed
uniformly but at random over any set of finite
measure, the likelihood of any particular subset
with positive measure containing at least one
point tends to unity as the number of points tends
to infinity. Mathematically if A is a subset of S
with a measure m such that

$$\frac{m(A)}{m(S)} \geqslant \varepsilon > 0,$$

and $p(A,n)$ is the probability that at least one
point of the sequence of n random points lies in A

then $$\lim_{n\to\infty} p(A,n) = 1.$$

We note that a proof of convergence of this kind for any iterative process does not imply that the result of any one iteration of the process containing a finite number of points need yield a good approximation to the solution.

Brooks (1958) and Anderssen (1972) discuss the properties of the pure random search algorithm PRS in which $f(x)$ is evaluated at N uniformly distributed random points in S. Their aim is to determine a value V such that not more than a predetermined proportion, α, of the locations of x in S have $f(x) < V$. They show that if P is the probability that such a V has been found then N, P and α are related by

$$P = 1 - (1 - \alpha)^N, \qquad (3.1)$$

Anderssen discusses further properties of this approach and introduces the idea of hypothesis testing. An estimate of the global minimum is only accepted when additional testing confirms the estimate at a given confidence level. He stresses that when insufficient *a priori* information is available to prove that a given estimate is the global minimum, then it is necessary to collect sufficient evidence about the nature of the problem so that the uncertainty in either the size or the error in the location of the global optimum is minimal. Numerically if we require $A_F = \{x : f(x) - f(x^*) < \varepsilon_F\}$ then ε_F is likely to be small, say 0.0001. If we require a high probability of success, $P = 0.99$ say, then a simple calculation indicates that we would require 20 000 function evaluations.

Now let us consider the multistart method, combined with the GT Approach. If $x^{(0)} \in R_j$ then it converges to M_j. So effectively the probability of locating the global minimum with one GT minimization is

$$\alpha_R = \frac{\text{Measure } R_*}{\text{Measure } S.}$$

Typically we may well have $\alpha_R = 0.1$ and a similar sum indicates that 40 GT minimizations would be required for a probability of success at the 0.99 level. As the GT method is usually very inefficient there is no guarantee that this would require less than 20 000 function evaluations.

A similar calculation could be undertaken for the multistart method using an efficient descent algorithm. Here the value of α will depend on the area of the basin and so will be smaller than α_R hence a larger number of more efficient local minimizations would be required to achieve the same probability of success.

As all these techniques are quite expensive, heuristic methods aimed at reducing the number of function evaluations have been suggested. One of the most interesting of these is Hartman's Method S2 (1972).

Algorithm S2

STEP 1 $v^+ = + \infty$,

STEP 2 Select $x^{(0)}$ at random

STEP 3 If $f(x^{(0)}) > v^+$ return to 2

STEP 4 Perform local minimization from $x^{(0)}$ to M_j and set $V^+ = f(M_j)$.

STEP 5 Return to Step 2.

Note that although no stopping rule is specified one can easily be incorporated. In this algorithm the local minimization is only undertaken when there is a certainty of improvement. This modification can often be beneficial, but it is easy to construct a function on which it is unlikely to be successful. Assume we have two local minima M_1 and $M_2, f(M_1) > f(M_2)$, assume also we are using the GT algorithm and $R_1 \gg R_2$ so that there is a high probability of M_1 being located first. Then the second minimum will only be found if a point is obtained at random in the level set $f(M_1)$.

410

If the measure of this level set is small then the chance of success will be small.

A more sophisticated approach based on the calculation of the function value at random points was suggested by Chichinadze (1967). He introduced a probability function $P(v)$ where P is the probability of $f(x) < v$ if x is selected uniformly at random in the space S. We note that the global minimum function value occurs
when $P(v) = 0$,
and the global maximum
when $P(v) = 1$.
The function $P(f)$ is of course not available, but if we calculate $f(x)$ at n randomly distributed points and let $m_i(v_i, n)$ = number of points at which $f(x) < v_i$ then the ratio m_i/n is an approximation to $P(v_i)$ which improves as $n \to \infty$. Indeed Chichinadze states that

$$\left| \frac{m_i}{n} - P(v_i) \right| = O\left(\frac{1}{n}\right).$$

To approximate the solution of $P(v) = 1$ he suggests fitting a linear combination of orthogonal polynomials $P_j(v)$ to the ratios m_i/n, *i.e.*, λ_1 are determined to minimize some measure of the agreement between $\sum \lambda_j P_j(v_i)$ and m_i/n at the different levels v_i. The root f^* of the equation

$$\sum \lambda_j P_j(v) = 0,$$

is then determined to obtain an estimate of the global minimum value of $f(x)$.

Following Chichinadze's original approach F. Archetti (1975) presents the theoretical basis of this approach in a more rigorous manner and describes his experience using the agreement between v^* and the result of a local minimization from the best point as a test for acceptability. Archetti ran this routine on the six hump camel back function six times using different sequences of random points in S, he reported a successful location of the global minimum on each occasion,

411

and only required 105 ± 25 function evaluations to satisfy his stopping criteria.

At the conference A. Curtis pointed out that the choice of a set of polynomial functions $P_j(v)$, was unfortunate. Assume we have a local minimum M_i and in the neighbourhood of the minimum the contours are quadratic, then if $v > f(M_i)$ corresponds to a small linear measure h we have

$$v = f(M_i) + O(h^2)$$

and

$$P(v) = \frac{m(A(v))}{m(S)}, \quad A(v) = \{x : f(x) < v\}$$

therefore $m(A(v)) = m(A(f_{M_i})) + O(h^n)$.

and $P(v) \doteqdot \begin{cases} P(f(M_i)) & , v < f(M_i) \\ P(f(M_i)) + O(v - f(M_i))^{n/2}), & v > f(M_i). \end{cases}$

There is therefore nonpolynomial behaviour of $P(v)$ near every local minimum, including the global minimum, and hence the choice of a polynomial to approximate v^+ is perhaps unfortunate. Again if no points are found during the random search in some region associated with the global minimum, then any acceptability test based on his comparison can be satisfied by any local minimum. Further work would therefore be required on this approach before any confidence could be placed on the results.

A related algorithm, in which directions rather than points are chosen at random, is the random search algorithm.

Algorithm RS

    STEP 1 Select $x^{(0)}$, $k = 0$

    STEP 2 Select $p^{(k)}$ at random

    STEP 3 $x^{(k+1)} = x^{(k)} + \alpha p^{(k)}$ where α is
            determined by a line search.

STEP 4 $k = k+1$ and return to 2.

The theoretical performance of this routine depends upon the line search undertaken at Step 3. Gaviano (1975) has shown that if the line search locates the unidirectional global minimum then

$$\lim_{k \to \infty} P(x^{(k)} \in A_F) = 1.$$

On the other hand less exact line searches can cause the technique to become trapped by other local minima. The most well known implementation of this technique is due to Bremmermann (1972) who fits a quartic to five points along the line in each linear search. Bremmermann's proof of ulti-mate convergence to the global minimum on n-dimen-sional quartic functions is a particular instance of Gaviano's more general result.

Our numerical experience with this routine is that it can be very erratic even on quartic functions such as Rosenbrock's function. The results of three typical runs on the six hump camel back function are given below.

	Run 1	Run 2	Run 3
Starting Point	0,0	1.5, 0.75	2.5, 1.5
Iteration Number	Function Value	Function Value	Function Value
5	-0.9408	-0.0729	-0.15706
15	-1.03159	-0.9446	-0.21282
25	-1.03161	-1.0022	-0.21538
35		-1.0313	-0.21538
45		-1.03161	-1.0249
150	-1.031628	-1.031628	-1.031628

Note 1   Each iteration requires 5 function evalua-tions.

413

Note 2 Silvestrowicz (1974) has shown that the
rate of convergence of this method to a
local minimum depends upon the condition
number at the Hessian. As the ultimate
global Hessian of the six hump camel back
function is almost circular these results
display a far faster ultimate convergence
rate than is usual.

4. DESCENT FROM A LOCAL MINIMUM

The heuristic method due to Hartman described
above can also be considered as an example of the
next class of methods we will discuss, namely
those that attempt to descend from a local minimum.
You will recall that in that technique after a
local minimum had been found having a function
value v^+, then a further search was only undertaken when a point was located with $f < v^+$. In
this routine the search for a better point was
purely random. In contrast to this Goldstein and
Price (1972) suggested a deterministic method for
locating such a point. Their method has fast convergence properties on polynomial functions of one
variable, but little is known of its performance
on more general functions.

Bocharov (1962) suggested that after a local
minimum had been located further progress could be
made by undertaking global searches in random
directions from that minimum. On n-dimensional
polynomial functions the Goldstein and Price
approach would be ideal for this one-dimensional
subproblem. If we then undertake a local n-dimen-
sional search whenever a point $f < v^+$ is located,
then it is easy to show that the probability of
ultimately locating the global minimum is unity.
The number of random directions required to locate
a point such that $f < v^+$ will depend on the shape
of the level set $f = v^+$. If it presents a narrow
face to the known local minimum then the method is
predictably inefficient.

The methods of Hartman (1972) and Opacic

414

(1973) can be viewed as heuristic modifications of this method that are intended to be more efficient, but in theory lessen the probability of success.

5. THE EFFECT OF BOUNDED DERIVATIVES

If we can obtain Lipschitz bounds on the function, *i.e.*, if a constant L exists and its value is known s.t.

$$|f(x_1) - f(x_2)| < L \|x_1 - x_2\| \text{ all } x_1, x_2 \in S \quad (5.1)$$

then finite methods of obtaining global minima can be constructed.

For the six hump camel back function

$$\frac{\partial f}{\partial x_1} = 8x_1 - 8.4x_1^3 + 2x_1^5 + x_2$$

$$\frac{\partial f}{\partial x_2} = x_1 - 8x_2 + 16x_2^3$$

and it is simple to show that when S is defined by

$$- 3 \leqslant x_1 \leqslant + 3, \quad - 1.5 \leqslant x_2 \leqslant + 1.5$$

then $L = 290$ is the lowest possible bound. The simplest and earliest way of using this information is in the grid search technique.

In this routine $f(x)$ is evaluated at the corners of an equispaced grid.
For example:
If we adopt a grid of width $a = 0.1$ this entails 1891 function evaluations for our example problem. Let the function evaluations be f_j and let f_j^* be the greatest function value at a vertex of the square Q_j, and f_G^* be the lowest function value obtained at a point of the grid. Then if

$$x \in Q_j$$

$$f(x) > f_j^* - aL\sqrt{n}$$

and if this is greater than f_G^* then the global minimum cannot lie in Q_j. On this example $f_G^* \sim - 1$ and so any Q_j for which $f_j^* > 41$ can be ignored.

415

This region however consists of a few squares well outside the largest contour shown f = 3. To eliminate all squares outside that contour we must use a grid of approximately a = 0.01 {*i.e.*, $n \sim 180000$ function evaluations} and although this number can be reduced by decreasing a iteratively, the number of function evaluations required is still prohibitive. And even then we only know that the minimum lies somewhere inside the contour f=3 and we would not have located it.

More efficient algorithms based on the use of L have been constructed by Etvushenko (1970) and Shubert (1972) but both suffer from the "curse of dimensionality" and are only really practical in 2 dimensions.

As these algorithms are efficient in one dimension they provide a means of obtaining the global minimum along any line in n-dimensional space. As the concept of a linear search is important in local minimization, we may anticipate that it will play a part in global optimization. To illustrate the concept behind this technique Shubert's method will now be described.

Shubert considered functions defined on an interval $a \leqslant x \leqslant b$ and satisfying condition 5.1, with the bound L again assumed to be known. The basic idea behind his approach can easily be extended to functions of many variables although it would not be practical because of the amount of computation involved. Shubert does not assume any *a priori* accuracy ε. Instead a bound on accuracy is calculated at each iteration. A location of the minimum points can also be estimated at any iteration.

An initial point $x \varepsilon < a, b > $ (e.g., $x^{(0)}$ = $(a + b)/2$ is chosen and $M^{(0)}$ = $f(x^{(0)})$ computed. A piecewise linear function $\phi^{(0)}(x)$ is built going down from $(x^{(0)}, f(x^{(0)}))$ in each direction with the constant slope L. It follows from 5.1 that

$$\phi^{(0)}(x) \leqslant f(x)$$

and also

$$\phi^{(0)}(x^{(0)}) = f(x^{(0)}).$$

Any point of absolute minimum of $\phi^{(0)}(x)$ is chosen as the next evaluation point $x^{(1)}$. The new estimate of the global minimum value is given by $M^{(1)} = \min(f(x^{(1)}), M^{(0)})$. A new piecewise linear function $\phi^{(1)}(x)$ is obtained by the following modification to $\phi^{(0)}(x)$. From the point $(x^{(1)}, f(x^{(1)}))$ a line of constant slope L is drawn down in each direction until it crosses $\phi^{(0)}(x)$ (or reaches an end of $< a, b >$). Clearly,

$$\phi^{(0)}(x) \leqslant \phi^{(1)}(x) \leqslant f(x)$$

$$\phi^{(1)}(x^{(i)}) = f(x^{(i)}) \quad i = 0,1.$$

Generally at the start of the kth iteration we have a continuous function $\phi^{(k-1)}(x)$ consisting of linear pieces with slope L. This function may be represented by the sequence of its (local) minima

$$(t_1, z_1), \ldots, (t_{r_{k-1}}, z_{r_{k-1}})$$

where

$$z_s = \phi^{(k-1)}(t_s), \quad s = 1, \ldots, r_{k-1},$$

ordered by decreasing $z_s : z_1 \geqslant z_2 \geqslant \ldots \geqslant z_{r_{k-1}}$. The smallest function value found until now,

$$M^{(k-1)} = \min_{0 \leqslant i \leqslant k-1} f(x^{(i)})$$

is also known. A typical situation after 4 steps is shown in Fig. 6. The algorithm therefore consists of the following iterative process.

kth iteration

$$1° \quad x^{(k)} = t_{r_{k-1}}; \quad f^k = f(x^{(k)})$$

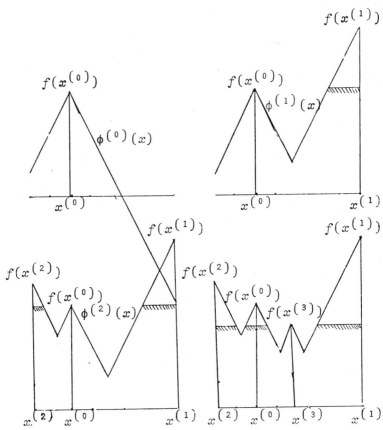

Fig. 6. Four iterations of Shubert's technique

$2°$ $M^{(k)} = \min (M^{(k-1)}, f^k)$

$3°$ To build $\phi^{(k)}$

(*i*) drop $(t_{r_{k-1}}, z_{r_{k-1}})$;

(*ii*) add $(t_1, z_1), (t_r, z_r)$, where

$$z_1 = z_r = (z_{r_{k-1}} + f^k)/2$$

418

$$t_1 = t_{r_{k-1}} - (f^k - z_{r_{k-1}})/2L$$

$$t_r = t_{r_{k-1}} + (f^t - z_{r_{k-1}})/2L$$

(*iii*) reorder a new sequence by decreasing z's.

Because

$$\phi^{(k)}(x) \leqslant f(x)$$

it is clear that

$$z_{r_k} \leqslant f(x^*) \leqslant M^{(k)}.$$

Therefore a minimum point x^* must be contained in one or more of the intervals

$$< t_s - (z_s - M^{(k)})/L, \; t_s + (z_s - M^{(k)})/L >$$

such that $z_s \leqslant M^{(k)}.$

It may be noticed that no points (t_s, z_s) with $z_s \geqslant M^{(k)}$ need to be stored as they do not influence further calculations.

This is an efficient linear search process when the bound L is known. Similar routines based on bounds on the second derivatives have been proposed by Brent (1973) and a special technique suitable for a particular situation is discussed by Beale (p.459).

All these techniques build up a set $E^{(k)}$ which has been eliminated from further consideration by iteration k as it has been shown that the the global minimum cannot occur in this region of space. The concept of building up such a set $E^{(k)}$ is important as it provides a deterministic basis for global optimization.

In the next section another space covering

technique is described which builds up sets $E^{(k)}$ by considering the regions of attraction of the GT Algorithm. This approach like those described above is only efficient for small dimensional problems. A heuristic technique based on these concepts has been suggested by Hartman (1973). This consists of a sequence of local minimizations from different starting points but instead of choosing random points as in the MS Algorithm the following strategy is used. The space is covered by a grid and if the search enters a hypercube of a grid this hypercube is placed in $E^{(k)}$, the starting point is then chosen by maximizing the distance from $E^{(k)}$. If the GT Algorithm were used, and the grid chosen consistently with a Lipschitz constant, then the set $E^{(k)}$ found in one search would form a subset of the Region of Attraction of that minimum. The method of choosing $x^{(0)}$ might well locate a different region of attraction and hence this could be an efficient way of locating most of the minima. There seems however to be no guarantee of success until the set $E^{(k)}$ covers S and this would apparently imply calculating the function value at least once in each grid cube, which was definitely not Hartman's intention.

6, THE GROWING ELLIPSES METHOD

This approach was originally suggested by Trecanni, Trabattoni and Szegö (1972). It is based upon the Liapunov Stability theory applied to the GT Algorithm

$$\dot{x} = \frac{-g(x)}{1+\| g(x) \|}$$

The basic idea is that if a local minimum M_j is known then its region of attraction R_j could all be excluded from further consideration. However the region of attraction is not known and they suggest constructing an approximation of this region based on elliptical shapes. Once constructed their approximation covers the basin B_j,

i.e., $\qquad R_j \supset E_j \supset B_j.$

For ease we shall transform the local minimum to the origin. Then the appropriate stability theorems can be expressed as follows.

Let $v(x)$ be a positive definite function (*i.e.*, $v(x) > 0$ for $x \neq 0$, $v(0) = 0$). Consider the component of a level set $\{x : v(x) < v_0\}$ containing the origin and assume that

$$\dot{v}(x) < 0 \quad (\dot{v}(x) = (\text{grad } v)^T \dot{x})$$

at all points of this component except the origin. Then the GT system will converge to the origin from all such points.

Trecanni *et al.* consider the elliptical function

$$V(x) = \tfrac{1}{2} x^T H x$$

so that

$$\dot{V}(x) = x^T H \dot{x} = -\frac{x^T H g}{1 + \| g(x) \|}.$$

$\dot{V}(x)$ is therefore negative, whenever $x^T H g > 0$. As they initially choose H to be the Hessian of the function at the minimum, $x^T H g > 0$ for small values of V_0, and will continue to be positive until either (*i*) $g = 0$, *i.e.*, a stationary point is found, or (*ii*) g is tangential to the surface $V = \text{const.}$ Such ellipses are therefore a subset of R_j. The TTS method can be described in terms of four steps.

STEP 1 The external penalty function method is used to minimize

$$V_0 = x^T H_0 x$$

s.t. $\quad\quad\quad\quad x^T x \geqslant t$

and $\quad\quad\quad\quad \dot{V}_0 \geqslant 0.$

The first constraint removes the origin which would otherwise be the solution to the problem and this method therefore determines the smallest value of $V_0(x)$ on which a point x_c exists with

$$\dot{V}_0(x_c) = 0.$$

The use of the external penalty function method locates a local solution of the constrained problem described above, the global solution is theoretically required. If v_1 is value of V_0 at this local minimum, they prove that provided all local minima have distinct function values the global minima could be found by solving a sequence of problems of the form:

$$\min \quad V_0 \;=\; x^T H_0 \, x$$

$$\text{s.t.} \quad x^T x \;\geqslant\; t$$

$$\dot{V}_0 \;\geqslant\; 0$$

$$v_L \;-\; V_0(x) \;-\; E \;\geqslant\; 0$$

and terminating the sequence when the problem becomes infeasible. This in itself provides an interesting approach to the original problem which does not appear to have been considered in detail. The main difficulty appears to be in determining when the sequence becomes infeasible.

STEP 2

The next step is to find the largest level set of $f(x)$ contained in $V_0(x_c)$. Let us denote the solution to this problem as z_0, then if $g(z_0) = 0$, it can be shown that z_0 is the saddle point on the boundary of both R_j and B_j.

STEP 3

The approximation to the region of attraction R_j is extended by constructing another quadratic function

$$V_k(x) \;=\; x^T H_k \, x/2$$

H_k being defined to have an eigenvector along the line $0z_{k-1}$.

The minimization performed now is

$$\min \ V_k(x)$$

s.t. (a) $\dot{V}_k(x) \geqslant 0$

(b) $x^T z_{k-1} \geqslant 0$, this restricts the minimization to the half of the quadratic surface on the same side at z_{k-1}

(c) a constraint that restricts the minimization to points external to the surfaces $V_j(x_c)$, $j = 1, \ldots, k-1$

(d) a constraint that prevents the plane formed by (b) cutting the earlier surfaces.

If x_c is now the solution to this problem, it can be shown that that part of the level set $V_k(x_c)$ which satisfies (b), (c) and (d) can be added to E_{k-1} to form E_R, a better approximation to R_j.

STEP 4

This repeats STEP 2, but the minimization is now performed on the more complex surface formed by the boundaries of E_k. Again we let the solution of this problem be z_k and if $g(z_k) = 0$ we have located the desired saddle point.

They showed the following theoretical results for this algorithm.

Theorem 1

The sequence of steps outlined in step 1 will converge after a finite number of steps.

Theorem 2

The sets E_k are contained in the region of attraction R_j, and the components of the level set $f(z_k)$ in the basin B_j.

Theorem 3

If the sequence z_k is finite it terminates at a saddle point. The typical situation is shown in Fig. 7. If the sequence $f(z_k)$ is unbounded then the origin is the only minimum.

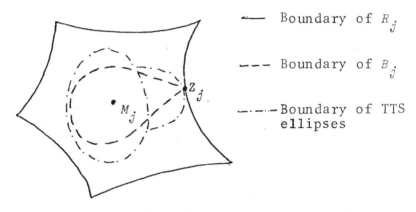

— Boundary of R_j

--- Boundary of B_j

—.—.—Boundary of TTS ellipses

Fig. 7. TTS method, typical situation

If the sequence z_k is infinite and certain conditions on the eigenvalues of H_k can be met then the limit $z_k \underset{k \to \infty}{=} Z^*$, the required saddle point.

C.R. Corles has, however, shown that situations arise in which the saddle point cannot be reached by the above method and hence on these functions it must be impossible to comply with the conditions on the eigenvalues of H_k. This situation will arise on any function that has points at which the gradient points towards the local minimum. One such point occurs on Rosenbrock's function at approximately (+0.01, +0.01) and a second example is illustrated in Fig. 8.

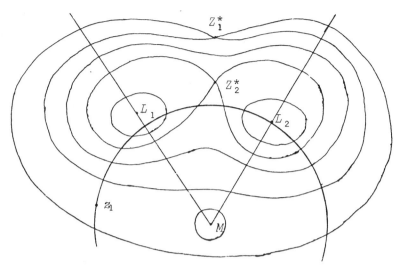

Fig. 8. TTS method, counter-example. No ellipse,
centred at M and growing from z_k, can enter
the cone L_1ML_2 and hence the sequence z_k cannot
tend to Z_1^* (or Z_2^*)

C.R. Corles (1975) has implemented an algorithm
for global optimization in 2 dimensions based on
the above approach. He solved the minimization
problem at steps 1, 2 and 4 in a very different way
from that envisaged by Trecanni *et al.*, choosing
to use a variable reduction technique in place of
the penalty function method. In addition as he
had found a counter-example to their policy of
keeping the centres of all the ellipses at the
local minimum, he chose to centre them at the
points z_k. Counter-examples to this policy are
also now available. His algorithm continued the
search after the saddle point had been located.
He allocated a high constant function value to
points inside the region already excluded and then
undertook a local minimization from a point near
the saddle point but outside the set E. This usually locates a different local minimum. This set
will either terminate at a new saddle point or
locate one previously obtained (Fig. 9.).

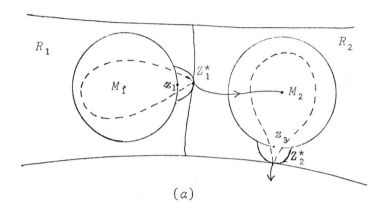

(a)

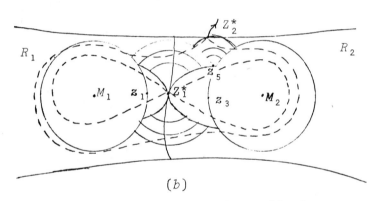

(b)

Fig. 9.   Treatment of saddle points (Corles
method):   (a) disjoint basins; (b) combined basins

In the former case we can continue as before,
in the latter the local minimization algorithm is
inappropriate as the saddle point is the minimum
of the modified function.   This is the circumstance
in which a combined basin was defined earlier and
by starting growing a circle centred at the saddle
point a set can be constructed which is contained
in the combined region of attraction and contains
the combined basin.

426

   Corles successfully located all the minima
and saddle points of the 6 hump function (see Fig.
10) and also of a number of other functions, using
this technique.

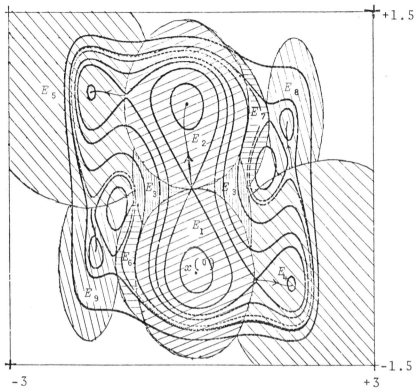

Fig. 10.   Six hump camel back function.   The 9
           ellipses grown by Corles   algorithm shown
           approximately to previous scale.
           "───► local minimization routine."

       The technique is not yet rigorous and diffi-
culties occur in the region of local maxima.   Further work on this technique is in progress.

7. BRANIN'S TRAJECTORY METHOD

   Consider the curves

$$g(x) = k \; \underset{\sim}{c} \qquad\qquad (7.1)$$

where $\underset{\sim}{c}$ is a constant vector and k a variable scalar. For every value of $\underset{\sim}{c}$, every critical point
x^* lies on each of these curves (with $k = 0$).
Moreover, if the Hessian matrix $G(x^*)$ is non-singu-
lar then there is an arc of each trajectory crossing x^*. Hence if the complete curve could be followed then all critical points of $f(x)$ could be
located. Difficulties arise when the curve consists of several disconnected pieces as they cannot
all be explored from a single starting point.

The curves are solutions of the differential
equation

$$\dot{g} = -g$$

or $\qquad\qquad \dot{Gx} = -g$

or $\qquad\qquad \dot{x} = -G^{-1}g.$

A change in our parameter t enables this to
be written as

$$\dot{x} = -(\text{adj } G).g.$$

Our trajectories therefore terminate both
(a) when $g = 0$, *i.e.*, stationary points of $f(x)$
(Branin (1972) termed these essential singularities)
and (b) when adj G is singular and g is parallel to
the eigenvector corresponding to the zero eigenvalue (extraneous singularities).

Arcs of the curve can therefore be obtained
by integrating the differential equations

$$\dot{x} = \pm \text{ adj } G.g$$

from any starting point. The sign must be changed
on crossing a singularity and some extrapolation
device adopted in that region.

On the basis of his experimental evidence
Branin made two conjectures (a) an extraneous singularity cannot be of a focus type, *i.e.*, trajectories cannot spiral into it; (b) in the absence

of extraneous singularities all the curves 7.1 are connected in the sense that an integration from any starting point will locate all the stationary points.

J. Gomulka (1975) has investigated the nature of extraneous singularities and confirmed the truth of the first of these conjectures. She showed that all trajectories are either closed curves or go from one singularity to another or to infinity (*i.e.*, leave *S*). The truth of the second conjecture is not yet determined, but Treccani (1975) has produced an example of a simple function whose contours are homeomorphic to a sphere which still possesses separate arcs of the same trajectory. It does, however, have an extraneous singularity. Examples of Branin's curves on the 6 hump camel back function are shown in Fig. 11, one passes through all stationary points in a connected trajectory, the other possesses three disconnected sections only two of which are shown.

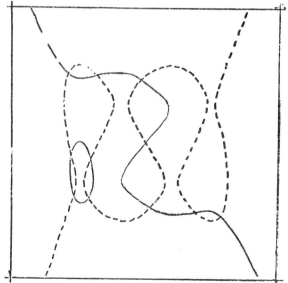

Fig. 11. Two Branin trajectories of the 6 hump camel back function: trajectory 1 -----
one continuous arc; trajectory 2 ———
three disjoint arcs (2 shown)

The problem of determining whether a function has trajectories with disconnected sections and locating these sections remains unsolved.

One way of attempting this was implemented by J.W. Hardy (1975). Restricting our discussion to the 2-dimensional problem, his algorithm undertook a further integration along the orthogonal trajectory at each saddle point. This heuristic extension was successful on fifteen of the sixteen functions tested and is illustrated in Fig. 12.

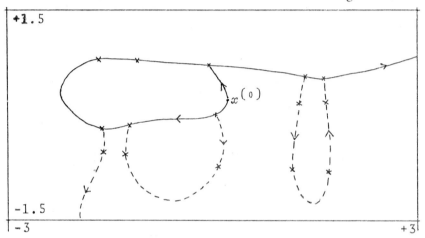

Fig. 12. An example of Hardy's method. 6 hump camel back function (scale changed):
——— initial trajectory and reflection;
---- successful orthogonal trajectories

The unsuccessful function was the first test function proposed by Goldstein and Price (1972). Two Branin trajectories of this function are shown in Fig. 13, each consists of three parallel straight lines. The global minimum lies at the central intersection and was never located.

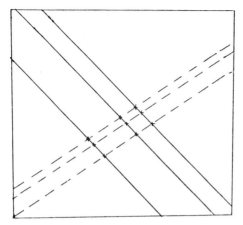

Fig. 13. Two sets of Branin's trajectories for the
Goldstein and Price function:
——— $c = \binom{-2}{3}$; ---- $c = \binom{1}{1}$

8. SUMMARY

The present situation in the problem of locating the global minimum of a non-convex function
is not all that satisfactory.

For problems in one dimension the Goldstein
and Price routine is ideal for polynomial functions and Shubert's or Evtushenko's routines can
be used on more general functions when the Lipschitz
constant is known.

The space covering techniques based on the
growing ellipses method of Trecanni *et al.* are
available for two-dimensional problems and
Evtushenko's routine can also be applied to these
when the Lipschitz constant is known.

There are no deterministic algorithms available for problems with more than 2 variables.
The concept of trajectories connecting all the
stationary points is possibly the most promising
approach to this problem but no extension has yet
been suggested which is guaranteed to locate the
global minimum. For this reason probabilistic

approaches are common and a number were discussed earlier which will theoretically locate the global minimum if continued for a long enough time.

In most practical situations the cost of the computation is limited, as otherwise the possible benefit of locating the global minimum would be less than the cost of the computation. Mockus (1971, 1975) has suggested that in these circumstances it is inappropriate to use a technique that only approaches the correct answer with infinite time. He suggests that it is then more appropriate to seek a technique that will obtain a good estimate in a given number of function evaluations. His method assumes that each function evaluation is very expensive and that it is therefore worthwhile undertaking a considerable overhead calculation between each function evaluation. Even on simpler less costly functions the change in emphasis may well lead to beneficial results and deserves further consideration.

ACKNOWLEDGEMENT

Dr. J. Gomulka, one of the authors, acknowledges the sponsorship of the SRC for the period in which this report was written.

All the authors attended the Workshop in Global Optimization, held at the University of Cagliari in October 1974, and acknowledge the influence of the many useful discussions they had there with the other participants. The proceedings of that Workshop are to be published by North Holland Press with the title "Towards Global Optimisation" and edited by Professor G.P. Szegö and L.C.W. Dixon.

REFERENCES

Anderssen, R.S. (1972) "Global Optimisation," *in* "Optimisation," *eds*. R.S. Anderssen, L.S. Jennings and D.M. Ryan, University of Queensland Press.

Archetti, F. (1975) "A Sampling Technique for Global Optimisation", *in* "Towards Global Optimisation," *eds.* L.C.W. Dixon and G.P. Szegö, North Holland Press.

Bocharov, N. and Feldbaum, A.A. (1962) *Automation and Remote Control,* **23**, No. 3.

Branin, Jr., F.H. (1971) "Solution of Nonlinear D.C. Network Problems via Differential Equations," Mem. Mexico 1971 Intemat. IEEE. Conference on Systems, Network and Computers, Oaxtepee, Mexico.

Branin, Jr., F.H. and Hoo, S.K. (1972) "A Method for Finding Multiple Extrema of a Function of *N* Variables," *in* "Numerical Methods of Nonlinear Optimization," *ed.* F.A. Lootsma, Academic Press.

Branin, Jr., F.H. (1972) "Widely Convergent Methods for Finding Multiple Solutions of Simultaneous Non-linear Equations," *IBM J. Res. Develop.,* 504-522.

Bremermann, H. (1970) *Mathematical Biosciences,* **9**, 1-15.

Brent, R.P. (1972) "Algorithms for Minimisation without Derivatives," Prentice Hall Press.

Brooks, S.H. (1958) *Op. Res.,* **6**, 244-251.

Chichinadze, V.K. (1967) *Eng. Cyb.,* No. 1, 115-123.

Corles, C.R. (1975) "The Use of Region of Attraction to Identify Global Minima," *in* "Towards Global Optimisation," *eds.* L.C.W. Dixon and G.P. Szegö, North Holland Press.

Dixon, L.C.W. (1974) "Nonlinear Optimisation: A Survey of the State of the Art," *in* "Software for Numerical Mathematics," *ed.* D.J. Evans, Academic Press.

Etvushenko, Y.G. (1971) *Zh. Vychisl. Mat. mat. Fiz.,* **11**, 6, 1390-1403.

Gaviano, M. (1975) "On the convergence of Random Search Algorithms for Minimisation Problems," *in* "Towards Global Optimisation," *eds*. L.C.W. Dixon and G.P. Szegö, North Holland Press.

Goldstein, A.A. and Price, J.F. (1971) *Maths. of Computation,* 25, 115.

Gomulka, J. (1975) "Remarks on Branin's Method for Solving Nonlinear Equations," *in* "Towards Global Optimisation," *eds*. L.C.W. Dixon and G.P. Szegö, North Holland Press.

Hardy, J.W. (1975) "An implemented extension of Branin's Method," *in* "Towards Global Optimisation," *eds*. L.C.W. Dixon and G.P. Szegö, North Holland Press.

Hartman, J.K. (1972) "Some Experiments in Global Optimisation," Naval Postgraduate School, Monterey, California NPS55HH72051A.

Hartman, J.K. (1973) "A New Method for Global Optimisation," Naval Postgraduate School, Monterey, California NPS55HH73041A.

McCormick, G.P. (1972) "Attempts to calculate Global Solutions that may have Local Minima," *in* "Numerical Methods for Nonlinear Optimisation," *ed*. F.A. Lootsma, Academic Press.

Mockus, J. (1975) "On Bayesian Methods of Optimisation," *in* "Towards Global Optimisation," *eds*. L.C.W. Dixon and G.P. Szegö, North Holland Press.

Nelder, J.A. and Mead, R. (1965) *Computer J.,* 5.

Opacic, J. (1973) *IEEE Trans. on Systems, Man and Cybernetics,* 102-107.

Shubert, B.O. (1972) *SIAM J. Numer. Anal.,* 9, 3, 379-388.

Silvestrowicz, J. (1974) "Numerical Optimisation of Nonlinear Functions Using Random Search Techniques," IBM, UKSC 0057.

Stoer, J. (1974) "On the convergence Behaviour of Some Minimisation Algorithms," *in* "Information Processing 1974," North Holland Press.

Trecanni, G., Trabattoni, L. and Szegö, G.P. (1972) "A Numerical Method for the Isolation of Minima", *in* "Minimisation Algorithms", *ed.* G.P. Szegö, Academic Press.

Trecanni, G. (1975) "On the critical points of continuously differentiable functions", *in* "Towards Global Minimisation", *eds.* L.C.W. Dixon and G.P. Szegö, Academic Press.

Trecanni, G. (1975b) "On the Convergence of Branin's Method: A Counterexample", *in* "Towards Global Minimisation", *eds.* L.C.W. Dixon and G.P. Szegö, Academic Press.

Wolfe, P. (1969) *SIAM Review*, **11**, 226-235; (1971) *SIAM Review*, **13**, 185-188.

MULTIPLE MINIMA IN A MODEL MATCHING PROBLEM

M.G. Brown

(British Aircraft Corporation)

SUMMARY

During the development of techniques for deriving estimates of the parameters of a system model so that the performance of the model should match the observed performance of the system, a series of control experiments was carried out to establish an upper limit to the "quality" of the match, in order to set realistic objectives for the ultimate matching exercise. The results of these control experiments were unsatisfactory, and it proved necessary to study the contours of the performance surfaces in detail. The results of this mapping exercise are presented and used to assess the benefits to be obtained by the application of optimization techniques.

During the research and development phase of a system development contract placed on the Guided Weapons Division of BAC, there was a contractual obligation to demonstrate that the performance observed during the system trials agreed with the predicted performance. This agreement could be demonstrated either by showing that outputs monitored during the trial were within fixed limits of the predicted signals, or that the system parameters were shown to be within their respective tolerance bands as specified by design. Only the second approach is discussed in this paper.

At the time this was a new obligation, and some disquiet was expressed about its viability.

The system in question had some unfriendly features.

1. The system was very nonlinear.

2. The system had time varying coefficients, whose rate of change could be of the same order as the frequency of the signals present.

3. The trials would be of short duration, in that each trial could compass only a few cycles of the frequencies of interest.

4. The system would not reach a steady state during the trial.

5. There would be cross-coupling present.

6. The system would be contaminated by noise, both generated in the system itself and in the monitoring equipment.

Because of these features, and because trials could not be carried out under perfectly controlled conditions, as they were subject to unrepeatability due to production tolerances, environmental parameters, etc., it was decided to conduct a series of controlled experiments to investigate the viability of the proposed technique for meeting the contractual obligation, and to establish a "yard stick of performance" to set an upper limit to the quality of the match to be expected with the trials data. If a good match could not be obtained on the controlled experiment in the laboratory, it would not be obtained with the results of a trial.

The technique chosen was to execute a model of the system with the input stimulus applied to the system during the trial, and to optimize the coefficients of the model to obtain the "best fit" between the output quantities observed on the trial and the same quantities as output from the system model. The system included 2 major loops, each with one input and three outputs, which were cross-coupled, but for the control experiments the cross-

437

coupling was ignored and only one major loop considered. This system is shown in Fig. 1, and it is
a measure of the complexity of this system that
only 4 of the unidentified blocks were linear with
constant coefficients, all the others were either
nonlinear or had time varying coefficients (or
both).

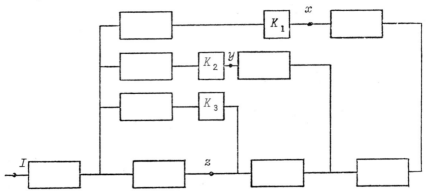

Fig. 1. System schematic

The optimization was to be carried out over 5 system parameters, and the performance criterion chosen was the integral of the error-squared, the
integration starting a short time after the initi-
ation of the experiment to allow transients to
settle.

The set of controlled experiments was carried
out on a simulation of the system. A simplified
model of the system was formulated, without the
cross-coupling but still with nonlinearities and
time varying coefficients, and 2 versions of this
model were simulated in parallel. One of these
was a reference model which had fixed value parameters, the other's parameters could be varied.
Both models were excited with the same stimulus,
and the appropriate outputs compared (see Fig. 2.)

The performance criterion was a weighted
integral of the error (squared) between respective
signals, and the experiment was first to fix the
parameters of the reference model, and to offset

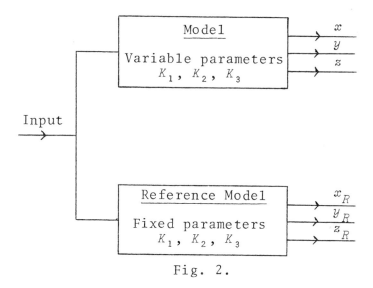

Fig. 2.

the values of the parameters of the variable model away from these reference values, then to carry out a series of dynamic runs on the simulation, changing the parameters of the variable model between runs under the control of an optimization algorithm.

The Nelder and Mead Simplex method was chosen because previous experience showed this to be the most robust method programmed for the available computer. The purpose of the reference model was to define a "target" point for the optimization, so that the deviations of the end results of the optimization process from this target could be compared with the required tolerances. The parameter space was normalized to cover the expected deviations of the parameters from nominal, and the target point was deliberately offset from the origin of the normalized space.

The original specification for the controlled experiments stated:

"Experiment 1: By fixing the values of K_4 and K_5

and setting the values (on the
variable model) of K_1, K_2 and K_3
away from the reference values, an
investigation will be made into
the validity of this technique in
optimizing the model variables to
the reference values.

Experiment 2: If experiment 1 is successful K_4
and K_5 will be introduced as fur-
ther variables and the same technique applied".

The results of experiment 1 were so unsatisfactory that experiment 2 was not attempted. The
final values for the 3 parameters were not consistent, had little dependence on the target values
but were very dependent on the starting values.
These results suggested that multiple minima were
present. At the same time, the suitability of the
input stimulus was being questioned, so work using
optimization algorithms was suspended and a new
investigation started, to establish the nature of
the performance surface and its variation with the
type of input stimulus.

The stimulus chosen for the controlled experiments was a series of "on-off" pulses, spread
through the test. The performance index was evaluated at each point of a 3-dimensional mesh of
1331 points covering the normalized parameter
space, 11 values of the 3 parameters K_1, K_2 and K_3.
A selection of these results is presented (Figs. 3,
4 and 5) as "three-dimensional" isometric plots of
the performance index, as ordinate, against points
in the K_1, K_2 space for fixed values of K_3.

In all figures, lines of constant K_1 and
lines of constant K_2 are shown, with the broken
lines indicating hidden lines. Fig. 3 shows the
contorted nature of the surface, with many multiple
minima, and little indication of a global minimum,
and Figs. 4 and 5, for values of K_3 which "straddle"
the target figure, are little better.

440

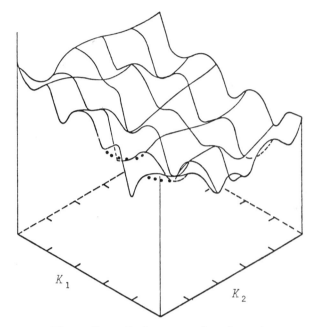

Fig. 3. Pulse train input

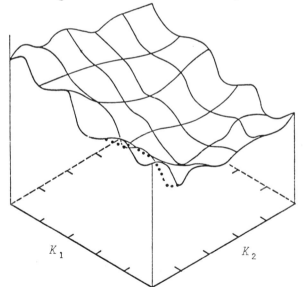

Fig. 4. Pulse train input

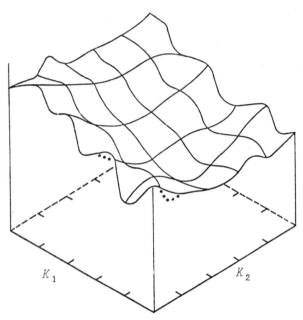

K_1 K_2

Fig. 5. Pulse train input

The performance surface for the pulse input showed that there was little hope of meeting the contract obligation with that input stimulus. On previous contracts, the stimulus used in experiments specially designed for system identification had been a simpler "on-off" step input, lasting for one quarter of the duration of the experiment. When the performance surface was mapped with this input, the resulting surface was much less contorted (Figs. 6 and 7), although it was still rather flat and had no pronounced minimum. The contours gave a minimum for a K_1 value near the target point, but the minimum was very flat in the K_2 direction.

The results to this point indicated that the poor results of the optimization sequence arose more from the paucity of information in the input stimulus than from the optimization. To maximize the information in the input, a pseudo-random binary sequence was used as the stimulus. The performance surface was again mapped (Fig. 8), and

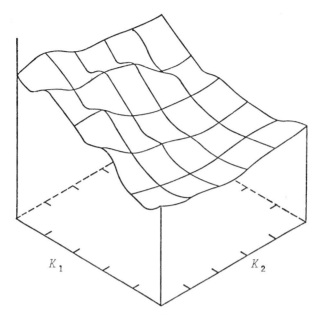

Fig. 6. Step input

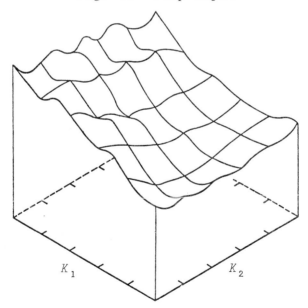

Fig. 7. Step input

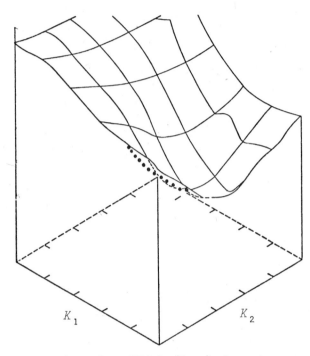

Fig. 8.   PRBS (low) input

showed much more reasonable characteristics, with little sign of multiple minima and a well defined minimum in both the K_1 and K_2 directions. However, the minimum for K_3 was not well defined, so the frequency of the pseudo-random binary sequence was increased, to increase still further the high frequency information in the input, and the mapping exercise repeated. The performance surface showed no sign of multiple minima (Figs. 9 and 10), with a sharply defined minimum in the K_2 direction, a flatter minimum in the K_1 direction, and a reasonable minimum in the K_3 direction.

The results of the experiments with the higher frequency pseudo-random binary sequence seemed very encouraging at first sight, but on closer inspection proved to be disappointing.

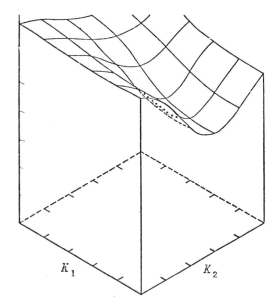

Fig. 9. PRBS (high) input

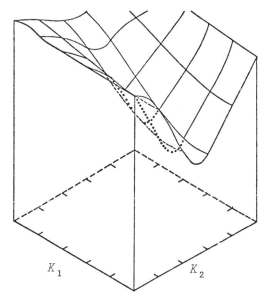

Fig. 10. PRBS (high) input

Comparison of the minimum with the target point showed that the value of K_1 at the minimum was further away from the value at the target point than it was for the other inputs. Since K_1 was the gain in the outermost feedback loop of the system (Fig. 1), it would be expected that this would be estimated with greater accuracy than would parameters K_2 and K_3, so the failure to be near the target values of K_1 was disturbing. If the signal levels at the higher frequencies in the sequence were great enough to give adequate signal level at the monitoring points, other parts of the system were saturating with consequent loss of stability margin, so this type of input was not appropriate to experiments with the real system. Finally, the aim had been to develop methods applicable to the analysis of all system trials, not just a few special trials with artificially injected inputs. Faced with this situation, the sponsors of the exercise decided that progress to date had not been encouraging, and withdrew their support.

The avowed purpose of this paper has been to describe a specific problem with multiple minima, but it could be argued that the multiple minima were the result of inadequate information content in the stimulus input as originally designed, and so this work is also an illustration of the fact that if the appropriate information is not present, an identification exercise will prove fruitless. The practical result has been that use of optimization methods in the G.W. Division for identification on trials matching work has been limited, with mapping methods being used far more widely (but see Cuming, Mullen and Ritchie (1973) and the paper by D.B. Crombie in this volume (p.574)). The application of optimization methods to problems at the system design stage has been far more successful.

REFERENCE

Cuming, D., Mullen, J. and Ritchie, A. (1973) "An Application of Simulation and Modelling Techniques to the Flight Proving of a Guided Missile Autopilot", Institute of Measurement and Control, UKAC 5th Control Convention.

Optimization Techniques Based on Linear Programming

E.M.L. Beale

(Scientific Control Systems Limited)

SUMMARY

Optimization techniques based on linear programming are used to solve large-scale problems. These methods are particularly appropriate because large problems are nearly always sparse. Modern implementations of the simplex method for linear programming exploit sparseness very efficiently: indeed new codes for large LP problems are over 5 times faster on the same computers as the best codes available 5 years ago. These improvements are mainly due to work in three areas: data layout, updating the basis and the choice of pivotal columns and rows.

Nonlinear problems can most easily be accommodated in this framework if the nonlinearities can all be expressed as sums and differences of nonlinear functions of single arguments. We then work with piecewise linear approximations to all nonlinear functions. In principle this is an old technique, but it has recently been made more efficient by using automatic interpolation based on the Dantzig-Wolfe decomposition principle, and also by allowing global optimization of nonconvex problems. The latter is done by Branch and Bound methods using the concepts of Special Ordered Sets. Integer variables can also be accommodated, again using Branch and Bound methods. In principle this approach provides very general facilities for global optimization: in practice large or even med-

ium-sized problems cannot always be solved by
these means, and those that can often require care
in formulation and in the selection of a solution
strategy.

Possibly local optima to more general nonlinear problems can be found by methods that simply
evaluate the objective function and constraints,
together with their first derivatives, at specific
points. A widely used approach to these problems
is the Generalized Reduced Gradient method. Another approach that is based even more directly on
linear programming is the Method of Approximation
Programming. There are versions of both these
approaches that use conjugate gradients to improve
convergence on smooth nonlinear problems. These
seem to have practical advantages over full quasi-
Newton methods when the effective dimension of the
problem may be large.

1. INTRODUCTION

Optimization techniques based on linear programming are used to solve large-scale problems.
These techniques are particularly appropriate
because large problems are nearly always sparse,
each variable occurring in only about 6 constraints
even when there are hundreds of constraints altogether. This is true whether the constraints are
linear or nonlinear, and whether the variables are
continuous or discrete. In these circumstances
the computing techniques used to exploit sparseness are far more significant than the precise
logic used to define the position of the next
trial solution. Modern implementations of the
simplex method exploit sparseness very efficiently
in the solution of linear programming problems.
This is useful, both because many practical optimization problems can be formulated naturally in
purely linear terms, and because linear programming methods can be extended to more general problems.

Section 2 outlines some recent advances in

methods of solving linear programming problems.

Section 3 is concerned with the most important reason for going beyond a purely linear programming model. This is the need to represent economies of scale. Then we cannot use a pure hill-climbing method, such as the simplex method, since this can only find a local optimum that may well not be a global optimum. This difficulty is usually resolved by introducing integer variables into the linear programming model, in particular variables that can only take the values zero or one. But recent developments in Special Ordered Sets allow us to use the same Branch and Bound methods when the problem also includes continuous nonlinear functions, provided that these are expressed as sums and differences of functions of single arguments.

Finally, Section 4 discusses developments in hill-climbing methods for nonlinear problems based on local linear approximations to the constraints and objective function.

2. DEVELOPMENTS IN LINEAR PROGRAMMING

The simplex method for linear programming, due to Dantzig (1951), is too well known to need describing here. But its implementation has recently been improved in several ways: indeed new codes for large LP problems are over 5 times faster on the same computers than the best codes available 5 years ago. This has been partly a matter of routine improvements in coding: analysing where the time is spent and where appropriate making better use of special facilities in the order-codes for the computer. But this is not the most important factor.

These new codes have been developed as in-core codes, with the possibility of going out-of-core if necessary, rather than as out-of-core codes that may sometimes fit in core. This approach has given the code designer more freedom over

449

data lay-out, but more substantial technical developments have been needed for such a significant
improvement in performance. In this paper we
review progress in three areas:

> Data Layout
> Updating the basis
> The choice of pivotal columns and rows.

First we need some notation. A general linear
programming problem can be written in matrix notation as

$$A x = b. \qquad (2.1)$$

This includes the objective function if the first
of the m rows relates the dummy variable x_0 to the
other variables. The problem is then to choose
non-negative values of x_1, x_2, ...x_n to maximize

x_0 subject to (2.1).

Now it has long been conventional to omit
the zero elements in A, so for each column one
just stores the row number and numerical value for
each non-zero element. But Kalan (1971) noted
that the same numerical value typically occurs
many times, and that the matrix can therefore be
stored more compactly by having a "pool" of distinct numerical values, and defining each non-zero
element by a row number and a pointer to the appropriate element in the pool.

Some more mathematical developments in the
storage and updating of the basis are at least
equally important. Let B denote the basis, that
is the matrix formed from the columns of A associated with the basic variables, and let $A^{(N)}$
denote the remaining, nonbasic, columns of A.
Then if we premultiply (2.1) by B^{-1} we deduce that

$$x^{(B)} = \beta - B^{-1} A^{(N)} x^{(N)}, \qquad (2.2)$$

450

where $x^{(B)}$ denotes the set of basic variables, $x^{(N)}$ the nonbasic variables, and

$$\beta = B^{-1} b.$$

The matrix $B^{-1} A^{(N)}$ is known as the tableau. Early implementations of the simplex method calculated the tableau as well as the vector β at each iteration. But this is inefficient on large problems because the tableau is generally far less sparse than the original matrix A. We therefore work with the original matrix A and some compact representation of B^{-1}. We rely on the fact that B^{-1} can be updated without re-inversion at each iteration, since B changes only in one column.

One can of course store B^{-1} as an explicit matrix. But this generally has far more non-zero elements than B when B is sparse, so it is unsuitable. The product form of the inverse was published by Dantzig and Orchard-Hays (1954), and many improvements were made over the years in the re-inversion routine, used to form a sparse product-form representation of the inverse of the current basis every 50 iterations or so. This re-inversion is equivalent to forming a matrix $\bar{B}$ from B by permuting the columns (explicitly) and the rows (implicitly) such that if elementary transformation matrices are then formed to represent pivoting on the diagonal elements of $\bar{B}$ in order, then these matrices are sparse.

It is now generally recognized that one can take even better advantage of sparseness by writing

$$\bar{B} = L\,U,$$

where L is lower triangular and U is upper triangular. This important idea was suggested by Markowitz (1957), although his implementation is not convenient for large problems. This approach can be related to the product form, since we can represent the columns of L, followed by the columns of U taken in reverse order, as ordinary elementary transformations defining B^{-1}.

After inverting, we must continue the iterative steps of the simplex method. One can represent the resulting changes of basis by elementary transformation matrices premultiplying the previous basis inverse. But each matrix is then formed from the coefficients in the current tableau of the variable becoming basic. And the columns of the tableau are often quite dense, so the number of non-zero elements defining B^{-1} grows rapidly with each iteration. So let us see how this can be avoided.

We write

$$B = GU,$$

where G is a general matrix whose inverse is represented in product form that may, but need not, be lower triangular, and U is upper triangular. In practice the rows of U may be permuted, but to simplify the notation we ignore this. If we now premultiply (2.1) by G^{-1}, and transfer the nonbasic variables to the right hand side, we have the equation

$$Ux^{(B)} = G^{-1}b - G^{-1}A^{(N)}x^{(N)}. \qquad (2.3)$$

After one iteration we must remove one variable from $x^{(B)}$, say the pth, and introduce another variable, say x_q. We can then rearrange the variables in (2.3) so that the left hand side refers to the new set of basic variables, taken in the order in which they entered the basis. The resulting scheme can be written

452

$$\overline{U}x^{(B)}{}' = \underset{\sim}{G}^{-1}\underset{\sim}{b} - \underset{\sim}{G}^{-1}\underset{\sim}{A}^{(N)}{}'\underset{\sim}{x}^{(N)}{}', \qquad (2.4)$$

where $x^{(B)}{}'$, $x^{(N)}{}'$ and $A^{(N)}{}'$ denote the new values of $x^{(B)}$, $x^{(N)}$ and $A^{(N)}$.

The columns of $\overline{U}$ are formed as follows:

the jth column of $\overline{\underset{\sim}{U}}$ is the jth column of $\underset{\sim}{U}$ for $j<p$.

The jth column of $\overline{\underset{\sim}{U}}$ is the $(j+1)$th column of U for $p\leq j<m$.

The mth column of $\overline{\underset{\sim}{U}}$ is $\underset{\sim}{G}^{-1}\underset{\sim}{a}_q$, if $\underset{\sim}{a}_q$ denotes the qth column of $\underset{\sim}{A}$.

The arithmetic required so far is similar to that required to update $\underset{\sim}{B}^{-1}$ in the conventional product form. We will have formed $\underset{\sim}{B}^{-1}\underset{\sim}{a}_q$ in order to find the pivotal row, and we can store $\underset{\sim}{G}^{-1}\underset{\sim}{a}_q$ as an intermediate quantity in this computation. But we must now consider what to do about the fact that, unless $p=m$, $\overline{\underset{\sim}{U}}$ is not upper triangular. Bartels and Golub (1969) suggested this approach as a means of ensuring numerical stability in the simplex method. And for $i=p$, $p+1$, ...$m-1$ they proposed the following:

If $|\overline{u}_{ii}| \geq |\overline{u}_{i+1,i}|$, subtract $\overline{u}_{i+1,i}/\overline{u}_{ii}$ times the ith row of (2.4) from the $(i+1)$th to make $\overline{u}_{i+1,i} = 0$.

If $|\overline{u}_{ii}| < |\overline{u}_{i+1,i}|$, do the same thing after exchanging the ith and $(i+1)$th rows.

At the end of this process $\overline{\underset{\sim}{U}}$ is upper triangular and the iteration is complete.

Forrest and Tomlin (1972) use a similar scheme which is more convenient for large problems,

since the new U is formed from $\bar{U}$ by simply deleting all non-zero elements from the pth row except
the last. This is achieved by premultiplying both
sides of (2.4) by an elementary row transformation,
i.e., a unit matrix with a modified pth row. The
new G^{-1} is therefore formed by premultiplying the
current G^{-1} by this row transformation. The new U
is formed by deleting the current pth column and
all entries in the pth row, and adding a new last
column as indicated.

The Forrest-Tomlin method is astonishingly
effective in reducing the build-up of non-zero coefficients in the total representation of B^{-1}.
Reid (1973) shows that a minor modification to the
Bartels-Golub method makes it competitive from
this point of view. The modification is to con-
sider both $|\bar{u}_{ii}/\bar{u}_{i+1,i}|$ and the numbers of non-
zero coefficients in the two rows when deciding
whether or not to exchange the rows. Reid also
comments on the fact that the Forrest-Tomlin method is theoretically no more stable than the original product-form method: he points out that this
theoretical defect can be overcome by making a
"panic invert" whenever the row transformation
added to A^{-1} contains an element that is an excessive multiple of the corresponding $\bar{u}_{i+1,i}$.

In practice the Forrest-Tomlin method works
well without a panic invert. This may be partly
because of recent developments in the selection of
pivotal rows and columns due to Harris (1973).
These reduce both the number of iterations and the
number of small pivots chosen. Her work on the
selection of pivotal columns is based on the fact
that the choice of the column with the most negative reduced cost, d_j, is arbitrary, since the
significance of the column is unaffected if all
its elements are divided by a scale factor s_j, in
which case the scaled reduced cost would be d_j/s_j.

454

She has therefore devised a method of computing dynamic scale factors s_j for each column representing estimates of the sizes of the numbers in the column of the tableau, which are updated from the coefficients in the previous pivotal row of the tableau.

Once the pivotal column is chosen, one has no choice of pivotal row unless there are near ties in the search for a row that minimizes the conventional ratio $\bar{a}_{io}/\bar{a}_{iq}$. When there are near ties she chooses the pivotal row to maximize the size of the pivot element, subject to the constraint that the trial values of all basic variables at the next iteration must exceed -TOLB (and be less than their upper bounds plus TOLB), where TOLB is the standard tolerance for zero on the value of a basic variable.

3. DEVELOPMENTS IN GLOBAL OPTIMIZATION BASED ON LINEAR PROGRAMMING

Global optimization problems are normally formulated in integer programming terms if they are to be solved by methods based on linear programming. They are then solved by Branch and Bound methods. Other approaches have been studied, and used to solve special types of problem; and it may be that they will produce ideas of practical importance, particularly as adjuncts to finding better bounds within the Branch and Bound framework. Fisher and Schapiro (1974) describe some interesting ideas on these lines.

The Branch and Bound methods now used are developments of those proposed by Land and Doig (1960). More recent work is summarized by Benichou *et al.* (1971), Geoffrion and Marsten (1972) and Forrest, Hirst and Tomlin (1974). There are two main differences between current methods and the early practical methods described in Beale (1968). One is that we no longer use the Last-In-First-Out method, sometimes known as simple backtracking, for choosing the next linear programming subproblem to be analysed. This method is rather more practical than the other simple-minded alternative of always choosing the subproblem with the best bound. But it now seems best to compromise between the desire to reach a completely feasible solution as quickly as possible and the desire to avoid postponing the analysis of promising subproblems for a long time while exploring less promising subproblems that are not logically hopeless.

The other feature that is now seen to be far less important than it seemed in 1968 is penalties in the sense of Driebeçk (1966). These are generally ineffective on large problems, even when strengthened to take advantage of the fact that some nonbasic variables are also required to take integer values. Instead we often fall back on a pre-assigned priority ordering of the variables to choose which one to branch on, and we use other externally generated heuristics to define the most promising subproblem to be explored next.

The current state of integer programming is that many practical problems have been solved successfully using integer programming models. On the other hand many large integer programming models have failed to produce useful results even after hours of time on large computers. It is therefore reasonable to be optimistic about applying integer programming to practical problems, but this optimism needs to be cautious. It is often best to start with a scaled-down version of the problem, and one must be prepared to change either the formulation or the solution strategy, or both, if the first runs on the computer are unsuccessful. Williams (1974) discusses a number of ways in which one can set about producing new formulations that are mathematically equivalent to the original formulation but are computationally more convenient. These often require additional constraints that are technically redundant when all the integer variables take integer values but which make the trial solution more realistic while these variables are not so restricted. Changes in the solution strategy simply require setting some parameters in the integer programming code.

The task of finding global optima to more general classes of nonconvex optimization problems is attracting increasing attention. McCormick (1972) points out that many such problems can conveniently be expressed in separable form, when they can be tackled by the special methods of Falk and Soland (1969) or Soland (1971), or by Special Ordered Sets.

Special Ordered Sets were introduced by Beale and Tomlin (1970), and have lived up to their early promise of being useful for a wide range of practical problems. They are of two types. Special Ordered Sets of Type 1, or S1 sets, are sets of variables of which not more than one member may be non-zero in the final solution. Special Ordered Sets of Type 2, or S2 sets, are sets of variables of which not more than two members may be non-zero in the final solution, with the further condition that if there are as many as two they must be adjacent.

S2 sets were introduced to make it easier to find global optimum solutions to problems containing piecewise linear approximations to nonlinear functions of single arguments in otherwise linear programming problems. We may want to include some function $a_{ik}(t_k)$ in the ith constraint, where the argument t_k is either a variable of the problem or else some linear function of other variables of the problem. If we can then find a finite set of values of t_k, say $t_k = T_{kj}$ for $j = 0, 1 \ldots$, such that $a_{ik}(t_k)$ is an approximately linear function of t_k between $t_k = T_{kj}$ and $t_k = T_{k,j+1}$ for all j, then we can proceed as follows.

Define a set of non-negative variables λ_{kj}

457

and the constraints

$$\sum_{j} \lambda_{kj} = 1 \qquad (3.1)$$

$$\sum_{j} T_{kj} \lambda_{kj} - t_k = 0, \qquad (3.2)$$

and then the nonlinear function $a_{ik}(t_k)$ is represented by the linear function

$$\sum_{j} \{a_{ik}(T_{kj})\} \cdot \lambda_{kj}, \qquad (3.3)$$

provided that not more than two of the λ_{kj} are non-zero and that if as many as two are non-zero they are adjacent.

If the function $a_{ik}(t_k)$ is convex, and occurs in a row for which the shadow price is known to be non-negative, then the proviso is automatically satisfied and the problem can be solved by ordinary linear programming. In particular this is so if the function is part of the objective function to be minimized, or if it occurs on the left hand side of a less-than-or-equal-to inequality. Otherwise we must take special precautions to avoid non-adjacent λ_{kj}. Miller (1963) introduced the so called λ-formulation of separable programming, which finds a local optimum solution to such problems. Various methods have been devised for finding global optimum solutions by adding further constraints and integer variables. But it is simpler and more efficient to work directly with the λ-variables as an S2 set.

Beale and Forrest (1973) have extended Special Ordered Set to solve problems of this type where the functions $a_{ik}(t_k)$ are defined by formulae, rather than by finite sets of values and linear interpolation. The method is essentially the same, but additional vectors λ_{kj} are fed into the linear programming problem as required, following the logic of the decomposition algorithm (due to Dantzig and Wolfe (1960)). As in other applications of Special Ordered Sets, branching is done on the upper and lower bounds on the permitted values of the arguments t_k. Bounding is done as in decomposition.

Most of the details of this scheme are rather technical, but one aspect may be of general interest. We need to be able to find the most negative reduced cost (within some small tolerance) of all possible new vectors λ_{kj} corresponding to values of t_k within its current bounds. This means finding a global minimum of some function

$$f = \Sigma_i \, \pi_i a_{ik}(t_k)$$

for the bounded scalar argument t_k.

In principle this is the simplest possible type of global optimization problem, but there is no finite algorithm for it unless we make some further assumptions.

In practice we solve this problem under the assumption that we can evaluate the $a_{ik}(t_k)$ and their first and second derivatives for any value of the argument t_k, and that we can divide the range of possible values of t_k *a priori* into a finite number of sub-intervals over which the second derivatives of all functions $a_{ik}(t_k)$ are monotonic.

We can evaluate f at the end points of all these sub-intervals, but we now need to find out whether there is a significantly smaller value within any sub-interval. From the values of the a_{ik} and their first and second derivatives at the ends of the sub-interval, we can calculate f and f' at the end points and also upper and lower bounds on f'' within the sub-interval. We can then find a lower bound on f within the sub-interval as follows.

Through each end point of the sub-interval draw a parabola representing a function with the same value and first derivatives as f at the end point, and with a constant second derivative equal to the lower bound on f''.

The true value of f must lie on or above both parabolae. And we can sharpen these bounds by drawing another parabola representing a function with a constant second derivative equal to the upper bound on f'', and placing this so that it touches both of the original parabolae. Between the points of common tangency the new parabola is an improved lower bound on f.

If the piecewise quadratic function fitted in this way does not have a lower bound significantly lower than the lowest value found so far, then we can eliminate this sub-interval. Otherwise we subdivide the sub-interval further at the point where the bound is lowest and repeat the argument on both sub-intervals separately.

Before leaving this formulation, three things are worth noting. One is that if the right hand side of (3.1) is replaced by a variable y, then the expression (3.3) represents $y\, a_{ik}(t_k/y)$. This type of expression arises in many applications: see for example McDonald *et al*. (1974). The second point is that the interpolation methods discussed here can be used in convex problems. They will then find the optimum solution without the need for any branching and bounding. The third point is that one may need to be cautious when applying these methods to non-convex problems, as with ordinary integer programming. In particular, logarithms can lead to difficulties. They are very useful in transforming products into sums which are therefore amenable to these methods.

But it seems to be generally unsatisfactory to allow some variable z and its logarithm to occur in a mathematical programming model if an underestimate of ln z causes the model to suggest things that are not really possible, and z can vary by more than about one order of magnitude.

4. HILL-CLIMBING METHODS BASED ON LINEAR PROGRAMMING

Methods based on a combination of hill-climbing and linear programming ideas can best be understood as variants of the Reduced Gradient Method due to Wolfe (1963). This is a generalization of one of the fundamental ideas of the simplex method so that it applies to nonlinear problems. The constraints are used to define locally linear approximations for some variables, the dependent or basic variables, in terms of the remaining independent, or nonbasic, variables. To avoid difficulties in defining feasible directions of change from the current trial solution, we arrange that any variable whose trial value is at its lower or upper bound is made independent. If the problem is linear, then, by following the logic of the simplex method, we need only consider trial solutions at which all independent variables are at their lower or upper bounds. For nonlinear problems this is not so, but the problem can usefully be considered as an unconstrained optimization problem in the space of those independent variables that are not up against their bounds, embedded in a locally linear programming problem. We may need to change the set of independent variables in the embedded problem as the iterative solution proceeds. This causes no real difficulties, although we have to accept the fact that the number of independent variables in the embedded problem, which represents the effective dimension of the problem, may be very variable.

The first large-scale implementation of this

approach was called the Method of Approximation
Programming by Griffith and Stewart (1961). It
predated Wolfe's paper, and also the development
of modern methods of unconstrained optimization.

It is therefore hardly surprising that it
used a mathematically unsophisticated approach to
the embedded unconstrained optimization problem:
namely a small-step gradient method. Nevertheless
Buzby (1974) demonstrates that the approach remains effective for practical large-scale nonlinear
programming.

Quasi-Newton methods are the most powerful
general methods for unconstrained optimization,
and Buckley (1975) has combined one with an efficient method for manipulating sparse linear constraints. But this can lead to awkward storage problems if the number of independent variables may
be large. The method of conjugate gradients is
therefore attractive, since it does not require
the storage of any square matrices with a linear
dimension equal to the number of independent variables. This was originally proposed for unconstrained optimization by Fletcher and Reeves (1964).
It is used in the Generalized Reduced Gradient
method of Abadie and Carpenter (1969), which has
proved a very successful general approach to nonlinear programming. Beale (1974) discusses a similar approach, but which follows the original
Method of Approximation Programming in that it
uses a production mathematical programming system
to solve a sequence of linear programming subproblems, with a "control program" generating the
matrix for each successive subproblem from the
results of previous subproblems.

Beale's paper discusses the theory of approximation programming. He generalizes the concept of nonlinear variables, defining them as a
set of variables such that, when they are all
assigned fixed values, the problem reduces to a
linear programming problem.

462

The effective dimension of the problem is then the number of independent nonlinear variables. The paper also proposes a specific method based on this theory. This method has been used successfully on problems significantly larger than those discussed in Beale (1974). But experience has shown that it can be improved. Its main fault is that it makes too many "exploratory steps". These are steps in which all the nonlinear variables are allowed to vary by up to their standard tolerances, as in the original Method of Approximation Programming.

These steps are used to select the independent variables for the embedded unconstrained optimization problem. We now recommend following each exploratory step by a possibly long sequence of line searches in the embedded problem, using both gradient and conjugate gradient search directions. We only make a new exploratory step to select new sets of independent variables when we fail to find a significantly better trial solution using the current set. Preliminary results with this new version are encouraging, but it seems best to wait for more experience before publishing the details.

REFERENCES

Abadie, J. and Carpenter, J. (1969) "Generalization of the Wolfe Reduced Gradient Method to the case of nonlinear constraints", *in* "Optimization" *Ed.* R. Fletcher, Academic Press, London and New York.

Bartels, R.H. and Golub, G.H. (1969) "The simplex method of linear programming using LU decomposition", *Communications of the ACM*, 12, pp 266-268 and 275-278.

Beale, E.M.L. (1968) "Mathematical Programming in Practice", Pitmans, London.

Beale, E.M.L. and Tomlin, J.A. (1970) "Special facilities in a general mathematical programming system for nonconvex problems using ordered sets of variables", *in* "Proceedings of the Fifth International Conference on Operational Research" *Ed.* J. Lawrence, Tavistock Publications, London.

Beale, E.M.L. and Forrest, J.J.H. (1973) "Global Optimization using Special Ordered Sets", Paper presented at the Eighth International Symposium on Mathematical Programming, Stanford,California, August 27th - 31st 1973.

Beale, E.M.L. (1974) "A Conjugate Gradient Method of Approximation Programming", *in* "Optimization Methods for Resource Allocation" *Eds*. R.W. Cottle and J. Krarup, English Universities Press, London.

Benichou, M.,Gauthier, J.M.,Girodet, P.,Hentges, G.,Ribiere, G. and Vincent, D. (1971) "Experiments in Mixed-Integer Linear Programming" *Math. Prog.*, 1, 76-94.

Buckley, A. (1975) "An alternative implementation of Goldfarb's minimization algorithm", *Math. Prog.* 8, 207-231.

Buzby, B.R. (1974) "Techniques and Experience Solving Really Big Nonlinear Programmes", *in* "Optimization Methods for Resource Allocation" *Eds*. R.W. Cottle and J. Krarup, English Universities Press, London.

Dantzig, G.B. (1951) "Maximization of a linear function of variables subject to linear inequalities", *in* "Activity Analysis of Production and Allocation" Chapter XXI *Ed.* T.C. Koopmans, Wiley, New York.

Dantzig, G.B. and Orchard Hays, W. (1954) "The product form for the inverse in the simplex method", *Maths. of Comp.*, 8, 64-67.

Dantzig, G.B. and Wolfe, P. (1960) "Decomposition principle for linear programming", *Operations Research,* 8, 101-111.

Driebeck, N.J. (1966) "An algorithm for the solution of mixed integer programming problems", *Management Sci.,* 12, 576-587.

Falk, J.E. and Soland, R.M. (1969) "An algorithm

for separable nonconvex programming problems", *Management Sci.*, 15, 550-569.

Fisher, M.L. and Schapiro, J.F. (1974) "Constructive duality in integer programming", *SIAM J. Appl. Math.*, 27, 31-52.

Fletcher, R. and Reeves, C.M. (1964) "Function Minimization by Conjugate Gradients", *Computer J.*, 7, 149-154.

Forrest, J.J.H. and Tomlin, J.A. (1972) "Updated triangular factors of the basis to maintain sparsity in the product form Simplex Method", *Math. Prog.*, 2, 263-278.

Forrest, J.J.H.,Hirst, J.P.H. and Tomlin, J.A. (1974) "Practical Solution of large mixed integer programming problems with Umpire", *Management Sci.*, 20, 736-773.

Geoffrion, A. and Marsten, R.E. (1972) "Integer programming algorithms: a framework and state-of-the-art survey", *Management Sci.*, 18, 465-491.

Griffith, R.E. and Stewart, R.A. (1961) "A nonlinear programming technique for the optimization of continuous processing systems", *Management Sci.*, 7, 379-392.

Harris, P.M.J. (1973) "Pivot Selection Methods of the Devex LP Code", *Math. Prog.*, 5, 1-28.

Kalan, J.E. (1971) "Aspects of Large-Scale In-core Linear Programming" *in* "Proceedings of the 1971 Annual Conference of the ACM", Chicago, Illinois, August 3rd - 5th 1971, 304-313.

Land, A.H. and Doig, A.G. (1960) "An automatic method of solving discrete programming problems", *Econometrica*, 28, 497-520.

McCormick, G.P. (1972) "Attempts to calculate Global Solutions of Problems that may have local minima", *in* "Numerical Methods for Nonlinear Optimization" *Ed.* F.A. Lootsma, Academic Press, London and New York.

McDonald, A.G.,Cuddeford, G.C. and Beale, E.M.L.

(1974) "Balance of care: Some mathematical models of the National Health Service", *British Medical Bulletin*, **30**, 262-270.

Markowitz, H.M. (1957) "The elimination form of inverse and its application to linear programming", *Management Sci.*, **3**, 255-269.

Miller, C.E. (1963) "The simplex method for local separable programming", *in* "Recent Advances in Mathematical Programming" *Eds*. R.L. Graves and P. Wolfe, McGraw Hill, New York.

Reid, J.K. (1973) "Sparse Linear Programming using the Bartels-Golub Decomposition", Paper presented at the Eighth International Symposium on Mathematical Programming, Stanford, California, August 27th - 31st 1973.

Soland, R.M. (1971) "An algorithm for separable nonconvex programming problems II: Nonconvex Constraints", *Management Sci.*, **17**, 759-773.

Williams, H.P. (1974) "Experience in the Formulation of Integer Programming Problems", *Math. Prog. Study* **2**, 180-197.

Wolfe, P. (1963) "Methods of Nonlinear Programming", *in* "Recent Advances in Mathematical Programming" *Eds*. R. L. Graves and P. Wolfe, McGraw Hill, New York.

Power System Scheduling Using Integer Programming

D.W. Wells

(Central Electricity Research Laboratories)

SUMMARY

If power is to be supplied by a power system as cheaply as possible while operating all its components within their normal limits the output of the generators must be coordinated. The coordinator has to decide which generators are needed when they should be started up or shut down and what share of the load each should supply. For complicated systems it is helpful to use a computer to evaluate suitable schedules of generator outputs.

When the problem of determining these schedules is formulated mathematically it is found that each generator is characterized by its power output, its temperature and an additional state variable. This latter variable can take two values corresponding to the generator being on or off but may not take intermediate values. Such discrete variables make the scheduling problem very complex and if they are not handled carefully can result in the computer performing an excessive amount of calculation. Fortunately the branch and bound method of integer programming can be combined with the simplex method of linear programming to solve this scheduling problem reasonably efficiently. It can make good use of any information that is available based on the operators experience of the power system or the results of previous calculations in similar conditions. Even with these techniques the best possible power schedule is unlikely to be found in a reasonable time so it is better to find a good schedule quickly and then improve it as much as possible in the time available.

1. INTRODUCTION

In an electric power system a number of generators with different characteristics supply power
to the loads through the transmission system. Control engineers coordinate these generators to produce this power as cheaply as possible while ensuring that all the generating and transmission equipment is loaded within its normal operating limits.
The engineers rely mainly on their experience but
in recent years computers have been used to advise
them. The power system is so complex that the
computers that were available in the past were not
able to take account of all the operating restrictions and find the best schedule in a reasonable
time. They could evaluate good schedules for a
single time instant but were less satisfactory
when evaluating schedules for a period of time
during which the demand was changing rapidly. As
computers improve it is worth investigating methods
of producing schedules under these conditions.

The method chosen must always find a useful
solution and should usually find a solution close
to the cheapest one possible together with an
estimate of the amount by which it differs from
this solution. In the following sections it is
shown that these objectives can be achieved by
formulating the problem as an integer program and
solving it by a suitable version of the branch and
bound method.

2. POWER SYSTEM SCHEDULING

The schedule of generator outputs for a power system is usually calculated in two stages.
Some hours in advance the plant ordering calculation decides which generators should run and when
they should start-up and shut-down. Later, when
the load can be predicted more accurately, the
plant loading calculation decides how much power
each generator should produce.

The loading calculation must produce a schedule that meets the total demand without overloading any part of the transmission system. It must also ensure that even if any one generator or transmission line was unexpectedly taken out of service the total demand would still be met without overloading anything. However because it does not decide which generators should run it needs to consider only the incremental cost of each generator. This is the rate of change of the total cost of operating a generator with respect to its output. For most units on the British system it is approximately constant although for a few generators it increases near maximum output so that a larger value must be used at high outputs. The loading calculation can be formulated as a linear program and solved by the simplex method (Wells (1968)).

The ordering calculation must take account of the start-up and no-load costs of each unit as well as its incremental cost. The no-load cost is a constant expense incurred all the time a generator is running and includes the cost of the heat that is lost through the casing of the machine. The start-up cost includes the cost of heating the unit to its operating temperature. This depends on its initial temperature which in turn depends on the time since it was last shut-down. Linear programming cannot be used to solve problems with this type of cost function so that a more complicated procedure such as integer programming must be used.

It is necessary to include only the most important constraints on operating the transmission system and these can be represented by simplified approximations. Restrictions on the rate of change of generator output must be included together with restrictions on the time between successive starts and shut-downs of each unit or different units in the same station. The resulting ordering calculation is much more complicated than the loading calculation but is not normally needed as often.

469

3. FORMULATION IN TERMS OF INTEGER PROGRAMMING

The state of each generator at each time instant is specified by a discrete variable (n) that indicates whether or not it is running, the power (p) that it is producing and the reserve of power(s) that is available to meet an unexpected increase in demand. The temperature of the turbine may also be included although this is an imprecise quantity since the turbine is not at a uniform temperature. However it is better to use a single value for the temperature of each turbine than to ignore temperature effects completely while it is doubtful if a more realistic but more complex model of a turbine is justified.

Consider initially one generator at a single time instant. If the generator is off the discrete variable is zero while if it is producing power this variable has the value one. This variable can have no other values

$$n = 0 \text{ or } n = 1 . \qquad (1)$$

The cost of running the generator (c) is zero while it is off but when it is on the cost is the sum of the no-load cost (c_1) and the product of the incremental cost (c_2) and the power generated

$$c = c_1 \, n + c_2 \, p . \qquad (2)$$

The power is zero when the generator is off while when it is on the power must be sufficient to keep the generator stable (p_{msg}) and less than the maximum possible value (p_{max})

$$n \times p_{msg} \leqslant p \leqslant n \times p_{max} \qquad (3)$$

The spare capacity is also zero when the generator is off while when it is on it must be less than the maximum value (s_{max}) and in addition the sum of the power and the spare capacity must be less than the maximum output of the generator

$$0 \leqslant s \leqslant n \times s_{max} \tag{4}$$

$$p + s \leqslant n \times p_{max} . \tag{5}$$

This cost and these constraints are linear functions of the state variables even though one of these variables can take only two possible values. The generator could be described more precisely by including the turbine temperature but the additional constraints and the modified cost would still be linear.

When a sequence of time steps is considered there are restrictions on the way in which the state of a generator can be changed. For instance there is a maximum rate at which the output of the generator can be increased or decreased. There may be constraints on the discrete state variable. For instance it may not be possible to start a generator within a certain time of it being shut-down or it may not be possible to start-up two generators in the same station at the same time. This type of restriction gives rise to linear con-straints on the discrete variables but does not introduce any additional discrete variables.

The generators are connected to the loads and with each other by the transmission system. The power flowing in the transmission lines is a nonlinear function of the power produced by the generators and taken by the loads but it can be replaced by an approximate linear function without introducing any serious errors. The most important transmission constraints are that the total power generated must equal the load and the total spare capacity must exceed a certain specified value. None of the transmission lines should carry more than its rated power even if any other transmiss-ion line trips out unexpectedly. This last requ-irement generates so many constraints that it may make the ordering calculation very long. It can

471

be reduced by selecting only the most important constraints or by simplifying the constraints even further. One useful simplification is to divide the network into regions that are usually referred to as groups and restrict the nett flow of power into or out of a group to a proportion of the total capacity of the lines delimiting the group.

4. THE BRANCH AND BOUND METHOD

The problem whose formulation has just been described may be summarized as follows. Find the cheapest value of a cost function (g) that is a linear function of a number of variables (x_i, y_j) subject to linear constraints where some of the variables (y) must take one of two discrete values.

$$g = \min_{x_i, y_j} c_i x_i + c_j y_j + d \qquad (6)$$

$$a_{ki} x_i + a_{kj} y_j + b_k \geqslant 0$$

$$y_j = 0 \text{ or } y_j = 1 \ .$$

This is an integer linear program and is best solved by the branch and bound method (McMillan (1970)). This consists of the following recursive procedure whose convergence is checked by monitoring upper and lower bounds on the cost function. Each two valued variable (y_j) is temporarily replaced by a continuous variable (z_j) taking all values from the lower to the higher discrete value. This generates an ordinary linear programming problem that is solved in the normal way by the simplex method.

$$h = \min_{x_i, y_j} c_i x_i + c_j z_j + d \qquad (7)$$

$$a_{ki} x_i + a_{kj} z_j + b_k \geqslant 0$$

$$0 \leqslant z_j \leqslant 1 \ .$$

If the solution of the linear program is dearer than the upper bound to the cost then this integer program has no useful solution and must be rejected.

If it has a solution in which each of the two valued variables is at one of its extreme values then this is also the solution of the integer programming problem. Its cost becomes the new upper bound to the cost.

If, as is usually the case, neither of these conditions holds then two smaller integer programs must be generated and solved (8). One of the two valued variables (y_m) is selected and the integer program is set up as before except that this variable is kept constant at one of its discrete values (9). This smaller integer program is solved by the branch and bound method. This is then repeated with the variable at its other discrete value (10). The solution of the large integer program is the solution of the cheaper of the two smaller integer programs except when they are both rejected as having no useful solution in which case it too must be rejected.

$$g = \min (g_0, g_1) \qquad (8)$$

$$g_0 = \operatorname*{Min}_{x_i, y_j (j \neq m)} c_i x_i + c_j y_j + d \qquad (9)$$

$$a_{ki} x_i + a_{kj} y_j + b_k \geq 0$$

$$y_j = 0 \text{ or } y_j = 1$$

$$g_1 = \operatorname*{Min}_{x_i, y_j (j \neq m)} c_i x_i + c_j y_j + (d + c_m) \ (10)$$

$$a_{ki} x_i + a_{kj} y_j + (b_k + a_{km}) \geq 0$$

$$y_j = 0 \text{ or } y_j = 1 \ .$$

There is rarely sufficient time available to complete an integer programming calculation so that it is usually necessary to use the best solution that can be found in the time available without being sure that it is the cheapest possible solution.

The efficiency of the branch and bound method depends critically on the order in which the discrete values are selected and so it must make full use of the information available to it when choosing each value. The most important information is that returned by the linear program. This gives a value for each temporary variable and a lower bound to the cost of setting the corresponding discrete variable to each of its possible values.

$$g \geqslant h + \min (\text{dec}_j \, z_j, \, \text{inc}_j \, (1-z_j)) \, V_j \, . \quad (11)$$

In addition the computer may know what the schedule was under similar circumstances on a previous occasion or the control engineer might have told the computer what he thinks the schedule should be. This information is less reliable and so it should only be used when the information from the linear program does not give a clear cut choice.

One of the best strategies is to try to avoid expensive schedules by finding the variable that would increase the cost by the greatest amount if set to a particular discrete value and then setting this variable to its other value

$$\text{if inc}_m \, (1-z_m) \geqslant \text{inc}_j \, (1-z_j), \, \text{dec}_j \, z_j \, V_j \quad (12)$$

$$\text{select } y_m = 0 \, .$$

If there are several variables with similar large costs it is best to select one that the linear program set to its cheaper discrete value because this generates the same linear program as the

previous stage of the calculation so the result can
be used again without doing another linear program.
This saves a lot of time on this stage of the calculation but could result in extra calculation being needed later

$$\text{if } \text{inc}_m \ (1-z_m) \geqslant \text{inc}_j \ (1-z_j), \ \text{dec}_j \ z_j \ V_j \qquad (13)$$
$$\text{but } z_m \neq 0$$

$$\text{and } \text{inc}_n \ (1-z_n) \simeq \text{inc}_m \ (1-z_m)$$

$$\text{while } z_n = 0$$

$$\text{select } y_n = 0 \ .$$

When searching back to check the alternative
to decisions that were taken earlier it is better
to select the cheapest case rather than the one
with the fewest variables not at permitted values.

5. CONCLUSION

Rigorous well defined algorithms should be
used to calculate schedules for generators in a
power system. The simplex method of linear programming and the branch and bound algorithm are
two such procedures. They are certain to find the
best schedule eventually and usually find it as
rapidly as any other method. They can also take
advantage of any *ad hoc* information to help them
find the result rapidly. However their main advantage is that they not only find an answer but
can if necessary provide information to show why
they chose that answer and how it would change if
any system parameters were changed. This can be
used to pinpoint data errors or to find any weak
point in the system. In the extreme case when
no schedule satisfies the restrictions on the system these rigorous methods find the constraints
that would have to be relaxed to make the system
soluble and it is even possible to extend them so
that they relax these constraints and thus find
a solution while warning the operator about what

they have done.

6. ACKNOWLEDGEMENT

This work was carried out at the Central Electricity Research Laboratories and is published by permission of the Central Electricity Generating Board.

REFERENCES

McMillan, C. (1970) "Mathematical Programming", John Wiley and Sons Inc.

Wells, D.W. (1968) "Method for economic secure loading of a power system", *Proc. IEE*, **115**, (8), p.1190-1194.

NOTATION

a_{ij}, a_{jk}	constraint coefficients
b_k	constraint constants
c_1	no-load cost of generator
c_2	incremental cost of generator
c_i, c_j	cost coefficients
d	constant expense
dec_j	decremental cost of z_j
g	cost of integer program
h	cost of linear program
i	index of continuous variables
inc_j	incremental cost of z_j
j	index of discrete variables
k	index of constraints
m	index of discrete variables
n	discrete variable

p	power
s	spare capacity
x_i	continuous variables
y_j	discrete variables
z_j	continuous variable replacing y_j

APPLICATIONS OF GEOMETRIC PROGRAMMING TO BUILDING DESIGN PROBLEMS

John Bradley and H.M. Clyne

(Institute for Industrial Research and Standards, Dublin)

SUMMARY

The duality theory of geometric (or polynomial) programming provides one method for solving a certain class of nonlinear optimization problems which arises commonly in engineering design. The results of solving a variety of structural engineering problems by this technique are discussed.

Problems considered include the comparison of different structural forms for industrialized buildings, the evaluation of various building components and processes, and the study of different thermal insulation strategies.

An evaluation of the usefulness and economy of the geometric programming approach for such work is made, an efficient numerical algorithm and computer code are presented, and some of the many geometric programming algorithms are compared from a computational viewpoint.

1. GENERAL INTRODUCTION

In many areas of structural analysis and design the engineer can rely on the availability of excellent computer software packages, such as the ICES, GENESYS, NASTRAN and ECI systems. Such programs are efficient, thoroughly tested, and have sophisticated problem oriented languages which make their use quite straightforward. When one tackles the optimization problems which arise from work on structures one is in a situation where suitable computer codes with problem oriented languages are either not available or are in an early stage of development. Other contributors to this conference have considered aspects of this problem in greater detail(*cf*. Brown *et al.*, p. 185 and Lootsma, p.252 of this volume).

In our work in the IIRS Building Division, we have found that the optimization problems which occur in structural engineering often share many of the following characteristics.

(1)   They are nonlinear and of medium size (say, up to 100 variables and constraints).

(2)   The nonlinearities consist in generalized polynomial terms, *i.e.*, terms of the form

$$c_i x_1^{a_{i1}} x_2^{a_{i2}} \ldots \ldots x_m^{a_{im}} \tag{1.1}$$

where the c_i's and the a_{ij}'s are constants. Transcendental functions are generally absent.

(3)   One is often required to solve the same basic problem, varying only the "technological" constants (*i.e.*, the c's and a's).

(4)   Models of structures are being continually changed, updated, etc. This must be easy to do, preferably without any computer programming language compilations.

(5)   A sensitivity analysis is usually required.

Since our resources in computer facilities and manpower are limited, practical considerations impose the two following further requirements.

(6)   The computer costs incurred in solving any optimization problem should be kept as low as pos-

sible since our computer work is done on a remote batch entry terminal and teletype linked to commercial service bureaux.

(7) The model design, data formulation and preparation, and computer execution stages should be as straightforward as possible in order to encourage widespread use of optimization methodology in our organization.

Consideration of these points led us to investigate a restricted form of nonlinear programming, namely geometric programming (GP), also called polynomial programming. This technique was first studied by Duffin *et al.*, (1967), and has been widely used in the US on chemical, electrical and industrial engineering problems (*cf*. Rijckaert (1973)). The first major structural and civil engineering applications were made by Templeman (1970) at Liverpool University, UK.

In section 2 we describe typical problems solved using the GP technique and outline the type of methodology one has to develop when solving practical optimization problems with semi-empirical functional forms. In section 3 we formulate the general polynomial programming problem and a restricted "posynomial" subproblem which is both interesting in its own right and useful in solving the general problem. Aspects of the GP duality theory are also presented. In section 4 we outline an efficient QP-based algorithm for solving polynomial programming problems and discuss its performance. In section 5 we evaluate some of the published GP algorithms, while in section 6 we present some conclusions of a general nature.

ACKNOWLEDGEMENTS

The work of L. Brown and W. Crowe of IIRS and M. Feeney of T.A. Garlands on the practical applications reported here is acknowledged. The authors' interest in GP was first stimulated by the work of Andrew Templeman of Liverpool University, and we thank him for a very fruitful contact and exchange of information and ideas. The work described here forms part of the first author's research programme for the degree of PhD at Trinity College, Dublin, and the helpful guidance of Dr. N.D. Francis is acknowledged.

2. STRUCTURAL ENGINEERING APPLICATIONS

We describe three structural engineering applications where the optimization problems can be formulated as, or approximated by, polynomial programs. These three problems are typical of the following general areas.

(1)   Optimum design of a single structural component (*i.e.*, a sandwich panel).

(2)   Optimum design of a composite structure (*i.e.*, an industrialized factory building).

(3)   Study of the optimum balance between construction and heating costs in housing.

Optimum design of sandwich panels (Crowe (1971))

This study is typical of the small optimization problems which arise in component design. Similar problems could include the design of a plywood box beam, a cylindrical tank, a roof truss or a portal frame.

Structural sandwich panels comprised of thin stiff facings bonded to both surfaces of a thick lightweight core can provide highly efficient constructions for partitioning, external walling, floor and roof deckings. A typical example is shown in Fig. 1. The earliest large scale use of such panels was for aircraft structures.

The purpose for undertaking this study was to establish a design procedure for a range of lightweight building components employing fibreboards as the principal stress carrying materials,

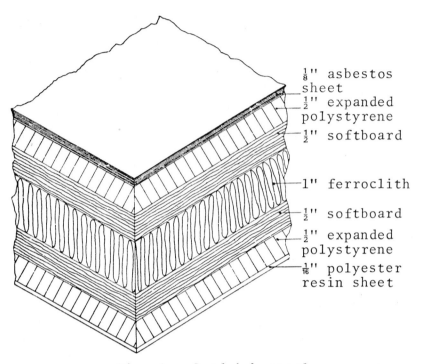

$\frac{1}{8}$" asbestos sheet

$\frac{1}{2}$" expanded polystyrene

$\frac{1}{2}$" softboard

1" ferroclith

$\frac{1}{2}$" softboard

$\frac{1}{2}$" expanded polystyrene

$\frac{1}{16}$" polyester resin sheet

Fig. 1. Sandwich panel

in order to rationalize the design and production of building panels. While the optimal aircraft panel is that with minimum weight, the constraints of the building industry are such that minimum cost is the governing criterion. Also, this cost factor is well represented (though scaled down) by the cost of materials alone, since the assembly and fabrication costs will be of a constant nature for a panel with a given number of constituents.

Facing materials considered covered the range of fibreboard sheet materials while core materials are available in a wide range of physical properties and cost. A trend can be established when a rational relationship between cost and per-formance is assumed. Cost data are available in tabular form, giving, for fibreboard, cost as a function of thickness (T_s) and elastic modulus (E_s)

for unit area, and, for core material, giving cost as a function of thickness (T_c) and shear modulus (G_c). An empirical relationship was sought, and a log-linear form gave an adequate fit to the data. Hence, the cost per unit width for a two component panel may be written as

$$g_0 = [2L\alpha_1] T_s^{\beta_1} E_s^{\gamma_1} + [L\alpha_2] T_c^{\beta_2} G_c^{\gamma_2} \qquad (2.1)$$

where L is the span, and the α's, β's and γ's are constants.

The consideration of structural response of panels is concerned with limitations on deflections and stresses, both of which are governed by codes of practice. They may be formulated as follows:

$$g_1 = \left[\frac{10W_1 L^3 d}{384}\right] T_s^{-1} T_c^{-2} E_s^{-1} + \left[\frac{K_1 W_1 L d}{8}\right] T_c^{-1} G_c^{-1} \leqslant 1 \quad (2.2)$$

$$g_2 = \left[\frac{W_2 L^2}{8}\right] T_s^{-1} T_c^{-1} E_s^{-1} \leqslant 1 \qquad (2.3)$$

$$g_3 = \left[\frac{W_2 L K_1}{2 F_{sh}}\right] T_c^{-1} \leqslant 1 \qquad (2.4)$$

$$g_4 = \left[\frac{W_2 L^2}{8 K_2}\right] T_s^{-1} T_c^{-1} E_s^{-\frac{1}{3}} G_c^{-\frac{2}{3}} \leqslant 1 \qquad (2.5)$$

$$g_5 = \left[\frac{2}{t_{max}}\right] T_s + \left[\frac{1}{t_{max}}\right] T_c \leqslant 1 \qquad (2.6)$$

where W_1 and W_2 are working loads, K_1 is a shape factor, K_2 is a buckling strength factor, F_{sh} is a limiting shear stress, d is a limiting deflection and t_{max} is the maximum total panel thickness allowed.

The terms in functions g_0 to g_5 are seen to be of form (1.1), *i.e.*, generalized polynomials. A composite graph showing some results of a series of optimizations is given in Fig. 2 and further results are given in Crowe (1971). The formulation given above does not take into account the discrete thicknesses of fibreboards actually manufactured. In practice rounding procedures are needed and can

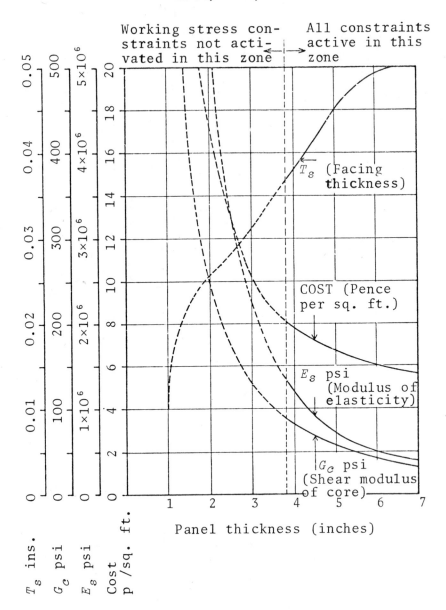

Fig. 2. Optimum design of composite panels. Characteristic curves established through geometric programming

be incorporated systematically into the optimization method (*cf*. Templeman (1970)). Also, while
the continuous designs found above are the cheapest
possible, but are impractical, alternative designs
can be compared with the optimal continuous design
and the precise penalty incurred by using existing
modular sizes can be calculated.

Optimization and factory structures (Bradley et al.
(1974))

      Structural design has been defined as the
process of determining the form and proportions of
a structure, subject to certain performance criteria
and based on an accepted measure of efficiency or
merit. A clear distinction is usually made between
determination of form (shape and relative arrangement of components) and of proportions (quantities
which define size for the component elements and
for the entire structure). The former is basically
a creative process which is open ended unless a
restriction is made to a specific class of alternative forms, while the latter may be reduced to a
rational procedure once a particular form has been
chosen. The theory of optimal structural design,
therefore, is concerned with the determination of
the optimum proportions of a variety of form types,
subject to given performance criteria. Some recent
progress has been made on aspects of form optimization (Gero (1973)).

      In the study summarized here, we formulate
and solve a cost minimization problem in the design
of building envelopes. The mathematical model is
organized to allow easy investigation of the changes
in cost and design variables arising from variations
in the basic structural form and materials and components. The cost of the building envelope forms
about 40 to 50% of the total cost of certain simple
factory structures of the type built on industrial
estates in Ireland. One such structure, with a
double portal frame roof, is illustrated in Fig. 3.

      The cost of the structural envelope can be
subdivided as follows.

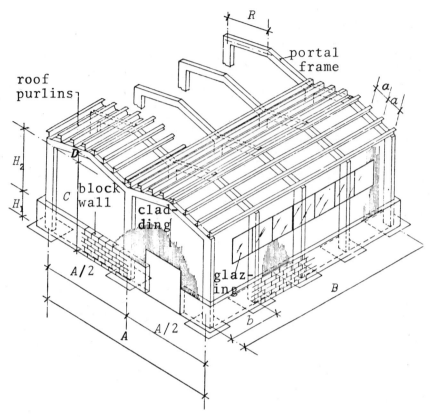

Fig. 3.   Double bay portal design

(a)   Structural frame component

Three basic types of structural frame were chosen for examination - portal frames, grid designs and truss designs. The derivation of a costing relationship for the roof frame is a complex problem, and it is possible to construct a "micro" model wherein the design of the frame is considered in great detail (*cf*. Lipson and Russell (1971)). We were satisfied with a semi-empirical "macro" model, a common situation in the building industry, where a major input to a problem is the result of experience and this can often be best expressed in empirical form.

486

For example, the portals were designed manually and, using the plastic analysis method, the portal weights (W_p) were calculated for a range of spans (A) and spacings (b), assuming a certain total roof loading. We sought a nonlinear regression of the form $W_p = f(A,b)$ and a second order polynomial gave an adequate fit to the data. For the purposes of costing, the structural frame weights were defined by formulae of this type. The unit cost information was in £ per lb of fabricated steel, this being the fashion in which contractors usually quote. Other more detailed approaches have also been used (*cf*. Lee and Knapton (1973)).

(b)   Remaining components

These include the columns and pads, roof purlins, decking, waterproofing, eaves beam, flashings, the wall components (*i.e.*, brickwork, panels, cladding), glazing, and the concrete floor slab, all of which can easily be expressed in terms of the geometric dimensions. A certain amount of difficulty arises when one tries to obtain accurate unit cost information for, what are often, manual tasks. We adopted compromise figures. We return to this point again in section 6.

Constraints

The design constraints on such a structure include restrictions on stresses and deflections in the frames and bounds on various geometric quantities and dimensions, all of a fairly standard type.

The aspect of this study, on which we wish to focus, concerns the methodology used in its practical formulation, solution and the treatment of the results. For example, the reduction of the data for the factory problem to the special form required by the polynomial programming code of section 4 is a simple but tedious matter of arithmetic. For such problems it is more convenient to

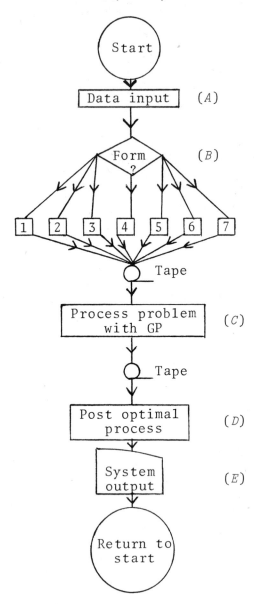

Fig. 4. Flow chart - optimization process (factory optimization)

separate the data formulation process from the optimization package, and to prepare a special post-optimal report generator. Fig. 4 illustrates this sequence where the code explanation is as follows.

(*A*)   Input data consist of simple unit cost information, geometric parameters and a code which allows the selection of one of a range of structural forms (portal, truss, grid, etc.).

(*B*)   Depending on the structural form selected, a simple Fortran subroutine is executed and the special data required by the GP code are prepared on magnetic tape.

(*C*) (*D*)The GP code performs the optimization and outputs on tape the minimum cost, optimal design variables, constraint activation details and information useful for a sensitivity analysis.

(*E*)   A final subroutine takes the rather cryptic GP results and produces a more readable report, an example of which is given in Fig. 5.

A summary of some of the results of the study is given in Table I. Clearly, for a given material specification, the interdesign cost variation is small, being of the order of £0.10 per square foot. The very flat cost function makes the model insensitive to all but the major span variation. This was one of the reasons why a regression approach for the roof costs was adopted rather than the construction of an accurate design model. The orders of accuracy of each segment of a model must be in proportion, since otherwise a gross approximation necessitated in one segment will render meaningless a highly accurate approximation made elsewhere.

We have found that the separation of the various processes associated with the optimization to be required for all but the simplest problems. Because of the fairly extensive data processing often required before the final form of the

Optimum cost = 24504.4   Cost per sq. ft. = 1.63

	IN.	FT.
Width	= 1034.3	86.19
Length	= 2088.2	174.01
Purlin spacing	= 94.0	7.83
Purlin span	= 298.3	24.86
Wall cladding	= 42.0	3.50
Walls	= 150.0	12.50

Truss weight	= 4482.0 lbs
Trussed purlin	= 167.3 lbs

	Percentage	Cost per sq. ft.
Weather + insul	= 11.4	0.187
Decking	= 11.1	0.182
Trussed purlin	= 4.8	0.078
Main truss	= 12.2	0.200
Walls	= 21.4	0.349
Wall cladding	= 3.4	0.055
Cladding trim	= 0.6	0.010
Glazing trim	= 0.6	0.010
Eaves flashing	= 1.0	0.017
Eaves beam	= 2.1	0.035
Floor slab	= 22.0	0.360
Columns + bases	= 2.2	0.037
Wall footing	= 6.5	0.107

Fig. 5. Sample post optimal program output.
Optimum Factory Design, truss with
trussed purlin

Table I
Cost per square foot for building envelopes

Roof			Steel Deck		Plastic Coated Steel Deck		Asbestos	
Wall			9" Block	PC Panel	9" Block	PC Panel	9" Block	PC Panel
Single Bay Portal	Purlins	Angle	1.65	1.79	1.72	1.90	1.56	1.70
		RSJ	1.55	1.69	1.63	1.77	1.46	1.59
		Zed	1.57	1.69	1.63	1.78	1.47	1.60
Double Bay Portal	Purlins RSJ		1.57	1.73	1.65	1.80	1.47	1.61
Grid with Purlins	Purlins RSJ		1.48	1.64	1.57	1.69	1.40	1.53
Grid without Purlins	no Purlins		1.50	1.64	1.58	1.72	1.40	1.55
Truss with Purlins	Purlins RSJ		1.57	1.72	1.65	1.79	1.49	1.63
Close-spaced Truss	no Purlins		1.56	1.71	1.64	1.79	1.56	1.71
Trusses in Two Directions	no Purlins		1.47	1.61	1.55	1.69	1.40	1.54

optimization problem is prepared, we feel that
even the advent of sophisticated problem-oriented-
languages will not entirely do away with this sub-
division.

Optimum balance between construction and heating
costs for housing (Clyne and Bradley (1974))

With the world's attention focused on energy
problems, and possible shortage of fossil fuels, it
is very appropriate to consider the demands placed
on the energy supply industries by domestic heating
requirements and how these demands can be moderated.

The nature and arrangement of the materials
and components in walls and roofs determine the
cost and thermal insulation level of any construction. Furthermore, the insulation level will affect
the heat loss by conduction through the building
fabric and, therefore, the capital cost of the
heating installation and the annual cost of the
heat loss. Hence, construction and heating costs
are intimately related, and one would surmize that
an optimum balance exists between the two. In
such a comparison, we use the net present value of
the annual costs over the life of the building.

The case of the optimal thermal design of
large office blocks has been considered by Wilson
(1973). Essentially, the problem is that of
obtaining comfortable thermal conditions in an
office building, at minimum cost. The initial
cost of the heat supply system and the annual cost
of the fuel used are dependent on the heat lost
through the building fabric and by ventilation.
At the stage of the building design for which
Wilson's analysis is tailored, the building shell
has been designed and may, or may not, have insulating material on its construction. Hence, the
structural form is not a variable or a cost source
which may be controlled. This reflects the reali-
ties of the situation in speculative building, and
the thermal cost model includes only fuel, heating
plant and insulating costs.

The thermal analysis of office buildings is complicated by the problem of intermittent heating. Wilson's model takes this into account, and he formulates constraints to deal with intermittency, fabric loss and condensation. The optimization problems which he solves are of polynomial form and he derives many optimal thermal policies.

Our concern was with domestic housing, and our model reflects the purpose of the study, *i.e.*, to advise Local Government Authorities on the formulation of minimum heating and insulation standards in the area of house construction. A schematic diagram illustrating the heat loss problem for a small house is given in Fig. 6 and a paper by Page (1974) uses simple methods of calculus to treat aspects of the general problem.

The problem we considered was that of building, say, a block of back-to-back terrace flats at a minimum economic cost of construction, *i.e.*, fabric cost plus net present value of energy losses. The variables to be identified were the building width (A), the depth (B), the number of storeys (N), the glazed areas on north, south, east and west (y_1, y_2, y_3, y_4), the insulation level of roof and floor (u_1, u_2), the capacity of the central heating plant and the thermal response of the building (H).

The cost function is of the form

$$[C_1]S + [C_2]H\theta^\beta$$

where S is a sum of component fabric costs (*i.e.*, walls, roof, ground floor, intermediate floors, partitions, glazing), θ is the generated temperature and C_1, C_2, β are known constants.

The model constraints cover the following areas:

(*a*) constraint on the heat demand of the building,

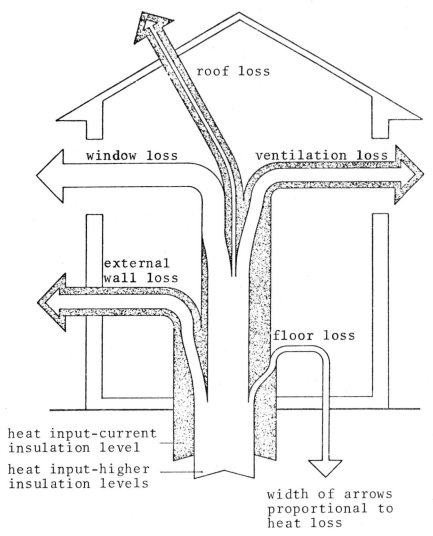

Fig. 6. Heat losses from a typical house

(*b*) relationship between the generated temperature and casual gains,

(*c*) constraint on composition of casual gains,

(*d*) constraint on glazing areas,

(*e*) insulation thickness constraints,

(*f*) various structural and geometric constraints.

The optimization problems which result are of polynomial form and are fairly large and complex. The formulation of optimal policies required the solution of many different programs so it was essential to have an efficient optimization code. The quadratic programming-based geometric programming computer code proved ideal for the purpose.

At present the model is being further extended to deal with the question of national energy demand for domestic heating taking into account both the existing underinsulated housing stock and the newer stock which is subject to more stringent regulations.

3. POLYNOMIAL PROGRAMMING - BASIC THEORY

The purpose of this section is to formulate the polynomial programming problem and to present aspects of duality theory which have been useful in practical computation.

The general polynomial program can be stated as follows:

Primal Program A - PP(A)

$\begin{Bmatrix} \text{minimize} \\ \text{maximize} \end{Bmatrix} g_0(\underset{\sim}{x})$

subject to $g_k(\underset{\sim}{x}) \begin{Bmatrix} \leq \\ \geq \end{Bmatrix} 1$, $k = 1,2,\ldots,p$

Here, $\underset{\sim}{x} = (x_1, x_2, \ldots, x_m) > 0$.

$$g_k(\underset{\sim}{x}) = \sum_{i \in J(k)} \sigma_i c_i \prod_{j=1}^{m} x_j^{a_{ij}}, \quad k = 0,1,\ldots,p \quad (3.1)$$

495

and the exponents a_{ij} are arbitrary real numbers but the coefficients c_i are positive. Also, $\sigma_i = \pm 1$.

The index sets $J(k)$, used throughout this section, are defined as follows:

$$J(k) = \{m_k, m_k+1, \ldots, n_k\} \quad k = 0, 1, \ldots, p$$

where

$$m_0 = 1, \ m_1 = n_0 + 1, \ldots, m_p = n_{p-1} + 1, \ n_p = n$$

i.e., the set of integers $J(k)$ contains the indices of the polynomial terms contained in the function g_k and n is the total number of such terms in PP(A) above.

For later use, one can transform PP(A) into the equivalent form below:

Transformed Primal Program A - TPP(A)

minimize $\quad x_0$

subject to $P_k(\underset{\sim}{x})/Q_k(\underset{\sim}{x}) \leqslant 1, \quad k = 0, 1, \ldots, p \quad$ (3.2)

$$\underset{\sim}{x} = (x_0, \ x_1, \ldots, x_m).$$

Here, $P_k(\underset{\sim}{x})$ and $Q_k(\underset{\sim}{x})$ are functions of the form (2.1), except that all the σ's are $+1$ (*i.e.*, "posynomials").

If one makes the restriction to functions consisting of polynomials with positive coefficients c_i and to minimization subject to "less-than-or-equal-to" inequality constraints, one obtains the following program:

Primal Program B - PP(B)

minimize $\quad g_0(\underset{\sim}{x})$

subject to $g_k(\underset{\sim}{x}) \leqslant 1, \ k = 1, 2, \ldots, p$

$$\underset{\sim}{x} = (x_1, x_2, \ldots, x_m) > 0.$$

496

Here,

$$g_k(\underset{\sim}{x}) = \sum_{i \in J(k)} c_i \prod_{j=1}^{m} x_j^{a_{ij}}, \quad k = 0, 1, \ldots, p \quad (3.3)$$

where the coefficients c_i are positive, $i.e.$, all the g_k's are "posynomials".

While the above program is not yet convex, a simple exponential transformation of the primal variables

$$x_j = \exp(z_j), \quad j = 1, 2, \ldots, m \quad (3.4)$$

leads to the following:

Transformed Primal Program B - TPP(B)

minimize $\quad g_0(\underset{\sim}{z})$

subject to $g_k(\underset{\sim}{z}) \leqslant 1, \quad k = 1, 2, \ldots, p.$

Here,

$$g_k(\underset{\sim}{z}) = \sum_{i \in J(k)} c_i \exp\left(\sum_{j=1}^{m} a_{ij} z_j \right) \quad (3.5)$$

where the c_i are positive and the a_{ij} unrestricted in sign.

This is clearly a convex programming problem, the g_k's being positive sums of exponentials.

While one can solve directly any of the above four programs by standard methods, the duality theory of GP allows one to formulate the following dual program corresponding to PP(B):

Dual Program B - DP(B)

maximize $\quad \ln V(\delta)$

subject to $\underset{\sim}{\delta} \equiv (\delta_1, \delta_2, \ldots, \delta_n) \geqslant 0$

$$\sum_{i \in J(0)} \delta_i = 1$$

$$\sum_{i=1}^{n} a_{ij}\delta_i = 0, \quad j = 1,2,\ldots,m. \qquad (3.6)$$

Here,

$$\ln V(\underset{\sim}{\delta}) = \ln\left\{\prod_{i=1}^{n}\left(\frac{c_i}{\delta_i}\right)^{\delta_i}\prod_{k=1}^{p}\lambda_k(\underset{\sim}{\delta})^{\lambda_k(\delta)}\right\} \qquad (3.7)$$

and $\lambda_k(\underset{\sim}{\delta}) = \sum_{i \in J(k)}\delta_i.$

This program involves the maximization of a concave nonlinear function of n variables $\underset{\sim}{\delta}$, subject to $(m + 1)$ *linear* equality constraints. It may be noted that, while $\ln V(\underset{\sim}{\delta})$ is continuous for $\underset{\sim}{\delta} \geqslant 0$, it is not differentiable if any component of $\underset{\sim}{\delta}$ is zero.

A further transformation, involving the solution of the underdetermined linear system (3.6), leads to the following:

Transformed Dual Program B - TDP(B)

maximize $\quad \ln V(\underset{\sim}{r})$

subject to $\delta_i(\underset{\sim}{r}) \geqslant 0, \quad i = 1,2,\ldots,n.$ $\qquad (3.8)$

Here,

$$\delta_i(\underset{\sim}{r}) = b_i^{(0)} + \sum_{j=1}^{d} r_j b_i^{(j)}, \quad i = 1,2,\ldots,n$$

$$\ln V(\underset{\sim}{r}) = \ln K_0 + \sum_{j=1}^{d} r_j \ln K_j - \sum_{i=1}^{n}\delta_i\ln\delta_i + \sum_{k=1}^{p}\lambda_k\ln\lambda_k \qquad (3.9)$$

$$K_j = \prod_{i=1}^{n} c_i^{b_i^{(j)}}, \quad j = 1,2,\ldots,d$$

$$\lambda_k = \sum_{i \in J(k)}\delta_i$$

and $\underset{\sim}{b}^{(0)}$ satisfies

498

$$\sum_{i \in J(0)} b_i^{(0)} = 1$$

$$\sum_{i=1}^{n} a_{ij} b_i^{(0)} = 0, \quad j = 1,2,\ldots,m$$

while the vectors $b^{(j)}$, $j = 1,2,\ldots,d$ satisfy the corresponding homogeneous systems of equations.

Here we have to maximize a nonlinear, concave function $\ln V(r)$ of $d = n - (m + 1)$ variables $r = (r_1, r_2, \ldots, r_d)$, subject to n linear, inequality constraints (3.8).

The duality theory for PP(B) and DP(B) states, in essence, that under certain weak conditions, the constrained maximum value of the dual function $\ln V(r)$ equals the constrained minimum value of the primal function $g_0(x)$, and that the following linear relations hold at the dual optimizing point:

$$\sum_{j=1}^{m} a_{ij} \ln x_j^* = \begin{cases} \ln \delta_i^* + \ln V(\delta^*) - \ln c_i \,, & i \in J(0) \\ \ln \delta_i^* - \ln \lambda_k(\delta^*) - \ln c_i, & i \in J(k), k > 0 \end{cases} \quad (3.10)$$

These linear relations (referred to as the "log-linear" system) can be solved for the optimum primal vector x^*, having once determined the (unique) optimum dual vector r^*.

This situation should be compared with the linear programming case, where the dual of one LP is just another LP, and the number of variables in a primal LP equals the number of constraints in the dual. The problem of duality for convex programs, leading to linearly constrained dual programs, is treated in a more general setting by Rockafellar (1970).

Hence, to solve the restricted posynomial B programs, one has a choice of three equivalent formulations, PP(B), TPP(B) and TDP(B), each of which has advantages in given circumstances. For example, when the number of primal variables (m) is approximately equal to the number of terms (n), the resultant number of dual variables $d = n - (m + 1)$ in TDP(B) will be small. Indeed, when, in the rare case, $d = 0$, the dual programs can be solved by simply inverting one square matrix.

There are various methods of solving the general negative term problem PP(A), one of which is to approximate it by a sequence of programs of form PP(B), *cf.* Duffin and Peterson (1973), Avriel and Williams (1970), Passy (1972). Morris (1972) has also considered the solution of general NLP's via a sequence of approximating posynomials. Passy and Wilde have also solved PP(A) via a non-convex "quasi-dual" program.

The most straightforward solution method is that of Avriel and Williams (1970) where the posynomials $Q_k(x)$ in TPP(A), *i.e.*,

$$Q_k(\underset{\sim}{x}) = \sum_j q_j(\underset{\sim}{x}) = \sum_j d_j \prod_{i=1}^{m} x_i^{b_{ij}} \qquad (3.11)$$

are approximated conservatively by a *single* term posynomial of the form

$$Q_k(\underset{\sim}{x}) \geqslant \bar{Q}(\underset{\sim}{x}; \underset{\sim}{x}^{(0)}) \equiv$$

$$Q(\underset{\sim}{x}^{(0)}) \prod_{i=1}^{m} \left[\frac{x_i}{x_i^{(0)}} \right]^{\sum_j (b_{ij} q_j(\underset{\sim}{x}^{(0)})) / Q(\underset{\sim}{x}^{(0)})} \qquad (3.12)$$

One starts the process with any feasible primal point $x^{(0)}$, replaces each $Q_j(x)$ in (3.2) by its single term approximation at $x^{(0)}$, namely $\bar{Q}(x; x^{(0)})$, and solves the resultant posynomial program to obtain the next point $x^{(1)}$. Continuing in this way, we generate a sequence $\{x^{(\alpha)}\}$, where $x^{(\alpha+1)}$ is the solution to the problem below.

Approximating Posynomial Program "α"

minimize x_0

subject to $P_j(\underset{\sim}{x})/\bar{Q}_j(\underset{\sim}{x};\underset{\sim}{x}^{(\alpha)}) \leqslant 1, \quad j = 0,1,\ldots,p.$

$\underset{\sim}{x} = (x_0,x_1,x_2,\ldots,x_m) > 0.$

Under certain mild restrictions, it may be shown that the sequence $\{\underset{\sim}{x}^{(\alpha)}\}$ converges to a "quasi-minimum", which, for most real problems, will be a local minimum. In fact, if $\underset{\sim}{x}^*$ is a q-min, then it is a minimum over any convex subset containing $\underset{\sim}{x}^*$, of the feasible set.

Finally we can summarize the relationship between the above six programs as follows; reading from left to right,

$\{PP(A) \equiv TPP(A)\} \Longrightarrow \{PP(B) \equiv TPP(B)\} \equiv \{DP(B) \equiv TDP(B)\}$
          approximates

4. A QUADRATIC PROGRAMMING BASED NUMERICAL ALGORITHM FOR POLYNOMIAL PROGRAMMING

A QP-based algorithm for solving PP(B) via the dual program TDP(B) will now be described. This algorithm will then be extended to deal with the more general problem PP(A), as outlined in section 3.

A logic diagram of the main features of the algorithm is given in Fig. 7, to which the numbered sections below refer.

(1)   The form of data input to the code can be standardized, as is already the case with the better known LP codes. We have adopted the format used by Templeman (1972) and recommend it for general use.

(2)   The setting up of the dual program TDP(B) involves essentially the solution of a system of $(m + 1)$ linear equations in n unknowns.

501

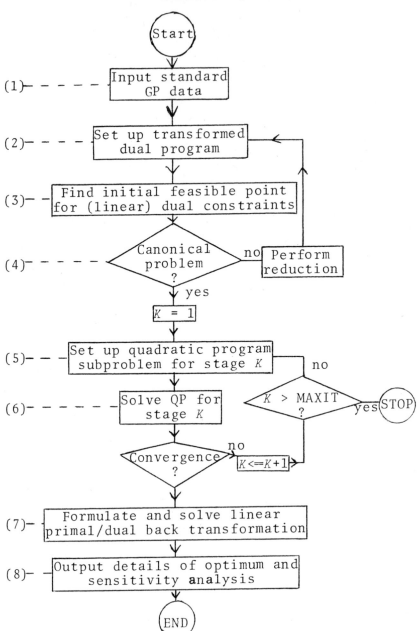

Fig. 7. Flow diagram for solution of prototype posynomial programs

(3)   To find an initial feasible interior point
of the set of linear inequalities (3.8) is a standard LP problem. We use the subroutine LAO2A of
the Harwell Subroutine Library, due to Fletcher
(1970).

(4)   Should no such interior feasible point exist,
the problem being solved is degenerate (in a sense
defined by Duffin *et al.* (1967)), and may be
reduced to "canonical" form. Such non-canonical
problems are rare (we have never met one in practice) and we have not included a reduction algorithm
in our production code.

(5)   From equation (3.9), the gradient and hessian
of $\ln V(\underset{\sim}{r})$ can be derived analytically as follows:

$$\frac{\partial}{\partial r_j}(\ln V(\underset{\sim}{r})) = \ln K_j - \sum_{i=1}^{n} b_i^{(j)}\delta_i + \sum_{k=1}^{p} \lambda_k^{(j)}\ln\lambda_k$$

$$\frac{\partial^2}{\partial r_j \partial r_l}(\ln V(\underset{\sim}{r})) = - \sum_{i=1}^{n} \frac{b_i^{(j)}b_i^{(l)}}{\delta_i} + \sum_{k=1}^{} \frac{\lambda_k^{(j)}\lambda_k^{(l)}}{\lambda_k}$$

where $\underset{\sim}{b} > 0$. Since the dual constraints are linear,
a local quadratic approximation to $\ln V(\underset{\sim}{r})$ at $\underset{\sim}{r}_k$
can be made which agrees with $V(\underset{\sim}{r})$ at $\underset{\sim}{r}_k$ in function, gradient and hessian, initially for $k = 0$,
where $\underset{\sim}{r}_0$ is the point from (3) above.

In the context of our GP algorithm, Fletcher's
linearly constrained algorithm (Fletcher (1972)),
which we follow closely, can be seen to have a
wider application to posynomial programs. Indeed,
large linearly constrained NLP's seem to arise in
practice as the duals of posynomial programs.

(6)   Of the many available quadratic programming
algorithms, we have found the Harwell Subroutine
VEO2A (Fletcher (1971)) both convenient and efficient, and it is the one which we use in our code.
Any other QP algorithm may be used and, indeed,
our geometric programming algorithm provides a
good environment in which to compare the efficien-

cies of alternative QP solution methods.

To avoid the problem of singular derivatives of ln $V(\underset{\sim}{x})$, if at any stage we detect a Lagrange multiplier $\lambda_h(\underset{\sim}{x})$ tending to zero (*i.e.*, primal constraint $g_h(\underset{\sim}{x})$ is going slack as the minimum is approached), we eliminate the relevant primal constraint and continue the solution procedure. This means that, in the final stages of the process when a very accurate dual optimum is being sought, we are dealing with an unconstrained QP problem. All the original primal constraints are checked for feasibility at stage (8) below.

(7) The primal/dual log-linear transformation equations (3.10) are solved by the least squares method.

(8) A detailed report is finally generated.

This algorithm has been extensively tested both on our own problems and on many in the literature. An interesting comparison of the GP/QP-algorithm's performance on a problem due to Beck and Ecker (1972) is given below. Problems 9A to 9F differ only in the exponent of one variable of one term of the cost function. This problem has 18 terms, 4 constraints (2 of which are slack at the optimum) and 7 primal variables, and is specified in Appendix I.

Convex-Simplex Method v GP/QP-algorithm (IBM 360/65)

Problem 9	Convex-Simplex Method (Beck and Ecker)	GP/QP-algorithm (Bradley)
A	6.37 seconds	3.84 seconds
B	6.23 seconds	3.14 seconds
C	43.70 seconds	3.16 seconds
D	38.62 seconds	2.46 seconds
E	54.32 seconds	2.59 seconds
F	27.72 seconds	2.74 seconds

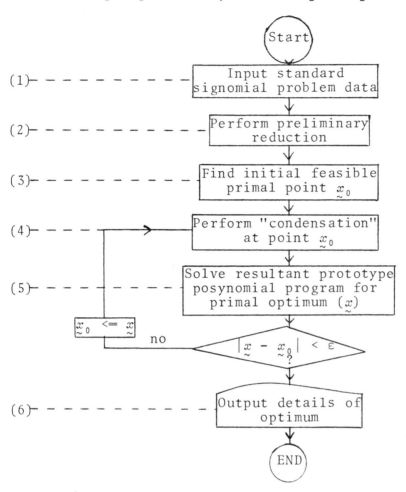

Fig. 8. Flow diagram for solution of general polynomial programming problems

505

The extreme sensitivity of the Convex-Simplex method is apparent together with the robustness of the GP/QP-algorithm.

Having constructed an efficient algorithm for the posynomial case, this algorithm may be used iteratively to solve the general polynomial program PP(A). The process is outlined in the numbered paragraphs below, which refer to Fig. 8.

(1)   The standard Templeman input format for PP(A) is used.

(2)   A preliminary reduction puts the problem into the form of TPP(A).

(3)   At present an initial primal feasible point, needed to start the process off, is input as data. Various attempts to automate completely the selection of a starting point, without implementing one of the time-consuming penalty function methods due to Fiacco and McCormick (1968) have not been entirely successful.

(4)   The single-term posynomial "condensation" of (3.12) is now performed. Any of the other condensation methods could have been used, but we use the Avriel and Williams (1970) method. Passy (1972) evaluates the various other methods, some of which have attractive convergence properties.

(5)   The QP-based posynomial algorithm GP/QP now solves the approximating posynomial program given in section 3.

(6)   If the method has converged, output the results. If not, pass control back to (4) for a further iteration.

From a computational viewpoint, three aspects of the above approximating scheme are interesting.

(a)   The number of dual variables, d, in the approximating posynomial programs equals $N_p - (m+1)$, where N_p is the total number of terms in the numerators of (3.2).

(b) It is easy to see that the feasible set, and hence the optimal set, for any of the approximating posynomial programs is contained in the (assumed compact) set of the original polynomial program. Consequently, every point in the sequence $\{x^{(\alpha)}\}$ is feasible.

(c) The optimal dual vector δ^* at stage α can be used to ensure that an initial feasible dual point near to the current optimum is found to start stage $(\alpha + 1)$.

The result of applying this condensation procedure to one of the polynomial problems arising from the factory optimization study, is given in the table below.

Condensation iteration	x_1	x_2	x_3	x_4	x_5	x_6	x_7	$g_0(x^*)$
0	1500	3000	65	200	1.0	0.0030	5000	
1	1009	2139	78	313	9.4	0.0019	4329	23474
2	1005	2147	78	316	9.0	0.0019	4248	23449
3	1002	2155	78	309	9.2	0.0019	4223	23440
4	1001	2155	78	310	9.2	0.0019	4220	23440
5	1001	2155	78	309	9.3	0.0019	4218	23440
6	1001	2155	78	309	9.3	0.0019	4218	23440

This type of convergence behaviour is fairly typical of the method.

5. AN EVALUATION OF GEOMETRIC PROGRAMMING ALGORITHMS

Algorithms used to solve polynomial programming problems can be considered under three headings.

(A)   Algorithms of general applicability which have been applied directly to solve the primal programs A and B of section 3.

(B)   Algorithms which have exploited the special properties of the primal programs B of section 3.

(C)   Algorithms which have exploited the special properties of the dual programs B and the duality theory.

(A) General methods

Any of the standard methods can be used to solve PP(A) or PP(B) directly. Himmelblau (1973) has evaluated some of the better methods in a recent paper. Some comparative results on polynomial problems solved by general methods have been given by Asaadi (1973). The QP-based algorithm GP/QP described in section 4 performed as well as any general method, although a comparison on any criterion other than "CPU time for a given accuracy", is difficult because of the nature of the iterations.

(B) Primal methods

The special properties and convexity of PP(B) can be exploited in many different ways. Three examples are given in the following:

(1)   Dawkins *et al.* (1974): Use is made of the method of convex programming by tangential approximation due to Hartley and Hocking (1963).

(2)   Zoracki (1973): Use is made of a result of Duffin and Peterson (1972) to reduce PP(B) to a posynomial program consisting of a single-term objective function and constraints with at most two terms. The cutting-plane method of Kelly is then used to solve the resulting "posybinomial" program.

(3)   Dembo *et al.* (1973):   The condensation procedure of section 3 is used to reduce PP(B) to a
program consisting entirely of single-term constraints. Linear programming can then be applied
directly to solve a sequence of such condensed programs.

(C) Dual methods

Methods in this **class** are most numerous.
Some of the better known are listed.

(1)   Templeman *et al.* (1972):   Applies a modified
Fletcher-Powell method to solve DP(B).

(2)   Westley (1971):   Applies the method of
Murtagh and Sargent (1969) to solve TDP(B).   A
similar method is **due** to Morris (1973).

(3)   Kochenberger *et al.* (1972):   Applies separable
programming to the (separable) DP(B).

(4)   Kochenberger *et al.* (1974):   Applies a modified Newton method to solve TDP(B).   A similar
method has been used by Bradley (1973) and
Jefferson (1973).

(5)   Beck and Ecker (1972):   Use is made of a
modified version of Zangwill's Convex-Simplex
method to solve DP(B).   The method incorporates an
interesting way of avoiding the non-differentiability issue mentioned in section 3.

(6)   Staats (1970):   The technique of surrogate
constraints is used and the **dual** surrogate program
is solved.

The non-differentiability issue raised in
section 3, which is associated with the dual programs B, has been tackled in many ways by the various authors.

(1)   Lower bounds can be imposed on the dual
variables to avoid singularity in the gradient and
hessian. This often leads to high CPU times and

inaccurate answers.

(2)   Slack variables can be used, Duffin and Peterson (1972). Any of the computationally stable methods involve the addition of terms, which increases d, the number of dual variables, thereby making the dual program more difficult to solve.

(3)   Methods based on the sequential isolation and elimination of slack primal constraints by means of the Lagrange multipliers associated with the corresponding dual constraints are very efficient in practice. In our experience, this method has never failed, and has the added advantage of reducing the size of the dual program progressively. It remains to give the method as sound a theoretical base as the slack variable method.

It would not be appropriate to quote CPU times for the various GP codes which have been kindly made available to the authors, since all such codes are experimental in nature. However, the following general remarks can be usefully made. (1)   No one of the above methods is obviously superior to all others. Only the constant use of such codes on the solution of a wide range of practical problems will guide us to the better methods.

(2)   The classic problem of whether to use LP based methods or higher order approximations for NLP problems is posed, in geometric programming, in the unique context where first and second derivatives are equally easy to obtain. It is our experience that the saving in CPU time when using second order approximations more than counter-balances the extra work involved. We also feel that, with good QP codes available, there is no virtue *per se* in reduction to LP. For example, in the method of Kochenberger *et al.* (1972), a CPU time of about 1 minute is quoted to obtain a slightly inaccurate solution to a small problem on an IBM 360/67, using separable programming. The same problem was solved in 1.5 seconds (IBM 360/65) to a high accuracy, using a second order method.

Here, the linear convergence of the SP method seemed to cause the trouble.

(3)   Whether it is more efficient to solve the primal or the dual programs remains an open question.

6, CONCLUSIONS

In this final section we isolate some points of interest which arose from our studies and comment on further possible extensions.

In civil engineering, one is often faced with optimization problems which contain discrete and discontinuous factors. Various methods are available to deal directly with such problems, but one is usually restricted to small examples containing few variables if computer time is to be kept within reasonable bounds. We have found that an initial use of continuous approximations helps to locate the area in which the discrete optimum lies and shows up those variables which have the largest effect on the problem, and those which can be effectively ignored. Hence the continuous approach is an essential preliminary to the discrete, simply because it can be used economically as an investigating tool to examine the nature of a complex problem.

Since optimization applications in civil engineering seldom, if ever, produce spectacular savings over time honoured traditional methods, one is often called upon to give a cost-benefit justification of optimization studies. We have found such benefits to be greatest in situations such as the following:

(a)   if the component or structure optimized is to be mass-produced, even marginal savings in resources can be important;

(b)   if many different structures are to be evaluated and compared, a systematic and unbiased

511

approach is necessary (e.g., the factory study);

(c) if the system being modelled is dependent in a nonlinear fashion on parameters which are likely to change over a period of time (e.g., effects of inflation, fuel price changes) a nonlinear study is required.

In some of the problems which we considered, we were faced with conditions of randomness in such quantities as unit costs. For example, the unit cost of bricklaying is not a constant c_b £ per square foot, but is a statistical distribution with a certain mean value. Avriel and Wilde (1970) have incorporated this type of situation into the framework of geometric programming for engineering applications, while Sengupta (1972) has developed methods applicable to econometric studies.

Finally, while the search for global solutions to non-convex problems is a refinement still beyond the reach of most civil engineering applications, nevertheless, Falk (1973) has applied the Falk-Soland algorithm to the negative term polynomial programming problem and has proved convergence to a global solution. However, the method is still restricted to very small examples.

APPENDIX I

The following test problem due to Beck and Ecker (1972) is referred to in section 4.

minimize

$$(10.0)\,x_1 x_2^{-1} x_4^2 x_6^{-3} x_7^{\alpha} \; + \; (15.0)\,x_1^{-1} x_2^{-2} x_3 x_4 x_5^{-1} x_7^{-0.5}$$
$$+ \; (20.0)\,x_1^{-2} x_2 x_4^{-1} x_5^{-2} x_6 \; + \; (25.0)\,x_1^2 x_2^2 x_3^{-1} x_5^{0.5} x_6^{-2} x_7$$

subject to

$$(0.5)x_1^{0.5}x_3^{-1}x_6^{-2}x_7 + (0.7)x_1^3 x_2 x_3^{-2} x_6 x_7^{0.5}$$
$$+ (0.2)x_2^{-1}x_3 x_4^{-0.5}x_6^{0.6667}x_7^{0.25} \leqslant 1$$

$$(1.3)x_1^{-0.5}x_2 x_3^{-1}x_5^{-1}x_6 + (0.8)x_3 x_4^{-1}x_5^{-1}x_6^2$$
$$+ (3.1)x_1^{-1}x_2^{0.5}x_4^{-2}x_5^{-1}x_6^{0.3333} \leqslant 1$$

$$(2.0)x_1 x_3^{-1.5}x_5 x_6^{-1}x_7^{0.3333} + (0.1)x_2 x_3^{-0.5}x_5 x_6^{-1}x_7^{-0.5}$$
$$+ (1.0)x_1^{-1}x_2 x_3^{0.5}x_5 + (0.65)x_2^{-2}x_3 x_5 x_6^{-1}x_7 \leqslant 1$$

$$(0.2)x_1^{-2}x_2 x_4^{-1}x_5^{0.5}x_7^{0.3333}$$
$$+ (0.3)x_1^{0.5}x_2^2 x_3 x_4^{0.3333}x_5^{-0.6667}x_7^{0.25}$$
$$+ (0.4)x_1^{-3}x_2^{-2}x_3 x_5 x_7^{0.75} + (0.5)x_3^{-2}x_4 x_7^{0.5} \leqslant 1$$

In A to F (p.504),the parameter "α" in the first term of the cost function takes the values $\alpha = -\frac{1}{4}, -\frac{1}{8}, \frac{1}{8}, \frac{1}{4}, \frac{1}{2}$ and 1. The activation or slackness of constraints one and four is highly sensitive to the value of α.

REFERENCES

Asaadi, J. (1973) "A Computational Comparison of Some Nonlinear Programmes", *Math. Prog.* **4**, 144-154.

Avriel, M. and Wilde, D.J. (1970) "Stochastic Geometric Programming", *in* "Nonlinear Programming", *ed.* H. Kuhn, Princeton.

Avriel, M. and Williams, A.C. (1970) "Complementary Geometric Programming", *SIAM J. Appl. Math.*, **19**, 125-141.

Beck, P.A. and Ecker, J.G. (1972) "A Modified Concave Simplex Algorithm for Geometric Programming", O.R. Research Paper, No. 37-72-P6, Rensselaer Polytechnic Inst.

Bradley, J. (1973) "NEWTGP - An Algorithm for the Numerical Solution of Prototype Geometric Programmes", Rept., IIRS.

Bradley, J., Brown, L. and Feeney, M. (1974) "Cost Optimisation in Relation to Factory Structures", *Eng. Optimisation*, **1**, 125-138.

Clyne, H. and Bradley, J. (1974) "A Study of the Optimum Balance Between Construction and Heating Costs for Domestic Dwellings", Internal IIRS Rept. No. 3446/74, Building Division.

Crowe, W. (1971) "Optimum Design of Composite Panels Employing Fibreboard and Other Facing Materials", *in* "Proc. Symp. on Low-Rise Lightweight Constructions", Budapest, April, 1971.

Dawkins, G.S., McInnis, B.C. and Moonat, S.K. (1974) "Solution to Geometric Programming Problems by Transformation to Convex Programming Problems", *Int. J. Solids Structures*, 10, 135-136.

Dembo, R., Avriel, M. and Passy, U. (1973) "Solution of Generalised Geometric Programmes", O.R. Mimeograph Series No. 140, Technion-Israel.

Dinkel, J., Kochenberger, G. and McCarl, B. (1974) "An Approach to the Numerical Solutions of Geometric Programmes", *Math. Prog.*, 7, 181-190.

Duffin, R., Peterson, E. and Zener, C. (1967) "Geometric Programming - Theory and Applications", John Wiley, New York.

Duffin, R. and Peterson, E. (1972a) "Geometric Programs Treated With Slack Variables", *Appl. Analysis*, 2, 255-267.

Duffin, R. and Peterson, E. (1972b) "The Proximity of (Algebraic) Geometric Programming to Linear Programming", *Math. Prog.*, 3, 250-253.

Duffin, R. and Peterson, E. (1973) "Geometric Programming With Signomials", *J. Opt. Theory and Appl.*, 11, 3-35.

Falk, J.E. (1973) "Global Solutions of Signomial Programmes", Rept., School of Eng. Sci., George Washington Univ., No. T-274.

Fiacco, A.V. and McCormick, G.P. (1968) "Nonlinear Programming: Sequential Unconstrained Minimisation Techniques", John Wiley, New York.

Fletcher, R. (1970) "The Calculation of Feasible Points for Linearly Constrained Optimisation Problems", UKAEA-Harwell Rept., No. R.6354.

Fletcher, R. (1971) "A FORTRAN Subroutine for General Quadratic Programming", UKAEA-Harwell Rept., No. R.6370.

Fletcher, R. (1972) "An Algorithm for Solving Linearly Constrained Optimisation Problems", *Math. Prog.*, 2, 133-165.

Gero, J. (1973) "Architectural Optimisation - A Review", presented at the Conference on Optimisation in Civil Engineering, Liverpool, 1973.

Hartley, H. and Hocking, R. (1963) "Convex Programming by Tangential Approximation", *Mangt. Sci.*, 9, 600-612.

Himmelblau, D.M. (1973) "Evaluation of Constrained Nonlinear Programming Techniques", Rept., Dept. of Chemical Eng., The Univ. of Texas, Austin.

Jefferson, T. (1972) "Geometric Programming with an Application to Transport Planning", PhD Thesis, Northwestern Univ., Illinois.

Kochenberger, G., Woolsey, R.E.D. and McCarl, B. (1972) "On the Solution of Geometric Programmes via Separable Programming", Mangt. Sci. Rept. Series, 72-9, Univ. of Colorado.

Lee, B.S. and Knapton, J. (1973) "Optimum Cost Design of a Steel Framed Building," presented at the Conference on Optimisation in Civil Engineering, Liverpool.

Lipson, S.L. and Russell, A.D. (1971) "Cost Optimisation of Structural Roof Systems", *J. Str. Div.*, Proc. ASCE, 97, 2057-2071.

Morris, A.J. (1972) "Structural Optimisation by Geometric Programming", *Int. J. Solids Structures*, 8, 847-864.

Morris, A.J. (1973) "Optimisation of Statically Indeterminate Structures by means of Approximate Geometric Programming", Proc. 2nd AGARD Symp. on Structural Optimisation.

Murtagh, B.A. and Sargent, R.W.H. (1969) "A Constrained Minimisation Method with Quadratic Convergence", *in* "Optimisation", *ed.* R. Fletcher, Academic Press.

Page, J.K. (1974) "The use of elementary techniques to optimise the shape of rectangular buildings to minimise conduction heat losses through the fabric", Dept. of Building Sci., Univ. of Sheffield, No. BS-15.

Passy, U. (1972) "Condensing Generalised Polynomials" *J. Opt. Theory and Appl.*, **9**, 221-237.

Rijckaert, M.J. (1973) "Engineering Applications of Geometric Programming", *in* "Optimisation and Design", *eds*. M. Avriel, M. Rijckaert and D. Wilde, Prentice-Hall.

Rockafellar, R.T. (1970) "Convex Analysis", Princeton Univ. Press.

Sengupta, J.K. (1972) "Stochastic Programming - Methods and Applications", North-Holland.

Staats, G.E. (1970) "Computational Aspects of Geometric Programming With Degrees of Difficulty", PhD Thesis, Univ. of Texas, Austin.

Templeman, A.B. (1970) "Structural Design for Minimum Cost Using the Method of Geometric Programming", *Proc. I.C.E.*, **46**, 459-472.

Templeman, A.B., Wilson, A.J. and Winterbottom, S.K. (1972) "SIGNOPT - A Computer Code for the Solution of Signomial Geometric Programming Problems", Rept., Dept. of Civil Engineering, Univ. of Liverpool.

Westley, G.W. (1971) "A Geometric Programming Algorithm", ORNL Report No. 4650.

Wilson, A.J. (1973) "Optimisation in Computer Aided Building Design", PhD Thesis, Dept. of Building Sci., University of Liverpool.

Zoracki, M.J. (1973) "A Primal Method for Geometric Programming", PhD Thesis, Rensselaer Polytechnic Institute, New York.

OPTIMIZATION IN THE PRESENCE OF NOISE
—A GUIDED TOUR

Peter Young

*(Control Division, Department of
Engineering, University of Cambridge)**

"The gradual discovery, that there
are certain classes of phenomena, in
which, though it is impossible to predict
what will happen in each individual case,
there is nevertheless a regularity of
occurrence if the phenomena be considered
together in successive sets, gives a clue
to the abstract enquiry upon which we
are about to embark."

John Maynard Keynes
A Treatise on Probability (1921)

1. INTRODUCTION

There are many methods of optimization available to the systems analyst; indeed, given the
veritable plethora of different procedures suggested in the technical literature over the past 20
years, he may well have great difficulty in deciding on the most appropriate procedure for any particular problem with which he is confronted. But
many of the best and most efficient techniques in
current use are totally deterministic in nature
and when applied to problems affected by "noise",
whether it be error in measurement or uncertainty

*Present address - Professorial Fellow at the
Centre for Resource and Environmental Studies,
Australian National University, Canberra.

in prediction, they are either unable to reach an optimum at all or they may reach a false optimum. As Wilde (1964) has said, "Efficiency in a search technique means nothing unless the procedure is certain eventually to reach the optimum sought; noise forces us to consider convergence before efficiency".

In this paper, we take a selective look at some of the methods available for overcoming the problems of both static and dynamic optimization in the presence of noise. In order to clarify the exposition we will also attempt to unify the various procedures by considering them within the context of the technique known to mathematicians as *stochastic approximation*. We will see that this technique, in one form or another, permeates many existing methods of statistical inference from the simple recursive procedure (Young (1974)) for computing the mean value of a random variable to the intricacies of optimally estimating the changing state of a dynamic system. And in discovering this unifying link between the various techniques we will, perhaps, be able to see how new deterministic optimization procedures, whether analytically or heuristically based, can be made more rugged in the presence of noise, more able to reach their pre-assigned goals despite the confusions introduced by the inevitable uncertainty that characterizes our observation of the real world.

Before initiating our tour through the mysteries of stochastic optimization, however, it is necessary to discuss briefly a minor semantical problem of the kind which always arises when one attempts to unify techniques that have been developed by people from different academic persuasions. The reader has already been exposed to one term "recursive" with which he may not be acquainted or, alternatively, may find an inappropriate description of the procedure involved. But we are all, to some extent, prisoners of common usage; unable to break away from the current vocabulary even though we may not personally consider it

518

satisfactory. In this sense, there can be no doubt
that a technique in which an estimate is updated
on receipt of fresh information is, in the vocabulary of the control and systems analyst, termed a
recursive algorithm. And, at the same time, it is
from the control and systems field that most work
on such recursive methods has emanated in the past
15 years (see, e.g., Kailath (1974), Young (1974,
1975a and b)). Of course, the statistician may not
agree with this terminology; indeed he may well
argue that he has the establishment on his side
for there is no apparent reference at all to the
term recursive in the concise Oxford English Dic-
tionary. To the statistician, such techniques
are probably better known as *sequential* or possibly
iterative methods of estimation. And even control
theorists may not all be convinced that recursive
is a good term to use: for instance, in the translation of Tsypkin's book on "Adaption and Learning in Automatic Systems" (1971) *iterative* is sometimes the chosen description - although that may
well echo more the mood of the translator than
that of Tsypkin himself. But here we will retain
recursive to mean either updating on receipt of
fresh information or, for example, working through
a block of data one item at a time as shown in
Fig. 1.

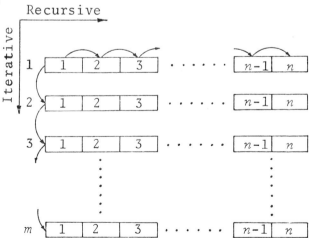

Fig. 1. Recursive and iterative data processing

519

The implication here is, of course, that the data
are often time-series data and the recursion is
with respect to time. In this sense, we will
retain *iterative* to mean the sequential processing
of a *complete* set of data as shown in Fig. 1 in
which, at each iterative step, the data base
remains the same and only some estimated variable,
say a control sequence or a parametric vector, is
modified; as when hill climbing is used to determine a set of parameters which optimize a given
performance index. Thus we might use a *recursive*
algorithm to estimate the parameters of a regression relationship which varies with time (Young
(1975b)) by passing through the time-series data
and attempting to track the parameter variations.
And subsequently, we might then process the data
iteratively, each time using the full data set, in
order to further *refine* the estimates and so obtain
better statistical efficiency*. It is realized,
of course, that these definitions are themselves
not mutually exclusive and that there may still be
situations in which it would be difficult to specify whether a procedure was recursive or iterative,
but they will suffice for our present purposes.

2. A SIMPLE EXAMPLE OF STOCHASTIC OPTIMIZATION: ESTIMATING THE MEAN VALUE OF A RANDOM VARIABLE

Working on the premise that most techniques
of general utility are, although the originator
may not always admit it, first developed by considering a simple example, let us look at the problem
of determining the mean value x of a random variable
y where

$$y = x + \varepsilon_y \qquad (1)$$

*With these definitions it will be seen that for a
given block data of finite size containing n data
elements, the recursive procedure terminates after
n steps whereas the iterative procedure continues,
as in the hill climbing method, until some satisfactory outcome is realized; for example the parameters no longer change with further iteration.

and ε_y is a "noise" term with zero mean and finite variance σ^2.

If we have k observations of y denoted by y_i, $i = 1,2,\ldots,k$, then it is well known that the "best" estimate $\bar{x}_k$ of x based on the k samples is the arithmetic mean, $i.e.$,

$$\bar{x}_k = \frac{1}{k}\sum_{i=1}^{k} y_i \, . \tag{2}$$

But what is the relationship between $\bar{x}_k$ based on k samples and $\bar{x}_{k-1}$ based only on previous $k-1$ samples? In other words, using the terminology discussed in the last section, what is the *recursive* algorithm for estimating the mean?

There are basically two ways in which we might obtain this recursive algorithm: the first by simple manipulation of the "block data" solution (2) and the second by considering the problem within the wider context of recursive least squares estimation (Young (1974)). Let us pursue both approaches in turn since this may better help to set the scene for the more general discussion which will follow in subsequent sections.

From (2) we see that

$$k\bar{x}_k = \sum_{i=1}^{k} y_i \quad \text{and} \quad (k-1)\bar{x}_{k-1} = \sum_{i=1}^{k-1} y_i \, ,$$

so that

$$\underset{\sim}{\bar{x}}_k = \frac{1}{k}[(k-1)\bar{x}_{k-1} + y_k]$$

or

$$\underset{\sim}{\bar{x}}_k = \frac{k-1}{k}\bar{x}_{k-1} + \frac{1}{k}y_k \, .$$

Multiplying by $k/k - 1$ we obtain

$$\frac{k}{k-1}\bar{x}_k = \bar{x}_{k-1} + \frac{1}{k-1}y_k$$

521

or

$$\frac{k}{k-1}\bar{x}_k = \bar{x}_{k-1} + \frac{1}{k-1}(y_k - \bar{x}_{k-1}) + \frac{1}{k-1}\bar{x}_{k-1} .$$

Finally, collecting terms, we see that

$$\left. \begin{array}{l} \bar{x}_k = \bar{x}_{k-1} + \frac{1}{k}(y_k - \bar{x}_{k-1}) \\[2em] \bar{x}_k = \bar{x}_{k-1} - \frac{1}{k}(\bar{x}_{k-1} - y_k) \end{array} \right\} \qquad (3)$$

or

which is the required recursive algorithm for $\bar{x}_k$.

While the non-recursive method of determining the mean value is well known, this recursive algorithm is comparatively little known. And yet the algorithm is significant in a number of ways: not only is it elegant and computationally attractive, but it also exposes in a most vivid manner the physical nature of the estimate for increasing sample size and so provides insight into a mechanism which, as we shall see, is useful in many more general problems.

Referring to equation (3), we see that the estimate of the mean after k samples is equal to the previous estimates $\bar{x}_{k-1}$ obtained after $k - 1$ samples plus a correction term which is the product of $1/k$ and the difference between the new sample observation y_k and $\bar{x}_{k-1}$. In effect, therefore, the previous estimate $\bar{x}_{k-1}$ is modified in proportion to the error between the observation of the random variable and the latest estimate of its mean value.

Another way of looking at the problem of estimating the mean value of a random variable is to consider it as a problem of least squares estimation (Young (1974)). Here y in equation (1) is viewed as an observation of an unknown constant x which is contaminated by the noise ε_y: the least squares estimate $\hat{x}$ of x is simply that value of $\hat{x}$ which minimizes a cost function J defined as the sum of the squares of the differences between y_i and $\hat{x}$ over the observation period, *i.e.*,

$$J \triangleq \sum_{i=1}^{k} [y_i - \hat{x}]^2 . \qquad (4)$$

The extremum (minimum in this case) of J is obtained in the usual manner by differentiating J with respect to $\hat{x}$ and equating the result to zero, $i.e.$,

$$\nabla J = \frac{\partial J}{\partial \hat{x}} = k\hat{x}_k - \sum_{i=1}^{k} y_i = 0 \qquad (5)$$

where $\hat{x}_k$ denotes the least squares estimate based on an observation size of k samples.

From equation (5) we see that

$$\hat{x}_k = \frac{1}{k} \sum_{i=1}^{k} y_i \qquad (6)$$

which is, of course, the arithmetic mean utilized previously. But continuing within the context of least squares estimation as developed, for example, in Young (1974), we can write equation (6) in the form

$$\hat{x}_k = p_k b_k \qquad (7)$$

where

$$p_k = \frac{1}{k} \quad \text{and} \quad b_k = \sum_{i=1}^{k} y_i . \qquad (8)$$

It is easily seen that p_k and b_k are related to their previous values p_{k-1} and b_{k-1} by the equations

$$\frac{1}{p_k} = \frac{1}{p_{k-1}} + 1 \qquad (9)$$

and

$$b_k = b_{k-1} + y_k . \qquad (10)$$

Multiplying throughout equation (9) by $p_k p_{k-1}$ we obtain

$$p_{k-1} = p_k + p_k p_{k-1} \qquad (11)$$

so that,

$$p_k = \frac{p_{k-1}}{1+p_{k-1}} \qquad (12)$$

and finally substituting (12) into (11)

$$p_k = p_{k-1} - \frac{p_{k-1}{}^2}{1+p_{k-1}} \; . \qquad (13)$$

This equation and equation (10) can now be substituted into equation (7) to yield

$$\hat{x}_k = [\, p_{k-1} - \frac{p_{k-1}{}^2}{1+p_{k-1}} \,][\, b_{k-1} + y_k]$$

which, on reduction and using the fact that $\hat{x}_{k-1} = p_{k-1}b_{k-1}$, can be written as

$$\hat{x}_k = \hat{x}_{k-1} - \frac{p_{k-1}}{1+p_{k-1}}[\, \hat{x}_{k-1} - y_k] \; . \qquad (14)$$

Equations (13) and (14) taken together constitute the recursive least squares algorithm for the estimation of the mean of the random variable y_k. In fact they are, not surprisingly, exactly equivalent to equation (3); as can be seen if we note that $p_{k-1} = 1/k-1$ and then substitute into equation (14) to yield

$$\hat{x}_k = \hat{x}_{k-1} - \frac{1}{k}[\, \hat{x}_{k-1} - y_k] = \hat{x}_{k-1} - p_k[\, \hat{x}_{k-1} - y_k] \, . \qquad (15)$$

The exercise of obtaining (15) in what appears to be a rather roundabout fashion is not wasted, however, for now we see how the algorithm fits within the context of optimization theory. Moreover, it is now clear that the error term $[\hat{x}_{k-1} - y_k]$ can be interpreted as a measure of the gradient of the *instantaneous* cost $[y_k - \hat{x}_{k-1}]^2$. In other words, equation (15) can be viewed as a gradient algorithm (Wilde (1964)) in which the estimate $\hat{x}_{k-1}$ is updated in a direction defined by the gradient of the instantaneous cost and with a magnitude or step size dictated by the weighting factor p_k.

But p_k is not constant; it is, in fact, in inverse proportion to the number of observations k. Thus as the algorithm proceeds and confidence in the estimate increases, so less and less notice is taken of the gradient measure, since it is more likely to arise from the noise ε_y than from error in the estimate of the mean value. And it is the harmonic sequence $1/k$, *i.e.*, $1,1/2,1/3,\ldots$, which provides the mechanism for the attenuation of the corrective action; indeed it can be seen that as the data base becomes very large, new observations have little effect since, under the basic assumption that the mean value is constant, they provide essentially no "new" information on which to base statistical inference*. In effect, the variable weighting factor p_k acts to smooth or "filter" out the inaccuracy injected by the observation noise ε_y. This interpretation of equation (15) as a filtering algorithm becomes even more transparent if a block diagram of the algorithm is constructed, as in Fig. 2. We now see that the algorithm can be considered as a discrete feedback system or filter characterized by a unity feedback loop together with a forward path which includes a time variable gain $p_k = 1/k$ and a discrete integrator $1/z-1$, where z is the forward shift operator and its inverse z^{-1} is the backward shift operator, *i.e.*, $z^{-1}\hat{x}_k = \hat{x}_{k-1}$. The feedback system of Fig. 2 is, in fact, a variable gain "low pass" filter mechanism of a type often used in communication and control systems design (Tsypkin (1971) and Takahashi *et al.* (1970)). Here the term "low pass" is applied because the filter in its fixed gain form (*i.e.*, p_k = constant) "passes" low frequency variations in

---

*Needless to say, if the assumption of a stationary mean value is not justified then such a procedure would be dangerous and could lead to poor estimation performance with, for example, heavily biased estimates. But we shall discuss this further in subsequent sections.

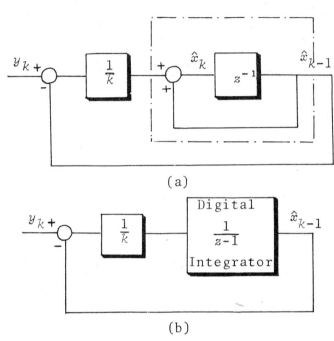

(a)

(b)

Fig. 2. Block diagram of the recursive algorithm
for estimating the mean of a random variable.

(a) basic diagram,
(b) reduced diagram showing the presence
of a digital integrator.

The diagram illustrates that the algorithm
is in the form of a time variable gain,
first order low pass filter

the input signal y_k but "filters off" any high frequency changes. When p_k is made a strictly decreasing function of time by defining it as $1/k$, the
effect is simply to reduce sequentially the "passband" of the filter until, with $k = \infty$, it passes
no signal at all not even dc (*i.e.*, zero frequency
components) and the output of the integrator
remains fixed at the final estimate of the mean $\hat{x}_\infty$.

With this interpretation in mind, it also
seems fairly clear by physical intuition, that the

526

averaging effect of the gain p_k will only work successfully if the "noise" ε_y is, in fact, a zero mean random variable with finite variance. And, as we shall see later, it is necessary to put even stricter requirements on the statistical properties of ε_y if the estimate is to be relatively efficient in a statistical sense, *i.e.*, if it is to possess low variance when compared with the theoretical minimum variance estimate.

3. STOCHASTIC APPROXIMATION

The recursive algorithm for estimating the mean value of a random variable as described by equation (15) of the previous section is an example of a technique known to mathematicians as a *stochastic approximation* (S.A.) or *stochastic gradient* procedure: the procedure is approximate in the sense that the noisy data are used recursively to generate better and better "approximations" or estimates of the true mean value; it is stochastic because of the random character of the additive noise effects; finally, it is sometimes referred to as a stochastic gradient procedure because, as we have seen, it represents the stochastic equivalent of the well known deterministic hill climbing procedures in which the search direction at any point is chosen by reference to the measured gradient of the cost function at that point. This latter remark is significant because such deterministic gradient techniques have been used both for the optimization of *functions* in a *parameter* space (Wilde (1964)) and the optimization of *functionals* in a *function* space (Lasdon *et al.* (1967)): they are, in other words, of wide practical significance. Much has been written of stochastic approximation and it would serve little useful purpose to review the topic in detail since several good essays on the subject are available elsewhere for the interested reader (see, e.g., Wilde (1964), Sakrison (1966), Mendel and Fu (1970) and Tsypkin (1971)). It will suffice here merely to point out the major characteristics of stochastic approximation in sufficient detail for the reader to appreciate its significance as a general tool for stochastic optimization.

Search techniques which are able to reach a pre-assigned goal despite the effects of noise or uncertainty were first termed stochastic approximation procedures in a seminal paper by Robbins and Munro (1951). They were concerned primarily with the problem of finding the root of a noisy function and it was left to Kiefer and Wolfowitz (1952) to point out that a similar approach could be applied to general problems of estimating the maximum of a unimodal function obscured by noise. Blum (1954) extended the approach to the multidimensional case and Dvoretzky (1956) considered in detail the conditions for the convergence of a stochastic approximation scheme; conditions which, as Wilde (1964) has pointed out, are of special significance in helping the practitioner to design his own S.A. procedure.

In order to generalize our treatment of stochastic approximation in terms of optimization theory, let us use the approach and nomenclature of Tsypkin (1971) who considers the problem of finding the value of a vector c which is optimal in the sense that it defines the extremum (maximum or minimum) of a performance index $J(c)$* which is given in the form of an expectation

$$J(c) = \int_x Q(x,c)\, p(x)\, dx$$

or, in shorthand form, as

$$J(c) = E_x\{Q(x,c)\}$$

where the realization $Q(x,c)$ is, in general, a functional of the vector $c = [c_1, c_2, \ldots, c_n]^T$ and also depends upon a vector of random sequences or processes $x = [x_1, x_2, \ldots x_n]^T$ with (usually unknown)

---

*Tsypkin refers to these general procedures of stochastic approximation as "probabilistic iterative methods".

probability density function $p(x)$. A simple form
of such a performance index or criterion of optimality is one based on the averaging of the realization $Q(x,c)$ with respect to time: if x is a random
sequence $\{x_i; \ i = 1,2,\ldots\}$ such a criterion takes
the form

$$J(c) = \lim_{k \to \infty} \frac{1}{k} \sum_{i=0}^{k} Q(x_i, c)$$

while if x is a random process $\{x(t); \ 0 \leqslant t < \infty\}$
it becomes

$$J(c) = \lim_{T \to \infty} \frac{1}{T} \int_{0}^{T} Q(x(t), c) \, dt .$$

Of course, the exact form of the performance
index will depend very much on the nature of the
problem under consideration and may well need to
be formulated bearing in mind constraints on the
various variables that characterize the problem.
We cannot deal in detail here with the question of
constraints, except to say that they may often
occur either as equality constraints, e.g.,

$$g(c) = E_x\{h(x,c)\}$$

or as inequality constraints, e.g.,

$$g(c) = E_x\{h(x,c)\} \leqslant 0$$

where the form of the vector
$g(c) = |g_1(c), g_2(c) \ldots g_m(c)|^T$ is, in general, unknown
and only the realization of the vector function
$h(x,c)$ is known. In the first case, the problem
may be solved by utilizing a Lagrange multiplier
type of approach, *i.e.*, by defining J in the form

$$J(c,\lambda) = E_x\{Q(x,c) + \lambda^T h(x,c)\}$$

while in the second case of inequality constraints,
it seems necessary to require a generalization of
the Kuhn-Tucker theorem which is itself a generalization of the Lagrange method for inequality constraints. For a more detailed and satisfying

treatment of this whole question, see Tsypkin
(1971) Chapters 2 and 3.

Posed in the above terms the solution to the
problem of optimality is basically one of finding
in some manner a vector $\underset{\sim}{c} = \underset{\sim}{c}^*$, which we shall
call the optimal vector, that defines the extremum
of the functional

$$J(\underset{\sim}{c}) = E_x\{Q(\underset{\sim}{x}.\underset{\sim}{c})\} \tag{16}$$

and which also satisfies any constraints on the
variables involved. In this general sense we may
be concerned not only with function optimization
in which the problem is to define a set of parameters which are optimal, but we could be confronted with the determination of optimal functions
over which certain functionals reach their extremal
values. A problem of functional optimization is,
of course, much more difficult to solve than a
parameter optimization problem, although it is sometimes possible to convert a functional optimization
problem into an equivalent parameter optimization
problem by expressing the functions as a linear
combination of other known linearly independent
functions characterized by a set of unknown coefficients or parameters.

For our present purposes, it is convenient
to consider only the unconstrained optimization
problem, realizing that it is possible, although
not easy, to handle such constraints using similar
algorithmic devices to those obtained in the unconstrained case. The condition of optimality for
the functional $J(\underset{\sim}{c})$ in equation (16) can be written
as

$$\nabla J(\underset{\sim}{c}) = E_x\{\nabla_c Q(\underset{\sim}{x},\underset{\sim}{c})\} = 0 \tag{17}$$

where as in the simple example considered in the
last section, ∇J represents the gradient of the
cost function and where

$$\nabla_c Q(\underset{\sim}{x},\underset{\sim}{c}) = \left[\frac{\partial Q(\underset{\sim}{x},\underset{\sim}{c})}{\partial c_1}, \frac{\partial Q(\underset{\sim}{x},\underset{\sim}{c})}{\partial c_2}, \ldots, \frac{\partial Q(\underset{\sim}{x},\underset{\sim}{c})}{\partial c_n}\right] \tag{18}$$

is the gradient of $Q(x.c)$ with respect to c.

The basic idea of solving equation (17) using stochastic approximation is to define a sequence of vectors c_k by the recursive algorithm

$$c_k = c_{k-1} - \gamma_k \{ \nabla_c Q(x_k, c_{k-1}) \} \qquad (I)$$

which can be compared directly with the simple recursive algorithm for estimating the mean value of a random variable given in equation (15) of the previous section. Unlike an equivalent deterministic gradient algorithm of this form, we know that even for $c = c^*$

$$\nabla_c Q(x, c) \neq 0$$

because of the random effects caused by the presence of x. From our experience in the last section, we have good reason to believe that in order to ensure convergence of the algorithm (I), *i.e.*, to ensure that c_k defined in (I) converges in some sense on the optimum value c^*, the weighting factor or gain γ_k must have rather special characteristics; indeed we might even guess that it should be a strictly decreasing function of k which reduces in some specific manner, perhaps as a harmonic sequence $1/k$.

In fact, such intuition is confirmed by theoretical analysis; as Tsypkin shows, the algorithm (I) will converge *almost surely* if the following conditions (which are basically derived from the conditions of Dvoretzky) are satisfied.

(a) (*i*) $\gamma_k \geq 0$; (*ii*) $\displaystyle\sum_{k=1}^{\infty} \gamma_k = \infty$; (*iii*) $\displaystyle\sum_{k=1}^{\infty} \gamma_k^2 < \infty$

(b) $\displaystyle\inf_{\varepsilon < \| c - c^* \| < 1/\varepsilon} E_x \{ [c - c^*]^T \nabla_c Q(x, c) \} > 0 \quad (\varepsilon > 0)$ $\qquad (19)$

(c) $E_x \{ \nabla_c^T Q(x, c) \nabla_c Q(x, c) \} \leq \alpha(1 + c c^T) \quad (\alpha > 0)$.

In physical terms, condition (a) requires that γ_k tends towards zero and the rate of decrease in γ_k is such that the variance of the estimate of $J(c)$

is also reduced to zero. In particular, although we want the corrective effect of γ_k to be small when c_k is close to c^*, we do not want it to vanish for $k < \infty$, for otherwise the procedure could stop before it had reached its goal; thus we require condition (a) (ii), $i.e.$, that the sum of the γ_k should be infinite so that if c_k tried to converge to other than the optimum, an infinite amount of corrective effort would be available. Of course this is not to say that convergence might not take an inordinately long time. For a good heuristic discussion on the need for condition (a) (iii) namely that the infinite sum of the variances is finite, see Wilde, chapter 6; basically, however, this property is required because, together with the requirement that the noise variance is finite, it ensures that the residual fluctuations caused by noise will die out in the long term, with the individual random errors caused by the noise at each step in the recursive algorithm tending to cancel each other out.

Condition (b) is required to define the behaviour of the surface $E_x\{\nabla_c Q(x,c)\}$ near zero, and thus the sign of the increments in c_k; while condition (c) ensures that the rate of increase of the expectation of quadratic form $\nabla_c^T Q(x,c)\nabla_c Q(x,c)$ with c is smaller than in a parabola of second degree. For further consideration of these conditions, however, see Tsypkin (1971) or Mendel and Fu (1970).

From the design standpoint, perhaps the most important of the conditions discussed above is (a) which defines the characteristics of the S.A. gain γ_k. The harmonic sequence $1/k:1,1/2,1/3,\ldots$ etc. is probably the best known example of a sequence which satisfies these properties and, as we have seen, is generated naturally in at least one estimation procedure. Other examples exist, of course, but it is necessary to be careful in choosing sequences to ensure that they do satisfy conditions (a). Some sequences, e.g., $1/k^2:1,1/4,1/9,\ldots$ are not divergent in the sense of condition (a) (iii); indeed for all $p > 1$

$$\sum_{k=1}^{\infty} \frac{1}{k^p} < \infty$$

and an S.A. procedure based on such a gain sequence might well not converge under all circumstances since the total correction effort is limited. On the other hand, the sequence $1/k^{\frac{1}{2}}:1,1/\sqrt{2},1/\sqrt{3}$, etc. would converge too slowly to dissipate the cumulative error, although a faster sequence with the power of $k > 1/2$ would be suitable.

One final comment on the basic principles of stochastic approximation concerns the nature of the realization $Q(x,c)$. Although Tsypkin, for example, gives little information on how the realization should be chosen, it is clearly extremely important from the point of view of the convergence properties and statistical efficiency of the S.A. algorithm. From purely physical considerations of the sort we have considered in this section, it seems reasonable that the realization $Q(x,c)$ should be chosen such that the noise on the *recursive residual* associated with the gradient term at each step in the recursive algorithm (I), should possess certain desirable statistical properties: for example, it should be zero mean and have finite variance so that the smoothing effect of the S.A. gain can be fully effective. In addition, and by consideration of other methods of statistical inference, we might also assume the recursive residuals should form a serially uncorrelated sequence of random variables since the problem becomes much simpler in statistical terms and it is much easier to make the solution statistically efficient.

In fact, it can be shown that all of these properties are required if the S.A. algorithm is going to perform well in a statistical sense; indeed similar properties appear to be required in all efficient recursive methods of statistical inference that function in a S.A. manner. In this sense, it could be considered that one of the primary functions of the S.A. algorithm is to ensure that the realization $Q(x,c)$ is such that the

recursive residuals constitute a zero mean, serially uncorrelated sequence of random variables; in
other words, they represent a discrete "white
noise" sequence.

In the control and systems literature(Kailath
(1968))this function of a recursive estimator in
yielding a white noise sequence of recursive residuals is termed the "innovations property" and the
sequence itself is termed the "innovations sequence"
The usual innovations approach is, therefore, first
to operate on the observed process by means of a
linear transformation (although it is possible to
conceive of analogous operations which might be
nonlinear) such that the resultant transformed or
"filtered" sequence has discrete white noise properties. The problem of statistical inference
with such white noise "observations" is, as we have
said, much easier to solve and the solution to the
simpler problem can then be expressed in terms of
the original observations by means of the inverse
of the original "whitening" filter. Similar arguments apply in the continuous-time case considered
briefly in a subsequent section.

As a concrete example of this innovations
approach, consider the mean value estimator: here
the innovations sequence is simply $\hat{x}_{k-1} - y_k$ and
it is clear that as $\hat{x}_k \to x$ so the sequence is, by
definition, serially uncorrelated if ε_y is itself
serially uncorrelated. What we are saying here is
that, in a good estimator, if ε_y is serially uncorrelated then $\hat{x}_{k-1} - y_k$ will also be serially uncorrelated *for all* k, although it will not necessarily
have constant variance because of the effects of
convergence (Kailath (1968)).

It is interesting to note that, in recent
years, a considerable amount of research effort
has been expended on the development of "adaptive"
procedures (e.g., Carew and Bélanger (1973) and
Neethling and Young (1974)) in which the recursive
algorithm (usually a state variable estimation
algorithm of the type discussed in section 5) is

adjusted until the sample statistical properties of the innovations sequence indicate that it is indeed zero mean and serially uncorrelated and that the algorithm is, therefore, performing optimally in this sense. It would be interesting to investigate whether similar adaptive or tuning procedures could be developed for more general S.A. algorithms.

4. STOCHASTIC APPROXIMATION: SOME EXTENSIONS

So far we have discussed the basic elements of S.A. applied to optimization problems and tried to provide a physical interpretation of the S.A. procedure. There are, however, a number of generalizations and extensions to the basic procedure which make it extremely flexible in application. In this section, we will consider some of these extensions, certain of which are heuristically based and others which, like the algorithm described in section 2, arise naturally when certain statistical estimation procedures are converted into recursive form.

4.1 *Multidimensional S.A. and optimum algorithms*

The first point to note is that the algorithm (I) although multidimensional in the sense that it updates the n vector c_k, is characterized by a scalar S.A. gain γ_k. In general, it seems reasonable that we might replace this by a matrix gain Γ_k, *i.e.*,

$$c_k = c_{k-1} - \Gamma_k \nabla_c Q(x_k, c_{k-1}) . \qquad (II)$$

The problem is, of course, how to determine the form of Γ_k, since it could have as many as n^2 elements. Heuristically, we might reduce the size of the problem by deciding to define Γ_k as a diagonal matrix with unequal diagonal elements, *i.e.*,

$$\Gamma_k = \begin{bmatrix} \gamma_{1k} & & & \bigcirc \\ & \gamma_{2k} & . & \\ & & . & . \\ \bigcirc & & & . \gamma_{nk} \end{bmatrix}$$

and then make each of the elements γ_{ik}, $i = 1,2,\ldots n$, satisfy the conditions (a) (i)*. But even so we do not have any rules for choosing the differences between the diagonal elements and it might be a hazardous trial and error business. Tsypkin (1971) discusses how we might choose the vector $\chi_k = [\gamma_{1k},\gamma_{2k},\ldots,\gamma_{nk}]^T$ of the diagonal elements which is optimal in the sense that it minimizes the performance index for the current values of c_k; $k = 1,2,\ldots,$ $i.e.$,

$$J(\underset{\sim}{c}_k) = E\{Q(\underset{\sim}{x},\underset{\sim}{c}_k)\}$$

He concludes that because the probability distributions of $\underset{\sim}{x}$ and $\underset{\sim}{c}_{k-1}$ are usually unknown it is, in general, necessary to simplify the formulation to the problem of minimizing the empirical or sample mean, $i.e.$,

$$J_e(\underset{\sim}{c}_k) = \frac{1}{k}\sum_{i=1}^{k} Q(\underset{\sim}{x}_i,\underset{\sim}{c}_k)$$

which corresponds to the substitution of the true probability density function by an empirical one. Even so the analysis is still not particularly simple and certain approximations become necessary before a solution is obtained; indeed Tsypkin appears to imply that, in general, it may be best to revert back to the scalar gain situation and find the optimal value for this scalar gain sequence $(\gamma_{best})_k$. In the simplest linear case he then shows that

$$(\gamma_{best})_k = \frac{1}{k + \dfrac{\sigma^2}{\bar{V}_0^2}}; \quad \bar{V}_k^2 = \frac{\sigma^2}{k + \dfrac{\sigma^2}{\bar{V}_0^2}}$$

where σ^2 is the variance of the additive noise and $\bar{V}_k^2$ is the mean square deviation of $\underset{\sim}{c}_k$ from $\underset{\sim}{c}^*$. And if, as will often be the case, there is no *a priori* information on the initial mean square deviation $\bar{V}_0^2$, then by setting $\bar{V}_0^2 = \infty$, we obtain

---

*Of course, if Γ_k had equal diagonal elements it would reduce to the scalar gain case.

$$(\gamma_{best})_k = \frac{1}{k}; \quad \bar{V}_k^2 = \frac{\sigma^2}{k}$$

which is the result obtained in the mean value estimation example considered in section 2.

There is perhaps a moral in this latter result: if we are dealing with a stochastic optimization problem in which a non-recursive analytic solution already exists, we may do better to look for a recursive version of this solution in which the stochastic approximation properties are implicit, rather than concocting an S.A. algorithm and then attempting to make it optimal in some manner. In this sense, one of the most useful algorithms that possesses S.A. properties is the recursive least squares algorithm (Young (1974)), the simplest example of which is, in fact, the mean value estimation algorithm discussed in section 2. The least square algorithm is basically concerned with a linear regression problem where the observation y is related to n linearly independent variables x_i, $i = 1,2,\ldots n$, by the relationship

$$y = a_1 x_1 + a_2 x_2 + \ldots + a_n x_n + \varepsilon_y \qquad (20)$$

where a_i, $i = 1,2,\ldots n$, are n unknown parameters and ε_y is the observation noise on y. If we have y such noisy observations y_i, $i = 1,2,\ldots,k$ and associated with the ith observation there is a set of variables x_j denoted by x_{ji}, $j = 1,2,\ldots,n$, then we can write

$$y_i = a_1 x_{1i} + a_2 x_{2i} + \ldots + a_n x_{ni} + \varepsilon_{yi};$$
$$i = 1,2,\ldots,k \qquad (21)$$

where the noise or error sequence ε_{yi}, $i = 1,2,\ldots.k$, is a zero mean serially uncorrelated sequence of random variables with variance σ^2 which is statistically independent of the variables x_{ji}. The least squares estimates $\hat{a}_j$, $j = 1,2,\ldots.n$, of the unknown parameters are those estimates which minimize the least squares cost function J, where

$$J \triangleq \sum_{i=1}^{k} [y_i - \sum_{j=1}^{n} x_{ji} \hat{a}_j]^2 \qquad (22)$$

or in vector terms

$$J \triangleq \sum_{i=1}^{k} [y_i - \underset{\sim}{x_i}^T \underset{\sim}{\hat{a}}]^2 \qquad (23)$$

where $\underset{\sim}{x_i}^T = [x_{1i}, x_{2i}, \ldots, x_{ni}]$ and $\hat{a} = [\hat{a}_1 \hat{a}_2, \ldots, \hat{a}_n]^T$.

The non-recursive solution to this problem is well known and is obtained, as in the single variable case, by setting the gradient of J with respect to the parameter estimates equal to zero in order to obtain the "normal equations" of linear regression analysis, *i.e.*,

$$[\sum_{i=1}^{k} \underset{\sim}{x_i} \underset{\sim}{x_i}^T] \underset{\sim}{\hat{a}} - \sum_{i=1}^{k} \underset{\sim}{x_i} y_i = 0. \qquad (24)$$

The solution of these equations can then be written in vector matrix form as

$$\underset{\sim}{\hat{a}}_k = P_k \underset{\sim}{b}_k \qquad (25)$$

where $\underset{\sim}{\hat{a}}_k$ is the least squares estimate based on k samples, while P_k and $\underset{\sim}{b}_k$ are defined as

$$P_k \triangleq [\sum_{i=1}^{k} \underset{\sim}{x_i} \underset{\sim}{x_i}^T]^{-1}; \quad \underset{\sim}{b}_k \triangleq \sum_{i=1}^{k} \underset{\sim}{x_i} y_i.$$

The analogy between this solution and the solution obtained in the single variable problem of section 2 is obvious on inspection and the equivalent recursive solution can be obtained by writing the recursive relationships for P_k and b_k, *i.e.*,

$$P_k^{-1} = P_{k-1}^{-1} + \underset{\sim}{x_k} \underset{\sim}{x_k}^T; \quad \underset{\sim}{b}_k = \underset{\sim}{b}_{k-1} + \underset{\sim}{x_k} y_k \qquad (26)$$

and continuing in a manner similar to that used in the scalar case, taking care, of course, to obey the rules of matrix algebra. The resulting algorithm takes the form (Young (1974))

$$\hat{a}_k = \hat{a}_{k-1} - P_{k-1}\underset{\sim}{x}_k[1 + \underset{\sim}{x}_k^T P_{k-1}\underset{\sim}{x}_k]^{-1}\{\underset{\sim}{x}_k^T \hat{a}_{k-1} - y_k\}$$
$$\text{(III}i\text{)}$$

or

$$\hat{a}_k = \hat{a}_{k-1} - P_k\{\underset{\sim}{x}_k\underset{\sim}{x}_k^T \hat{a}_{k-1} - \underset{\sim}{x}_k y_k\} \qquad \text{(III}ii\text{)}$$

where P_k is obtained from the second recursive relationship

$$P_k = P_{k-1} - P_{k-1}\underset{\sim}{x}_k[1 + \underset{\sim}{x}_k^T P_{k-1}\underset{\sim}{x}_k]^{-1}\underset{\sim}{x}_k^T P_{k-1} \text{(III}iii\text{)}$$

which can also be compared with the results in the scalar case.

Here again we see from (IIIiii) that the algorithm is of a gradient type and it can further be shown (e.g., Ho (1962)) that the matrix P_k possesses S.A. like properties. But the S.A. properties are, like those of the scalar problem, inherent in the formulation of the problem itself and we do not need to construct artificially a suitable matrix gain sequence since it is provided for us in equation (IIIiii). This same type of algorithm can be obtained directly using a S.A. argument based around a second order Newton algorithm, as shown by Kashyap *et al.* (Mendel and Fu (1970) Chapter 9), but it is felt that this is a little artificial; indeed it could be argued that this approach to the derivation of the recursive least squares algorithm was prompted by prior knowledge of the algorithm obtained in the manner outlined above*. This is not a criticism in any sense; indeed there is good evidence that the algorithm (III) was itself originated, at least in the control literature, by knowledge of the pioneering work of Rudolf Kalman on state variable estimation (a subject that, as we have said, will receive some attention in subsequent sections).

---

*Tsypkin also arrives at algorithm (III) from S.A. arguments but here again the approach is much less straightforward than that discussed here.

4.2 *Continuous-time algorithms*

Although in this age of the digital computer
we are often concerned with sampled data and discrete-time formulations, it may sometimes be advantageous to use a continuous-time approach. To
obtain a continuous-time version of the S.A. algorithm (II), we first note that it can be written as

$$\Delta \underset{\sim}{c}_k = - \Gamma_k \nabla_c Q(\underset{\sim}{x}_k, \underset{\sim}{c}_{k-1})$$

where $\Delta \underset{\sim}{c}_k = \underset{\sim}{c}_k - \underset{\sim}{c}_{k-1}$. It is then easy to see by
a limiting argument that a continuous version of
the algorithm will be of the form

$$\frac{d\underset{\sim}{c}(t)}{dt} = - \Gamma(t) \nabla_c(\underset{\sim}{x}(t), \underset{\sim}{c}(t)) \qquad \text{(IV)}$$

or in the scalar gain case

$$\frac{d\underset{\sim}{c}(t)}{dt} = - \gamma(t) \nabla_c(\underset{\sim}{x}(t), \underset{\sim}{c}(t)) \qquad \text{(V)}$$

Algorithms such as (IV) and (V) are analogous
to the continuous-time steep or steepest descent
algorithms that have been in use in the control and
computer field for some time (Young (1965)); for
example such equations with constant gains are often
used on analog computers to solve sets of algebraic
equations. The importance of the link between
these well known deterministic algorithms and the
lesser known continuous-time S.A. algorithms is
that it may often be advisable to incorporate such
a variable gain device to counter the inevitable
noise that affects any electronic mechanization;
indeed it is quite likely that variable gain modifications to continuous steepest descent solutions
are used in practice, although these modifications,
if they exist, will probably be based on some heuristic reasoning rather than a knowledge of stochastic approximation*.

---

*Although the theorist may not wish to admit it, often tech-
niques developed with great mathematical rigour and effort
have been used by a practitioner who has found by trial and
error and a detailed knowledge of his problem that they work.
His reasoning may have been completely heuristic and based
totally on intuition, but the solution is not necessarily
any the worse for that.   540

4.3 Search algorithms

If for some reason it is not possible to obtain the gradient of the realization $\nabla_c Q(x,c)$, but sample values of $Q(x,c)$ are available, then it is still possible to use S.A. by replacing the gradient in the algorithm by an estimate of the gradient obtained from the sample information; in fact this is the basis of the approach used by Kiefer and Wolfowitz (1952) mentioned earlier. How this estimate should be obtained is to a large degree open to the experimenter to choose; he may, for instance, choose one of the many methods that have been developed for deterministic hill climbing. To quote but one possible example, Tsypkin (1971) suggests perturbing c in a certain manner* and then computing the average gradient over the perturbation from the sample values of the realizations $Q(x,c)$ obtained during the perturbation experiments.

4.4 Acceleration of convergence

We have commented little up to this point on the rate of convergence of S.A. algorithms. If designed properly they are guaranteed to converge "almost surely" as $k \to \infty$. But such asymptotic convergence may be intolerably slow. Here, as is often the case in the design of hill climbing algorithms, there are two approaches open to the designer, one heuristic, the other analytic. Of the first type there are many examples but we will mention here only one, namely the method of Kesten and Lapidus *et al.* (see Wilde (1964)).

Kesten and later Lapidus *et al.* note that the ideal S.A. scheme will take large steps when far away from the optimum but should shorten its steps rapidly as it approaches the optimum. With this in mind, they monitor the gradient (actually the average gradient for they use a Kiefer-Wolfowitz scheme) and argue that when far from the optimum there will be few reversals of the sign of the

*Actually he uses a synchronous detector, but such detail is not important to the present discussion.

541

gradient but as the optimum is approached, the random effects of the noise will cause repeated reversals of sign. They then apply the simple rule that the S.A. gain will only be reduced when the direction of search changes: in this way there is a built-in "adaption" of the S.A. gain sequence and the scheme seems to be very effective in promoting more rapid convergence than that obtained using the standard S.A. gain sequence.

As far as analytic approaches are concerned, it would appear that if an optimal non-recursive analytic solution exists and is converted into an equivalent recursive solution, then the resulting implicitly generated S.A. gain sequence yields rapid convergence and it is not usually necessary to attempt any acceleration. Thus, for example, the recursive least squares solution is nearly always superior in convergence to a scalar gain S.A. algorithm applied to the same problem (see, e.g., Ho and Blaydon (1966)) simply because the implicitly generated matrix gain sequence P_k contains within it information gleaned from the sampled data which allows it to adjust the estimates in a mutually dependent fashion and so achieve faster convergence. In fact, it can be shown (Young(1974)) that if the noise sequence ε_{yi}, $i = 1, 2, \ldots, k$ in the regression model (20) has the statistical properties noted earlier, then P_k is directly proportional to the covariance of the estimation errors $P_k^* = E\{\tilde{a}\tilde{a}\}^T$, where $\tilde{a} = \hat{a} - a$ (actually $P_k^* = \sigma^2 P_k$) and the advantages of using such information in guiding the search are obvious.

This superiority of "naturally occuring" S.A. algorithms, such as recursive least squares, over their more heuristically designed S.A. competitors cannot be over-emphasized. As a general rule, if one can apply an analytically developed recursive algorithm to a given problem then one should do so, unless there are attendant problems, such as computational load, which might make the heuristically based algorithm, which is usually simpler in computational terms, more attractive:

one would then be trading speed of convergence for
low computer storage requirements.

Certainly some scalar gain S.A. algorithms
can be exceptionally slow to converge and the potential user should, more than anything else, realize
this. The point is, however, that heuristically
designed S.A. algorithms are, just because of their
simplicity, extremely flexible and widely applicable.
And therein lies their attraction: better an algorithm which takes a long time to obtain an answer
than no algorithm at all. After all there is at
least the possibility with an S.A. algorithm that,
having designed it and found it too slow to converge, the designer can modify it in some manner
in order to improve its convergence properties.
And perhaps his time might be better spent doing
this than looking for some non-existent or very
complex optimum solution: such problems are the
domain of the mathematician and the practitioner
is not often sufficiently well equipped to tackle
them.

5. AN EXAMPLE OF DYNAMIC OPTIMIZATION IN THE PRE-SENCE OF NOISE: THE LINEAR-QUADRATIC-GAUSSIAN PROBLEM

We have mentioned in previous sections the
difficulties of stochastic optimization in the
function space and the fact that S.A. offers one
approach to solving such problems, albeit one which
has not received much attention in the past and
whose practical utility is difficult to judge.
There is, however, one famous example of dynamic
optimization which has a particularly elegant
closed form solution and, lo and behold, has imbed-
ded within it a stochastic approximation-like mechanism.

The example is concerned with a linear dynamic
system of the form,

System: $\qquad x_{k+1} = A x_k + B u_k + \xi_k$ (26)

observation

equation $\quad : \qquad y_k = C x_k + \eta_k$ (27)

which is shown in block diagram form in Fig 3 (within dotted box). In these equations x_k is an n vector of state variables, u_k is a m-dimensional control input vector; y_k is a p-dimensional output or measurement vector while ξ_k and η_k are, respectively, n and p vectors of Gaussian random variables which can be considered as system disturbances (ξ_k) and measurement noise (η_k). These vectors of random variables are assumed mutually independent in statistical terms but the elements of each individual vector although assumed serially uncorrelated in time (*i.e.*, "discrete white noise") may be mutually correlated at the same instant of time with the associated covariance matrices denoted by Q and R, respectively. In mathematical terms, therefore,

$$E\{\underset{\sim}{\xi}_j \underset{\sim}{\xi}_k^T\} = Q\delta_{jk}; \quad E\{\underset{\sim}{\eta}_j \underset{\sim}{\eta}_k^T\} = R\delta_{jk}; \quad E\{\underset{\sim}{\xi}_j \underset{\sim}{\eta}_k\} = 0$$

where $\delta_{jk} = \begin{cases} 1 & j=k \\ 0 & j \neq k \end{cases}$ is the Kronecker delta function.

A mathematical description such as that given in equations (26) and (27) can be applied to many different physical and even socio-economic dynamic systems and is an extremely useful vehicle for general systems analysis (see, e.g., Takahashi *et al.* (1970)). An optimization problem for this type of system might be posed in the following manner: what is the control sequence u_k, $k = 1,2,\ldots,N$, which will minimize the following quadratic cost function J?

$$J = E\{\tfrac{1}{2} \sum_{i=1}^{N-1} \underset{\sim}{x}_i^T N \underset{\sim}{x}_i + \underset{\sim}{u}_i^T M \underset{\sim}{u}_i\} \tag{28}$$

where N and M are weighting matrices associated with the quadratic forms which are, respectively, positive semi-definite, and positive definite. This problem which is known to control and systems analysts as the Linear, Quadratic, Gaussian (Linear system; Quadratic cost function; Gaussian disturbances) or LQG problem has received a vast amount of attention in the last 15 years; indeed some might say with certain justification that it has

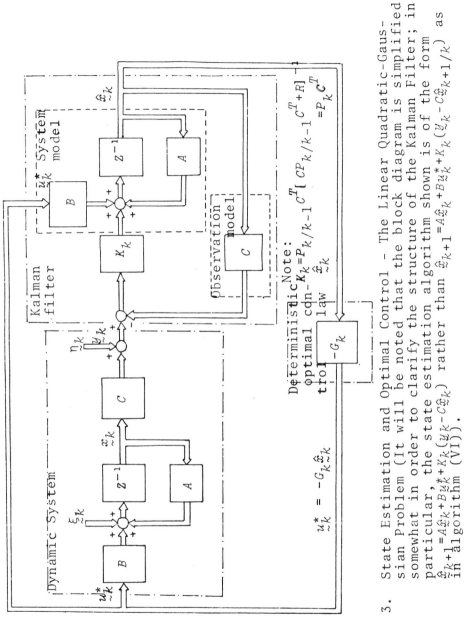

Fig. 3. State Estimation and Optimal Control – The Linear Quadratic-Gaussian Problem (It will be noted that the block diagram is simplified somewhat in order to clarify the structure of the Kalman Filter; in particular, the state estimation algorithm shown is of the form $\hat{\underset{\sim}{x}}_{k+1} = A\hat{\underset{\sim}{x}}_k + B\underset{\sim}{u}^*_k + K_k(\underset{\sim}{y}_k - C\hat{\underset{\sim}{x}}_k)$ rather than $\hat{\underset{\sim}{x}}_{k+1} = A\hat{\underset{\sim}{x}}_k + B\underset{\sim}{u}^*_k + K_k(\underset{\sim}{y}_k - C\hat{\underset{\sim}{x}}_{k+1/k})$ as in algorithm (VI)).

545

received more attention than it deserves. The reason for this attention is undoubtedly the elegance of the solution. It can be shown (see, e.g., Takahashi *et al.* (1970)) that the optimal control sequence u_k^* is obtained from a feedback control law of the form

$$u_k^* = - G_k \hat{x}_k \tag{29}$$

where G_k is an $n \times n$ time variable feedback gain matrix while $\hat{x}_k$ is an optimal estimate of the state vector x_k obtained from the following "Kalman filter" algorithm

Correction on receipt of latest data sample y_k

$$\begin{cases} \hat{x}_k = \hat{x}_{k/k-1} + P_{k/k-1} C^T (CP_{k/k-1} C^T + R)^{-1} \\ \qquad\qquad \{y_k - C\hat{x}_{k/k-1}\} \quad \text{(VI}i\text{)} \\ P_k = P_{k/k-1} - P_{k/k-1} C^T (CP_{k/k-1} C^T + R)^{-1} \\ \qquad\qquad CP_{k/k-1} \quad \text{(VI}ii\text{)} \end{cases}$$

Prediction between samples based on state equations

$$\begin{cases} \hat{x}_{k/k-1} = A\hat{x}_{k-1} + Bu_{k-1}^* \quad \text{(VI}iii\text{)} \\ P_{k/k-1} = AP_{k-1} A^T + Q. \quad \text{(VI}iv\text{)} \end{cases}$$

We will not discuss this algorithm in detail here as this has been done elsewhere (e.g., Young (1974) and Dorf (1965)), but the reader will no doubt note the similarity between equations (VIi) and (VIii), and the recursive least squares algorithm (III) discussed earlier*. This is no coincidence since the Kalman filter is, in fact, a general recursive least squares estimator and can be obtained by extending the recursive least squares regression algorithm to handle time variable unknowns (or state variables) (Young (1974)). The important point to note is that the gain matrix P_k can be interpreted in stochastic approximation terms. Not, it should be emphasized, that P_k still varies in the same simple manner observed in

---

*Here the "innovations" sequence referred to earlier is $y_k - C\hat{x}_{k/k-1}$.

the case of constant unknowns, for now the additional equation (VIiv) is present and this can have considerable effect on the P_k variation. In fact, as in the regression case, $\hat{P}_k$ can once more be interpreted as the covariance matrix of the estimation errors and the additional equations are required to account for the *a priori* information available about the variation of the states; information that is inherent in the state equations (26). But the "smoothing" effect of S.A. is still a predominant part of the algorithm - indeed it is essential to the estimation procedure; it is simply that the effect is, to some extent, masked by the incorporation of the *a priori* information on the variation of the states and we are not as aware of its presence as in the simpler examples discussed previously.

One final point about the LQG problem is that the solution makes use of a theorem known to control and systems analysts as the *Separation theorem* and to economists (who appear to have originally developed it) as the *Certainty Equivalence theorem*. This theorem, which is largely restricted to the LQG problem* gets its name because it allows the control and estimation parts of the optimal solution to be solved separately: thus the optimal control law (29) is, in fact, the optimal control law for the equivalent deterministic system without any stochastic disturbances, while the Kalman estimation algorithm is designed on the basis of the "closed loop" system equations with $\overset{*}{u}_k$ defined in (29). The entire optimal stochastic system obtained in this way is shown in Fig. 3.

6. SOME PRACTICAL EXAMPLES

Since the title of the Conference was Optimization in Action, presumably a prerequisite of any survey paper is some discussion on applications. In the case of optimization in the presence of

*Although it can be shown that some other types of optimization problem satisfy a form of separation theorem.

noise, the topic is so large that it is difficult to choose examples which are in any way representative of the total field of application and one must merely choose those which will provide some indication of the kind of problems that can be tackled using the techniques surveyed. And since any survey, particularly one which is as selective as this, tends to reflect the views of the author, it is perhaps reasonable that the applications should be those with which he has had personal contact.

6.1 A continuous-time example

One of the first practical encounters which the author had with the problems of optimization in the presence of noise (or, to be more precise, uncertainty) was in connection with the use of continuous steepest descent methods to obtain low order differential equation models of dynamic systems. In this particular case, noise in the sense of measurement noise or system disturbances was no problem since the data from the dynamic system were obtained during planned experiments where such stochastic effects were not present: the "noise" in this case was in fact the "uncertainty" associated with the use of a low order model to describe a possibly higher order system.

In fact, at the time these particular experiments were carried out, this particular aspect of the work was, perhaps, not given the prominence it deserved; indeed the publication which described the investigations (Young (1965)) did not discuss the topic at any length, although another contemporary paper concerned with hybrid methods of estimation (Young (1966)) did consider noise effects but utilized non-recursive digital algorithms. It was some years later in connection with another related problem that the importance of the earlier analog results became clear. This problem was one of designing an adaptive auto-stabilization system for a missile or aircraft. Here it was necessary to estimate the possible linear relationship between

548

low noise measurements of the vehicles' pitch angular acceleration $\dot{q}$, its pitch rate q obtained from a pitch rate gyro, its acceleration a_z obtained from an accelerometer and, finally, its rudder deflection δ obtained from a position sensor on the actuator, *i.e.*,

$$\dot{q} = a_1 q + a_2 a_z + a_3 \delta .$$

Several approaches both continuous and discrete-time were used, as described in Young and Yancey (1971), but we will discuss here an approach based on a S.A. modification of the continuous steepest descent equations developed in the earlier paper (Young (1965)). The steepest descent method was applied to a cost function which we could, in the present context, define as the mean square "equation" error*, *i.e.*,

$$J = E\{[\dot{q} - \hat{a}_1 q - \hat{a}_2 a_z - \hat{a}_3 \delta]^2\} = E\{[\dot{q} - \underset{\sim}{x}^T \underset{\sim}{\hat{a}}]^2\} \quad (30)$$

where

$$\underset{\sim}{x}^T = [q a_z \delta] \quad \text{and} \quad \underset{\sim}{\hat{a}} = [\hat{a}_1 \hat{a}_2 \hat{a}_3]^T .$$

The scalar gain S.A. algorithm for this problem takes the form of equation (V), *i.e.*,

$$\frac{d\underset{\sim}{\hat{a}}}{dt} = - \gamma(t)\{\underset{\sim}{x}\underset{\sim}{x}^T \underset{\sim}{\hat{a}} - \underset{\sim}{x}\dot{q}\} \quad (31)$$

where the expression in parentheses is the gradient of $Q = [\dot{q} - \underset{\sim}{x}^T \underset{\sim}{\hat{a}}]^2$. A schematic diagram of the analog mechanization of this scheme is shown in Fig. 4 where it will be seen that the S.A. gain is incorporated by means of a multiplier modulating the input signals to the integrator. This diagram

---

*Actually, at first an exponentially weighted past average (EWP) criterion function was used because of the possibility that the parameters might vary over time; and later, an alternative approach which utilized "air data" information to provide information on the parameter variations was found to be superior. Some details of these investigations can be found in Young and Yancey (1971).

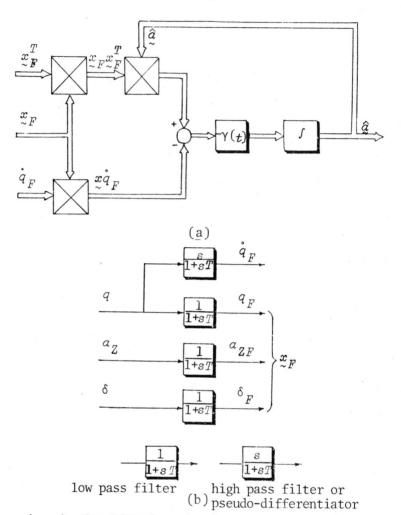

(a)

(b)

low pass filter   high pass filter or
                    pseudo-differentiator

Fig. 4. Analog Stochastic Approximation Applied
to Missile Model Estimation (a) basic
block diagram (b) generation of data from
measurements using Multiple Filters
(Young (1966))

is, of course, a vector-matrix block diagram in
the sense that the integrator is a "block" integrator which, in practice, is composed of three separate integrators each processing a separate line

of the vector-matrix equation (31) and yielding an estimate of a single parameter.

At the time of these experiments, it was not possible to mechanize a true S.A. gain generator which is, of course, not too easy using purely analog elements. Nevertheless the results obtained in simulation tests were quite good and indicated that the system could be a basis for practical mechanization. Unfortunately, as is often the case in practical situations, it was decided that a hybrid (analog-digital) mechanization had definite advantages and design studies continued using such an approach*. It is not possible, therefore, to show any realistic results obtained with the system shown in Fig. 4. It is interesting, however, to look at the results obtained from detailed simulation studies using the hybrid procedure which was based on a modified recursive least squares algorithm (Young and Yancey (1971)). Fig. 5 shows the estimate of the parameter a_3 as obtained during an 80 second simulation "mission" of the missile involved: it is clear that the parameter is estimated very well despite its large and rapid variations. This good "parameter tracking" performance is obtained by a special modification to the recursive least squares algorithm which allows for the use of "air data" information, such as dynamic pressure and altitude (Fig. 6) in order to provide an *a priori* stochastic model of the parameter variations which can then be incorporated directly into the recursive least squares algorithm (Young (1969), (1974)).

6.2 An application in pattern recognition

One of the principle areas in which stochastic approximation has been applied is in connection with pattern recognition, and many examples could be cited. A considerable amount of research in the Soviet Union, for example, has been carried out on the development of a technique known as the

*At this time (1968) solid state analog components like multipliers were not freely available.

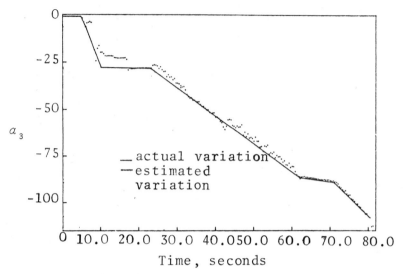

Fig. 5.   Estimation of time variable missile para-
          meter a_3

Method of Potential Functions, which was first
suggested by Aizerman *et al*. (1964) and is described
very nicely by Kashyap *et al*. (see Mendel and Fu
(1970)). This technique, which can be considered
within the context of stochastic approximation,
can involve iteration in the function space and
has, apparently, been used on many different problems.

A recent example of S.A. applied to pattern
recognition in the UK (Kittler and Young (1975))
concerns the synthesis of a discriminant function
that is initially assumed linear in a set of linearly independent functions $\phi_j(y)$ of the input
patterns y. However, in order to obtain good sensitivity in the vicinity of the decision surface,
as required in this sort of application, a nonlinear criterion function is considered of the form

$$J(\hat{a}) = \sum_{i=1}^{k} [c_i - q(\phi_i^T \hat{a})]^2 \qquad (32)$$

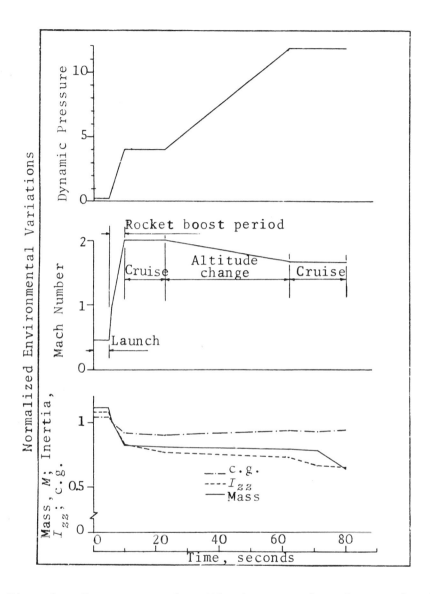

Fig. 6. Representative Mission Details (normalized)

where c_i is, effectively, a variable associated
with the known classification of y in a "training"
set of data that are to be used to determine the
discriminant function, while $\hat{a}$ and ϕ_i are, respectively, the vector of estimates of the unknown
parameters and the vector of linearly independent
functions ϕ_j that characterizes the function. The
function $g(n_i)$ is chosen to have a saturating property, *i.e.*, for large values of n_i, $g(n_i)$ approache
c_i. In particular, $g(n_i)$ was chosen as

$$g(n_i) = K_1 \tan^{-1}(n_i) + K_2 \qquad (33)$$

where K_1 and K_2 are chosen in a special manner
(Kittler and Young (1975)) which need not concern
us here.

Since the function $g(\phi_i^T\hat{a})$ is not linear in
the parameters $\hat{a}$ the problem of optimization is
not so straightforward as in the linear case and
it is necessary to resort to a linearization procedure. Here the gain matrix of the recursive
estimator is generated from the linearized quantitie
but the gradient measure (in this case simply the
error) is retained in nonlinear form. The linearization is recomputed at each recursive step on the
basis of the current estimates $\hat{a}$; in other words,
the function is linearized around a trajectory
defined by the recursive estimates*. The final
algorithm takes the form

$$\hat{a}_k = \hat{a}_{k-1} + P_k^*\phi_k^*[c_k - K_1 \tan^{-1}(\phi_k^T\hat{a}_{k-1}) + K_2] \quad (34)$$

where P_k^* and ϕ_k^* are obtained from the linearized
relationships, and P_k^* is generated as in the normal
least squares algorithm, by the recursive relationship

$$P_k^* = P_{k-1}^* - P_{k-1}^*\phi_k^*[1 + \phi_k^{*T}P_{k-1}^*\phi_k^*]^{-1}\phi_k^{*T}P_{k-1}^* . \quad (35)$$

---

*An approach often used in nonlinear filtering
(see Farison (1966)).

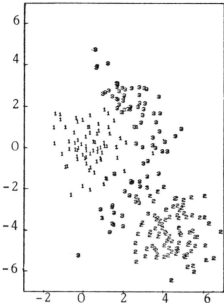

Fig. 7. Scatter diagrams of first two components of pattern vectors

Table I

		Machine Classification	
Class	1	2	3
1	97	0	6
2	0	92	3
3	1	4	97

Ordinary least squares algorithm

		Machine Classification	
Class	1	2	3
1	102	0	1
2	0	93	2
3	2	2	98

Modified least squares algorithm

(Actual Classification)

555

The efficacy of the above scheme is shown in Fig. 7 and Table I. Fig. 7 shows a scatter diagram for the first two components of a 10 dimensional pattern vector which has to be classified into 3 classes. Using a training set of 300 samples, the classification required the generation of 6 separating surfaces each described by discriminant functions defined by 11 unknown parameters (Kittler and Young (1975)). The classification results obtained using a testing set of 300 samples (different from but using the same parent statistical distributions as the training set) are shown in Table I which also compares the results with those obtained using an ordinary linear least squares algorithm: the improvement in classification is quite marked as can be seen. Similar techniques to these have been used successfully in the classification of vector electrocardiograms (Kittler and Young (1975)).

6.3 Recursive estimation applied to multivariable time-series models

As a final example, let us consider the development of recursive methods of time-series analysis and their application to the study of water quality in river systems. A paper at a previous IMA Conference (Young (1974)) introduced a particular recursive approach based on instrumental variable (IV) and approximate maximum likelihood (AML) modifications to the normal recursive least squares algorithm and showed how this IV-AML technique had been applied to the characterization of a relationship between rainfall and runoff flow in a river system. Here we shall discuss a multivariable (multi-input-multi-output) version of this recursive technique (MIVAML) and show how it is useful in obtaining a good stochastic multivariable model of water quality in a non-tidal river system.

The basic model of water quality, which is concerned with the changes in Dissolved Oxygen (DO) and Biochemical Oxygen Demand (BOD) in a reach of a non-tidal river, is a discrete-time version of

an earlier continuous-time model that was developed
using a method of state-parameter estimation known
as the Extended Kalman Filter (Young (1974)).
These later studies were aimed at improving the
statistical efficiency of the model parameters in
this example and, at the same time, developing a
general statistical technique that is both easy to
use, computationally efficient and able to produce
estimates with good relative statistical efficiency
(*i.e.*, with low variance albeit rarely minimum
variance parameter estimates).

An initial description of the MIV-AML method
of time-series analysis has appeared elsewhere
(Young and Whitehead (1975)) and we will not go
into too much detail here. Suffice it to say that,
in the language of the econometrician (Johnston
(1963)), the technique can be applied to "reduced
form" discrete dynamic models of the general form

$$y_k = \Pi v_k + w_k \qquad (36)$$

where y_k is a vector of "endogenous" or internally
generated variables, v_k is a vector of "predetermined" variables that can consist of both "exogenous" or deterministic input variables u_k and
lagged values of the endogenous variables, *i.e.*,
$y_{k-1}, y_{k-2}, \ldots$ etc., while w_k is a vector of stochastic disturbances that may be mutually or serially
correlated (*i.e.*, "coloured" noise).

The basic estimation problem posed by equation (36) is to use the noisy observations of y_k
and v_k to obtain consistent, relatively efficient,
estimates of the parameters characterizing the Π
matrix. In order to solve this problem we first
note that the ith elemental row of the vector matrix equation (36) can be written

$$y_{ik} = v_k^T \pi_i + w_{ik} \qquad (37)$$

where π_i^T is the ith row of the matrix Π and w_{ik} is
the ith element of w_k. Thus, one simple approach
to the problem of estimating the unknown parameters

557

is to decompose the over-all estimation problem into n separate sub-problems (where n is the order of y_k), each defined in terms of an estimation model such as (37) which is linear in the parameters composing the vector $\underline{\pi}_i$.

At first sight, it might appear that a consistent estimate of the parameter vector $\underline{\pi}_i$ in each of these elemental estimation models could be obtained by normal linear least squares regression analysis. But further examination reveals that this is not the case: because of the noise inherent on the variables associated with the unknown parameters, equation (37) is, in statistical terms, a *structural* rather than a *regression* model (see Kendall and Stuart (1961)): as a result, if a simple least squares estimate is obtained in the usual manner, then the estimate will, in most cases, be *asymptotically* biased and thus statistically inconsistent.

One of the simplest methods of obviating this problem is to use the IV-AML method (Young (1974)) outlined briefly in Appendix I. This method has the advantage that it is recursive, thus allowing for the identification and possible elimination of non-stationary parameters (Young (1975) and Whitehead and Young (1975)). In the present context, it is also interesting because the constituent recursive equations are, by their similarity to the ordinary least squares equation, clearly making use of a stochastic approximation-like mechanism.

The MIVAML technique has been applied to the following discrete-time model of BOD(x_1) and DO(x_2) in the reach of a river system.

$$
\underline{y}_k = \begin{bmatrix} k_1 \dfrac{V_m}{Q_{k-1}} & 0 & k_2 & 0 & k_3 & 0 & 0 \\[2ex] k_4 & k_5 \dfrac{V_m}{Q_{k-1}} & 0 & k_6 & k_7 & k_8 \dfrac{Q_{k-1}^{0.67}}{h^{2.85}} & k_9 \end{bmatrix} \underline{v}_k + \underline{w}_k \quad (38)
$$

where

$$\underset{\sim}{y}_k = [\, x_{1k} \; x_{2k} \,]^T \, ;$$

$$\underset{\sim}{v}_k = [\, x_{1,k-1} \, x_{2,k-1} \, u_{1,k-1} \, u_{2,k-1} \, S_{k-1} \, C_{s,k-1} \, w_{k-1} \,]^T \, ;$$

V_m is the mean volumetric "hold-up" in the reach (ft^3); Q_k is the volumetric flow rate ($ft^3 day^{-1}$) and h is the mean depth (ft). The undefined elements of $\underset{\sim}{v}_k$ are as follows.

u_1 is the upstream or input BOD (mg l^{-1})

u_2 is the upstream or input DO (mg l^{-1})

S is a term dependent upon sunlight hours and Chlorophyll A

C_s is the saturation concentration of DO (mg l^{-1})

w is the input of DO due to weirs or run-off at high flow (mg l^{-1}).

Thus, in terms of equation (37)

$$\underset{\sim}{\pi}_1 = \left[k_1 \frac{V_m}{Q_{k-1}} \quad 0 \quad k_2 \quad 0 \quad k_3 \quad 0 \quad 0 \right]^T$$

$$\underset{\sim}{\pi}_2 = \left[k_4 \quad k_5 \frac{V_m}{Q_{k-1}} \quad 0 \quad k_6 \quad k_7 \quad k_8 \frac{Q_{k-1}^{0.67}}{h^{2.85}} \quad k_9 \right]^T$$

where $k_1, k_2, \ldots, k_9$ are the unknown parameters to be estimated and the terms in the known or measured variables V_m, Q_{k-1} and h that occur in this particular case are, for the purposes of estimation, absorbed into the definition of the elements of $\underset{\sim}{v}_k$ for each elemental estimation problem. This example shows how quite complicated nonlinear relationships can be considered provided the model remains linear in the *unknown* parameters.

In this case, estimates of k_1 to k_9 were obtained using an iterative version of the multivariable IV procedure in which there are repeated recursive runs through the observed data with the *auxiliary model* parameters (see Appendix I) updated

after each run. This iterative procedure, which
refines the estimates and appears to produce greater statistical efficiency (Young and Whitehead
(1975)), is considered complete when further iteration yields negligible change in the estimates.
This approach is described elsewhere (Young (1974))
for the single input, single output case; the procedures are, however, identical in the multivariable
situation. Fig. 8 compares the model output of DO
and BOD with the observed values over a 228 day
period (with two short breaks).

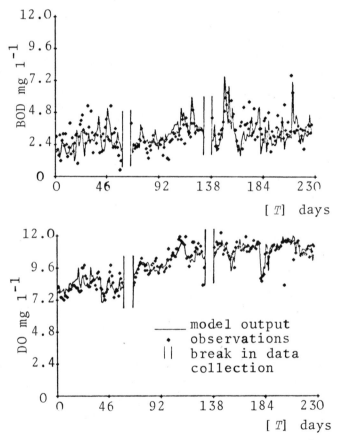

Fig. 8. Estimation results for water quality
         model

The parameter estimates for the model in this case are very little different from the estimates obtained over only the initial portion of the data (actually during summertime), thus indicating the efficacy of the model in describing the system over the whole year. The fit to the DO is better than for the BOD simply because the DO measurement is more accurate than the BOD measurement, which is notably bad, especially, as in this case, when the levels of BOD are low.

CONCLUSIONS

Any survey paper is, to a certain extent, a statement of the author's own view of the particular field of study. And this is, perhaps, more true in the present case simply because the topic "optimization in the presence of noise" is open to such a wide interpretation that any reasonable length survey must, of necessity, be rather selective in what it discusses: to attempt to cover the whole field would certainly be a mistake, particularly if the object is, as in the present case, to provide a reasonable introduction to a general audience. Here we have tried to unify the discussion on stochastic optimization by considering it within the context of stochastic approximation, a technique fairly well known to mathematicians but not, apparently, widely known or understood elsewhere except in certain specialist disciplines, such as pattern recognition. The object has been to give the reader a physical feel for the subject and to show that stochastic approximation, in one form or another, permeates many existing methods of statistical inference from least squares parameter estimation through techniques of pattern recognition and machine "learning" to state variable filter and estimation theory.

It should be emphasized that no attempt at all has been made to present a comprehensive picture of the state of the art of stochastic optimization in all its different forms. We have not, for example, considered optimization in the presence

of parametric uncertainty, a topic which, with the
increased interest in socio-economic and environmental systems analysis, will surely become very
important in the next few years. It is true that
techniques of the general type outlined here might
be applied to such problems (Kendrick and Majors
(1974)) but it is not clear whether it might not
be better to consider other approaches based either
on Monte-Carlo simulation (Barrett *et al.* (1973))
coupled with the use of non-parametric statistics,
or alternatively, purely analytic solutions.

In preparing the paper, I have drawn particularly heavily on four previous publications
(Wilde (1964), Tsypkin (1971), Sakrison (1966) and
Mendel and Fu (1970)) which have discussed stochastic approximation, and in particular Tsypkin's book
(1971) on "Adaption and Learning in Automatic Systems". In a sense, therefore, this paper constitutes in part a review of these previous studies
of the subject in which a conscious effort has
been made to simplify and so present the material
in a way which is acceptable to the general reader.
At the same time, it is hoped that the paper will
also serve to clarify certain points raised by
these previous publications and so help the reader
to understand the subject more fully in order that
he may be in a better position to design practically
useful procedures of his own.

In this latter sense, at least, the present
paper is within the spirit of this Conference for
it is attempting to encourage the practical application of stochastic optimization techniques. For
there is no doubt that any mathematical characterization of the real world must be stochastic if it
is to be of practical use and the analyst who
wishes to answer problems of the real world must
use techniques that recognize this fact and are,
above all else, rugged in the presence of stochastic phenomena. The method of least squares, for
instance, has survived as a powerful analytical
tool for approaching two centuries simply because
it possesses these qualities. And if we are to

generate new and more sophisticated analytical tools in the future, it is clear that they must also possess similar qualities of simplicity, combined with general application potential. It is the author's opinion that such future tools will, if they are to succeed, possess some if not all of the properties of stochastic approximation outlined in this paper.

We began the paper with a quotation from the great Cambridge economist John Maynard Keynes: while it is a quotation which provides the essential message of this paper, it is hoped that the reader will not conclude that our enquiry has been a predominantly abstract one. On the contrary, the paper will only have fulfilled its purpose if the physical interpretation provided here helps the reader to better understand the techniques that have been discussed and that, in this way, he is stimulated to make use of them in practical applications.

ACKNOWLEDGEMENTS

The author gratefully acknowledges the help of his colleagues and collaborators, Josef Kittler and Paul Whitehead in the preparation of results presented in sections 6.2 and 6.3.

APPENDIX I

THE MULTIVARIABLE IV-AML METHOD OF TIME-SERIES ANALYSIS (YOUNG AND WHITEHEAD (1975))

For convenience, let us consider a specific example of the reduced form model (36) in which the signal topology is more clearly defined. In particular, let us consider the state-space model (26) in which all of the state variables can be observed in the presence of noise, $i.e.$,

$$x_{k+1} = A\underset{\sim}{x}_k + B\underset{\sim}{u}_k + \underset{\sim}{\xi}_k \qquad (A1)$$

$$y_k = x_k + \eta_k \qquad \text{(A2)}$$

and where ξ_k and η_k are zero mean vectors of random variables with rational spectral density. Substituting (A2) into (A1), we obtain the following relationship between the measured variables

$$y_{k+1} = A y_k + B u_k + w_{k+1}$$

or

$$y_k = A y_{k-1} + B u_{k-1} + w_k \qquad \text{(A3)}$$

where

$$w_k = \eta_k - A \eta_{k-1} + \xi_{k-1} .$$

The basic estimation problem posed by equations (A1) and (A2) is to use the noisy observations of the state variables given in (A2) to obtain consistent estimates of the $n^2 + nm$ parameters that characterize the A and B matrices in (A1). In order to solve this problem we first note that the ith elemental row of the composite equation (A3) can be written as

$$y_{ik} = z_k^T a_i + w_{ik} \qquad \text{(A4)}$$

where

$$z_k^T = [\, y_{1,k-1} y_{2,k-1}, \cdots, y_{n,k-1}, u_{1,k-1}, \cdots, u_{m,k-1}\,]$$

$$a_i = [\, a_{i1}, a_{i2}, \cdots, a_{in}, b_{i1}, \cdots, b_{im}\,]^T$$

and

$$w_{ik} = \eta_{ik} - a_{i1} \eta_{1,k-1} - , \cdots, - a_{in} \eta_{n,k-1} + \xi_{i,k-1}$$

Thus one simple approach to the problem of estimating the unknown parameters is to decompose the over-all estimation problem into n separate subproblems, each defined in terms of an estimation model such as (A4) which is linear in an $n + m$ subset of the $n^2 + nm$ unknown parameters.

At first sight, it might appear that a consistent estimate of the parameter vector $\underset{\sim}{a}_i$ in each of these elemental estimation models could be obtained by normal linear least squares regression analysis. But further examination reveals that this is not the case: because of the inherent noise on the variables associated with the unknown parameters equation (A4) is, in statistical terms, a *structural* rather than a *regression* model; as a result, if a simple least squares estimate of $\underset{\sim}{a}_i$ is obtained in the usual manner, then the estimate will, in most cases, be *asymptotically* biased and thus statistically inconsistent (Young (1974)).

One of the simplest methods of obviating this problem is to use an Instrumental Variable (IV) modification to the simple least squares regression algorithm. This approach is well known in both the statistical and control literature (Young (1974)) and will not be described in detail here: it will suffice merely to point out that the IV estimate $\hat{\underset{\sim}{a}}_{ik}$ based on a data set of k samples is obtained by the solution of the following vector-matrix equation (which is simply a modification of the equivalent "normal equations" of linear regression analysis),

$$\hat{P}_{ik}\hat{\underset{\sim}{a}}_{ik} = \hat{\underset{\sim}{b}}_{ik}.$$ (A5)

Here $\hat{P}_{ik}$ is an $n \times n$ matrix and $\hat{\underset{\sim}{b}}_{ik}$ is an n vector, each defined as

$$\hat{P}_{ik} = \left[\sum_{j=1}^{k} \hat{\underset{\sim}{x}}_j \underset{\sim}{z}_j^{T}\right]^{-1}; \quad \hat{\underset{\sim}{b}}_{ik} = \sum_{j=1}^{k} \hat{\underset{\sim}{x}}_j y_{ij}$$ (A6)

while $\hat{\underset{\sim}{x}}_j$, $j = 1,2,\ldots,k$ is an *instrumental variable vector* chosen to be as highly correlated as possible with the hypothetical "noise free" vector

$$\underset{\sim}{x}_j = [\, x_{1,j-1}, x_{2,j-1}, \cdots, x_{n,j-1}, u_{,j-1}, \cdots, u_{m,j-1}]^{T}$$

but totally uncorrelated with the noise n_{ij}.

An alternative recursive IV solution to the problem, in which the estimate $\hat{\underset{\sim}{a}}_{ik}$ after k samples

is obtained as the linear sum of the previous estimate $\hat{a}_{i,k-1}$ plus a corrective term based on the new
information y_{ik}, z_k and $\hat{x}_k$ received at the kth sampling instant, can be obtained directly from the
non-recursive solution by simple matrix manipulation (Young (1974)) similar to that used in the
ordinary least squares case discussed in section
4.1, *i.e.*,

$$\hat{a}_{ik} = \hat{a}_{i,k-1} - k_k \{ z_k^T \hat{a}_{i,k-1} - y_{ik} \} \qquad (A7)$$

where k_k is a gain vector defined by

$$k_k = \hat{P}_{k-1} \hat{x}_k [1 + z_k^T P_{k-1} \hat{x}_k]^{-1}$$

and the matrix $\hat{P}_k$ is obtained by a second recursive
algorithm

$$\hat{P}_k = \hat{P}_{k-1} - \hat{P}_{k-1} x_k [1 + z_k^T P_{k-1} \hat{x}_k]^{-1} z_k^T P_{k-1} \qquad (A8)$$

It can be shown that the estimates obtained
either by the block data solution (A5) or its recursive equivalent (A7) and (A8) are consistent and
relatively efficient in the statistical sense
(*i.e.*, they have low variance) provided the IV vector $\hat{x}_k$ is highly correlated with x_k; indeed experimental evidence based on Monte-Carlo simulation
studies indicates that with a well chosen IV vector,
the IV estimation variances are near the Cramer-Rao
lower bound (Neethling (1974)) and compare well
with the variances of other estimators (see, e.g.,
Iserman *et al.* (1973)).

But how can the IV vector be selected so as
to ensure that comparatively low variance estimators are obtained? Here we can be guided by experience with IV methods developed previously for
single input, single output time-series models
(Young (1974)): noting the physical nature of the
problem, an *auxiliary model* of the system is first
constructed on the basis of *a priori* estimates of
the unknown parameters; this model is then used to
generate an output which can be considered as an
initial estimate of the noise free output of the

system, and which can, therefore, be used to define the IV vector. In the multivariable case considered here, the same approach can be employed, as shown diagrammatically in Fig. 9. Here the auxiliary model, like the basic system model, is multivariable and is activated by the deterministic input vector u_k: at each sampling instant its state vector $\hat{\underline{x}}_k^* = [\hat{x}_{1k}\hat{x}_{2k},\ldots,\hat{x}_{nk}]^T$ provides the source for the instrumental variables needed to define the IV vector $\hat{\underline{z}}_k$ required at the next sampling instant; in particular, the IV vector is defined as follows

$$\hat{\underline{z}}_k = [\hat{x}_{1,k-1},\hat{x}_{2,k-1},\ldots,\hat{x}_{n,k-1},u_{1,k-1},\ldots,u_{m,k-1}]^T$$

where it will be noted that since the deterministic input variables are assumed to be known exactly, they can be used directly in the definition of the IV vector and need not be estimated*.

The choice of the *a priori* estimates required by the auxiliary model can be based upon either physical knowledge of the system obtained from prior identification studies or from the biased least squares estimates obtained in an initial estimation run. As the estimation improves, however, it makes sense to update the auxiliary model in some manner in order to enhance the quality of the IV vector and so improve the statistical efficiency of the estimates. This updating can be carried out in one of two ways: first, for on-line purposes, the recursive estimates can be used as the basis for continuous updating of the auxiliary model parameters; alternatively, an off-line, iterative procedure can be used in which there are repeated runs through the block of data with the auxiliary model estimates updated only after each run is finished. This iterative procedure is considered complete when further iteration yields negligible change in the estimates. Both approaches

---

*If the system is enclosed within a feedback loop of known structure, then it will be necessary to estimate the input variables as well since these will then be contaminated by circulatory noise in the closed loop (see Young (1970)).

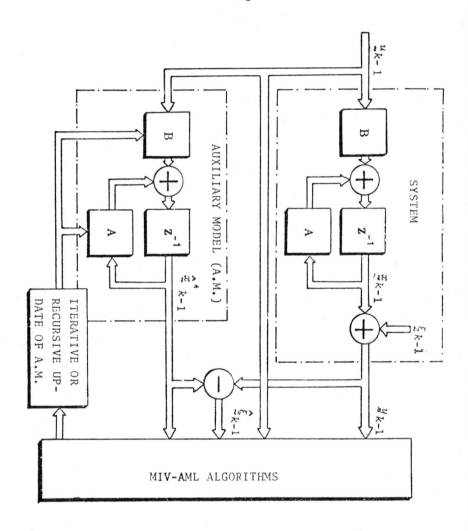

Fig. 9   Multivariable IV-AML Estimation Procedure

are described in detail elsewhere for the single input, single output case; the procedures are, however, identical in the multivariable situation.

Experience has shown that, not surprisingly, the iterative method is able to "refine" the estimates and yield superior estimation performance, particularly in applications such as macro-economic modelling where there is a relative paucity of data. It is also interesting to observe that this iterative IV procedure can be considered as an extension to the Two Stage Least Squares (TSLS) methods so popular in econometrics (see, for example, Johnson (1963)); indeed we might refer to it as Multi-Stage Least Squares (MSLS).

Having obtained relatively efficient estimates of the parameters in A and B, it remains to estimate the statistical characteristics of the noise vector $\underset{\sim}{w}_k$. This is carried out by defining an estimate of $\underset{\sim}{w}_k$ as

$$\hat{\underset{\sim}{w}}_k = \underset{\sim}{y}_k - \hat{\underset{\sim}{x}}_k$$

and then using a second recursive algorithm, the Approximate Maximum likelihood (AML) algorithm, to obtain estimates of the parameters in approximate multivariable model for $\underset{\sim}{w}_k$ (see Young and Whitehead (1975)).

REFERENCES

Aizerman, M.A., Braverman, F.M. and Rozonoer, L.E. (1964) "The Probability Problem in Pattern Recognition Learning and the Method of Potential Functions", *Automation and Remote Control*, **25**, No. 9.

Barrett, J.F., Coales, J.F., Ledwich, M.A., Naughton, J.J. and Young, P.C. (1973) "Macro-economic Modelling: A Critical Appraisal", Proc. IFAC/IFORS Conference on Dynamic Modelling and Control of National Economies, Warwick, England July, 1973.

Blum, J.R. (1954) "Multidimensional Stochastic Approximation Methods", *Ann. of Math. Stat.*, **25**, No. 4.

Carew, B. and Bélanger, P.R. (1973) "Identification of Optimum Filter Steady State Gains for Systems with Unknown Noise Covariances", *IEEE Trans. on Aut. Control*, **AC–18**, 582-587.

Dorf, R.C. (1965) "Time Domain Analysis and Design of Control Systems", Addison-Wesley.

Dvoretsky, A. (1956) "On Stochastic Approximation", Proc. Berkeley Symp. on Math. Statistics and Probability, Univ. of California Press.

Farison, J.B. (1966) "Statistical Linearisation and Variance Estimation of Products of Functions of Random Variables with Small Variations", *Proc. IEEE*, **54**, 1971-1972.

Ho, Y.C. (1962) "On Stochastic Approximation and Optimal filtering methods", *J. Math. Anal. Appl.*, **6**, No. 1.

Ho, Y.C. and Blaydon, C. (1966) "On the Abstraction Problem in Pattern Classification", Proceedings National Electronics Conference (USA).

Iserman, R., Baur, U., Bamberger, W., Kneppt, P. and Siebert, H. (1973) "Comparisons and Evaluations of Six On-line Parameter Estimation Methods with Three Simulated Processes", *in* "Identification and System Parameter Estimation" North Holland/American Elsevier.

Johnston, J. (1963) "Econometric Methods", McGraw Hill.

Kailath, T. (1968) "An Innovations Approach to Least Squares Estimation - Part I: Linear Filtering in Additive White Noise", *IEEE Trans. on Aut. Cont.*, **AC–13**, 646-655.

Kailath, T. (1974) "A View of Three Decades of Linear Filtering Theory", *IEEE Trans. on Information Theory*, **I T–20**, 145-181.

Kendall, M.C. and Stuart, A. (1961) "Advanced Theory of Statistics", Griffin.

Kendrick, D.A. and Majors, J. (1974) "Stochastic Control with Uncertain Macroeconomic Parameters", Dept. of Economics, University of Texas, Austin; Project on Control in Economics, Rep. No. 74-3.

Keynes, J.M. (1973)(originally published 1921) "A Treatise on Probability", Economic Society Edition, Macmillan.

Kiefer, J. and Wolfowitz, J. (1952) "Stochastic Estimation of the Maximum of a Regression Function", *Ann. of Math. Stat.*, 23, No. 3.

Kittler, J. and Young, P.C. (1973) "A New Approach to Feature Selection Based on the Karhunen-Loeve Expansion", *Pattern Recognition*, 5, 335-352.

Kittler, J. and Young, P.C. (1975) "Discriminant Function Implementation of a Minimum Risk Classifier", *Biological Cybernetics*, 18, No. 3/4, 169-179.

Lasdon, L.S., Mitter, S.K. and Waren, A.D. (1967) "The Conjugate Gradient Method for Optimal Control Problems", *IEEE Trans. on Automatic Control*, AC–12, No. 2, 132-138.

Mendel, J.M. and Fu, K.S. (1970) "Adaptive Learning and Pattern Recognition Systems", Academic Press.

Neethling, C.G. (1974) PhD Thesis, Control Division, Dept. of Engineering, Univ. of Cambridge, England.

Neethling, C.G. and Young, P.C. (1974) "Comments on Identification of Optimum Filter Steady State Gains for Systems with Unknown Noise Covariances", *IEEE Trans. on Aut. Cont.*, AC–19, 623-625.

Robbins, H. and Monro, S. (1951) "A Stochastic Approximation Method", *Ann. of Math. Stat.*, 22, No. 1.

Sakrison, D. (1966) "Stochastic Approximation: A recursive Method for Solving Regression Problems", *Advances in Communication Systems*, 2.

Takahashi, Y., Rabins, M.J. and Auslander, D.M. (1970) "Control and Dynamic Systems", Addison-Wesley.

Tsypkin, Ya.Z. (1971) "Adaption and Learning in Automatic Systems", Academic Press.

Whitehead, P.G. and Young, P.C. (1975) "A Dynamic-Stochastic Model for Water Quality in Part of the Bedford-Ouse River System", Proc. IFIP Working Conference on Modelling and Simulation of Water Resource Systems, Ghent, Belgium, July/August 1974, North Holland.

Wilde, D.J. (1964) "Optimum Seeking Methods", Prentice-Hall.

Young, P.C. (1965) "The Determination of the Parameters of a Dynamic Process", *The Radio and Electronics Engineer*, **29**, No. 6.

Young, P.C. (1966) "Process Parameters Estimation and Self Adaptive Control", *in* "Theory of Self Adaptive Control Systems", *ed.* P.H. Hammond, Plenum Press, 118-139.

Young, P.C. (1969) "Applying Parameter Estimation to Dynamic Systems - Parts I and II", *Control Engineering*, **16**, No. 10, 119-125; No. 11, 118-124.

Young, P.C. (1970) "An Instrumental Variable Method for Real-Time Identification of a Noisy Process", *Automatica*, **6**, 271-287.

Young, P.C. (1971) Discussion of the paper "Dynamic Equations for Economic Forecasting with GDP - Unemployment Relation and Growth of GDP in the UK as an Example", *J. Roy. Stat. Soc. A*, **134**, 220-223.

Young, P.C. (1974) "Recursive Approaches to Time-series Analysis", *Bulletin IMA.*, **10**, Nos. 5/6, 209-224.

Young, P.C. (1975) Discussion of the paper "Techniques for Testing the Constancy of Regression Relationships Over Time", *J. Roy. Stat. Soc. B*, **37**, No. 2, 149-192.

Young, P.C. "Recursive Estimation: A Practical Approach to Dynamic Systems Analysis", *in preparation.*

Young, P.C. and Whitehead, P.G. (1975) "A Recursive Approach to Time-Series Analysis for Multivariable Systems", Proc. IFIP Working Conference on Modelling and Simulation of Water Resource Systems, Ghent, Belgium, July/August, 1974, North Holland.

Young, P.C. and Yancey, D. (1971) "A Second Generation Adaptive Autostabilisation System for Aircraft and Missiles", Technical Publication TN 404-109, Naval Weapons Center, China Lake, California.

Identification of Polynomial Coefficients of a Missile System Using Flight Trials Data

D. Crombie

(British Aircraft Corporation)

SUMMARY

This report contains the results of the development of a technique for using flight trial data to identify the coefficients of a missile system during a test flight. The method was successfully used with specially prepared data, but the results obtained using the flight trial data did not give a satisfactory solution. The reasons for the unsatisfactory solution are discussed.

1. INTRODUCTION

The British Aircraft Corporation Guided Weapons division at Stevenage is concerned with the design, manufacture and testing of guided weapons systems. The work involves building mathematical models of the systems and using them to investigate many features of the total system, e.g., accuracy, reliability, sensitivity and lethality. The Corporation has both digital and hybrid computing facilities and often a model is established on both facilities so that any modelling done will be accomplished in the most economic manner.

The Hybrid Computing Office at Stevenage has performed twenty-three optimization studies using eight different problems; six different algorithms were available and some problems utilized more than one algorithm. The result obtained may be summarized as shown

successes	18
failures	5 .

Analysis of the failures showed three categories

(*a*) application of an algorithm not suited to the environment

(*b*) problems with multiple local minima in the cost function

(*c*) insufficient information content to enable the minimum to be identified

One of the failed problems is discussed in this paper to act as a guide to others contemplating using optimization methods for the first time.

The work described in this report took about two years to complete since there were long periods between the different sections. The objective was to investigate the technique of using an optimization routine, in association with some flight trials data, to determine the transfer function of the missile during different periods of its flight. Two different methods were used and both are discussed.

The various sections of the work were:

(1) definition of forcing function

(2) generating data from the detailed model

(3) performing the flight trial

(4) processing the flight trial data

(5) optimization using the simplified model with the data from the detailed model

(6) optimization using the simplified model with the data from the flight trial

(7) optimization using general polynomial transfer function.

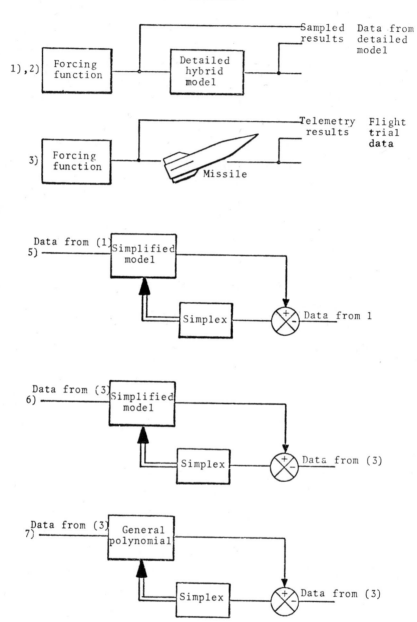

Fig. 1. Schematic diagram of sections of work

A schematic diagram of the experiment is shown in Fig. 1.

If the missile polynomial coefficients were identified by the technique then any malfunction during the flight would be easier to detect.

A detailed model of the system was already in existence on the hybrid computer and had been checked against flight trial results.

2. DEFINITION OF FORCING FUNCTION

2.1 *Requirement*

Before performing a flight trial to obtain the required "in flight" data the detailed hybrid model was used to determine a suitable forcing function to be used during the flight. Several criteria had to be satisfied.

The forcing function must:

(*a*) be reproducible at the firing range

(*b*) excite the missile to a measurable degree

(*c*) not cause the missile to operate for long periods in the nonlinear region

(*d*) not violate any range safety conditions.

2.2 *Pseudo-random binary source*

A pseudo-random binary sequence (PRBS) was decided as being satisfactory with regard to criteria (*a*) and (*b*). The number of bits in the register and the clock frequency were determined from consideration of the highest frequency the system was expected to deal with, and the lowest frequency likely to be encountered in use. The highest frequency likely to be demanded was 10Hz and a suitable low frequency was taken as 0.5Hz. These factors led to a six-bit shift register driven by a 20Hz clock being suggested as a forcing function.

2.3 Frequency of forcing function

The detailed model of the system was excited with the PRBS using different clock rates to ensure that the most suitable frequency was chosen. The system did not respond to the higher frequency components of the PRBS when the clock rate was above 25Hz, but at 20Hz the system was responding well to all the demands. The response to lower frequencies was good but it was decided to retain as much high frequency information as possible so the 20Hz clock rate was chosen.

2.4 Amplitude of forcing function

The amplitude of the forcing function was varied to determine which magnitude best satisfied the conflicting requirements of 2.1(*b*) and 2.1(*c*). A large amplitude caused the missile to operate in the nonlinear region for long periods of time, yet gave measurable values of the system parameters. A small amplitude satisfied the condition for operating within the linear region, yet the trajectory deviation became immeasurable. The trajectory, being the double integral of control surface angle, was very sensitive to any bias level in the demand; since all PRBS sequences contain a bias level an inverse repeat sequence was used. (An inverse repeat sequence attempts to minimize the effect of the bias by changing the polarity of the complete sequence each time the sequence restarts.)

The results of the amplitude investigation was that a 5° peak to peak signal gave satisfactory variations in the demanded control surface angle and the gyro angle, but the trajectory information was not suitable for matching purposes.

2.5 Required forcing function

The forcing function most suitable for the flight trial input was a 6 bit PRBS with a clock rate of 20Hz and a magnitude of 5° peak to peak.

3. GENERATION OF TEST DATA FROM DETAILED MODEL

The forcing function defined in 2.5 was generated and used with the detailed hybrid computer model. The model had been programmed to record several parameters at intervals of 0.01 second. The parameters recorded were:

(a) demand from ground

(b) demand received at missile

(c) gyro angle.

These parameters were filed on to magnetic tape for use in later parts of the experiment.

4. FLIGHT TRIAL

4.1 *Conditions of flight trial*

A flight trial was organized which was to use the conclusions of 2.5 as the forcing function. A telemetry set was installed in the missile to record the demand received at the missile PSIDM and the missile gyro angle PSIGM and there was a telemetry set on the ground to record the demand sent up the wire PSIDG. Due to practical limitation at the range, the amplitude of the forcing function used in the flight trial was 6.6° peak to peak.

4.2 *Results of flight trial*

The telemetry recordings produced during the flight were transcribed on to an ultraviolet recorder and the variables were examined for consistency. The examination showed that the three parameters of interest had reacted to the forcing function as expected. The data for each parameter were unscrambled and interpolated before being punched on to cards as a set of data at 0.01 second intervals. The two packs of telemetry data were read into a digital computer and the data were merged and filed on to magnetic tape for use later.

5. OPTIMIZATION METHOD

The optimization method selected for this experiment was the Nelder and Mead (1965) modified version of the simplex algorithm. The algorithm had been programmed and tested prior to this work, and since a financial restraint was in operation this method was chosen.

The performance index (cost function) used for the optimization algorithm was the integral of the squares of the difference between desired output (PSIDG) and achieved output (PSIGM) at every time point.

$$PI = \int_0^t (PSIDG - PSIGM)^2 dt$$

This performance index gave equal weighting to each data point and was independent of the polarity of the difference.

In the use of hybrid computers the parameters require "amplitude scaling" to operate within the voltage range of the equipment. The parameter space of an optimization problem should be scaled so that any gross differences in the parameter sensitivity should be eliminated. The coefficients of the polynomial were given ranges so that the analog scaling and the sensitivity scaling were both satisfied.

There were four conditions under which the algorithm had been programmed to terminate the search:

> (*i*) the difference between each element of the vector defining the present estimate of the minimum and the previous estimate being less than a chosen value;

> (*ii*) the modulus of the difference between the best estimate of the minimum and the function value at each other point of the simplex all being less than a chosen value;

(*iii*) number of function evaluations exceeding
a given value;

(*iv*) manual intervention by the operator.

The first criterion eliminates the possibility of the algorithm collapsing the simplex
points if the calculated performance index still
shows variations either real or noise induced.
The second criterion eliminates the situation
where the algorithm is seeking a minimum on a flat
surface. The third criterion places an upper boundary on the cost of the search and the fourth criterion is a safeguard against incorrect operation.

For this work the first two criteria were
the ones in use.

6. OPTIMIZATION USING DATA FROM THE DETAILED MODEL

6.1 Choice of polynomial transfer function

The form of the polynomial transfer function
used for the simplified model is shown in Fig. 2.
The detailed model used on the hybrid had provided
sufficient information about the various components
of the missile to enable several approximations to
be used. The kinematics of the missile had been
simplified to a double integral and a gain. The
coefficients to be determined by the optimization
routine were the seven shown in Fig. 2 with letters
A, B, C, D, E, T_1, T_2. The reason for selecting
this representation was that if the coefficients
were identified accurately, using the data generated from the detailed model, then the assumptions
concerning the kinematic representation would be
justified.

Crombie

Fig. 2. System block diagram of simplified model

582

6.2 Time sections used

The missile kinematics vary during the flight, hence the values obtained for the polynomial coefficients were expected to vary during the flight. The total simulated trial, performed with the detailed hybrid model, lasted for 20 seconds, the initial attempt at identifying the polynomial coefficients used three sections of the flight, each lasting for 6 seconds. The results shown in Fig. 3 show that the polynomial coefficients $A - E$ as identified by the optimization process are close to the nominal values and that some show a definite relationship with time.

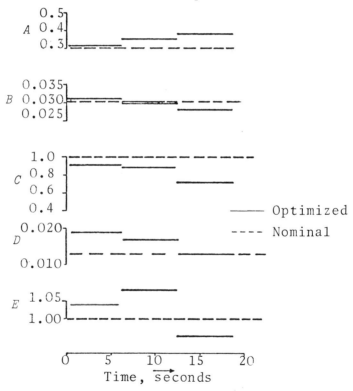

Fig. 3. Results of simplified model with data from detailed model

The optimized values of T_1 and T_2 are shown in Fig. 4, the effect that these values have on the frequency components of the input is shown in the Bode plot, gain *versus* frequency. The effect of the various optimized values was very similar to the nominal values. The values of T_1 and T_2 were not considered in further sections.

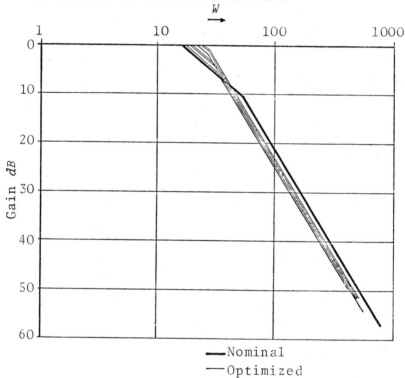

Fig. 4. Bode plots of T values optimized using data from detailed model

The length of the time sections used was reduced into 2.5 second sections and the coefficients were once again identified using the optimization algorithm. The results obtained are shown in Fig. 5 (the multiple lines for some time sections indicate the results of different optimizations), the variation of the coefficients with time shows a good agreement with the results from the longer time sections apart from $B(10 - 12.5)$ when the value chosen is much lower than expected.

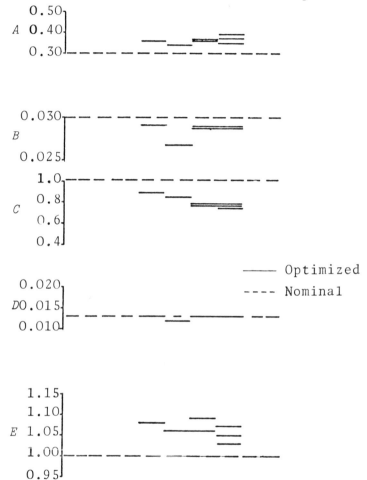

Fig. 5. Optimized coefficients of simplified model using data from detailed model

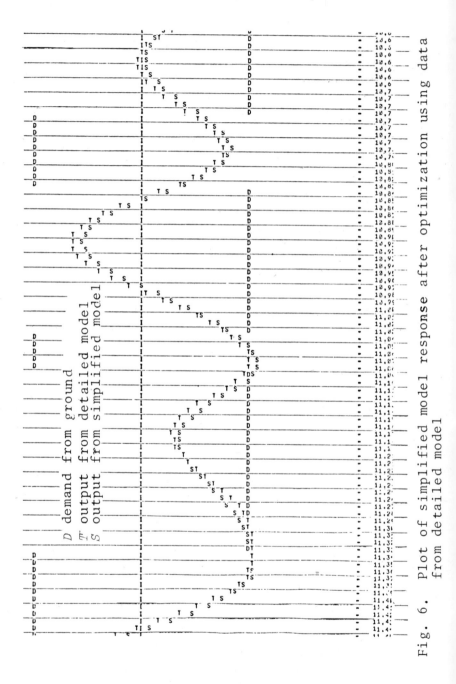

Fig. 6. Plot of simplified model response after optimization using data from detailed model

D demand from ground
T output from detailed model
S output from simplified model

6.3 Analysis of results

The results obtained, although showing a general agreement between the longer and shorter time sections, still had some variations between the coefficient values for the same time sections. The reasons for the variations will be discussed in detail later (section 9). The patterns established as a result of this work were:

(*a*) A increases with time

(*b*) B **decreases** with time

(*c*) C decreases with time

(*d*) D constant value

(*e*) E no definite pattern

The approximations made for the representation of the missile kinematics have been justified by the coefficient values being close to the nominal values for the detailed model. The result of one of the optimized runs is shown in Fig. 6.

The method was considered as being operational and the next part of the work was attempted.

7. OPTIMIZATION USING DATA FROM FLIGHT TRIAL

7.1 Transfer function and time sections used

The same procedure as described in section 5 was used, but with the data obtained from the flight trial replacing the data obtained from the detailed model. The time sections used were 2.5 seconds from 7.5 to 17.5 seconds.

7.2 Results and analysis

The results obtained are shown plotted in Fig. 7, these do not show any definite pattern for any of the coefficients. A section of the result of the optimization is shown in Fig. 8; D is the demand, S is the simplified model output and T is

587

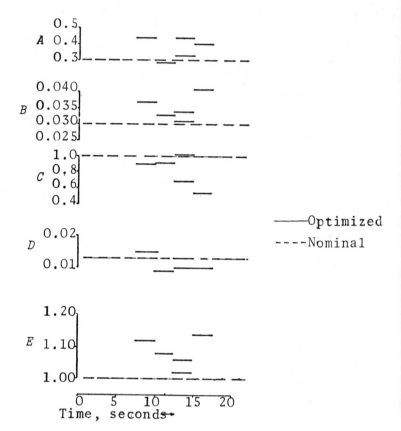

Fig. 7. Optimized coefficients of simplified
model using data from flight trial

the telemetry output. Examination of the traces
show that the two outputs (T and S) are not in
good agreement, there are certain portions of the
record where the T outputs do not follow the pattern expected from consideration of the demand, D.

The coefficients of the polynomial were not
accurately defined as a result of this work; there
were variations of up to 40% between the values
obtained using the data from the detailed model
and those obtained using the data from the flight
trial.

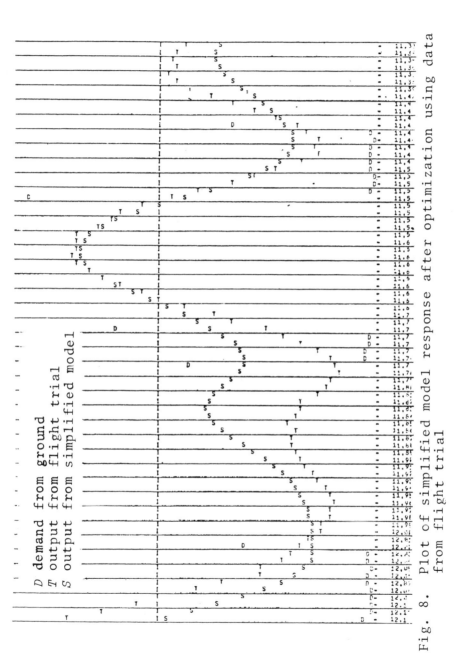

Fig. 8. Plot of simplified model response after optimization using data from flight trial

D demand from ground
T output from flight trial
S output from simplified model

The noise introduced by the recording and processing of the telemetry data has been sufficient to confuse the performance index calculation, hence the optimized coefficients are not in agreement with the expected values.

8. OPTIMIZATION USING GENERAL POLYNOMIAL TRANSFER FUNCTION WITH FLIGHT TRIALS DATA

8.1 Form of transfer function

An alternative approach to the problem of identification of the missile parameters is to use a general third order polynomial of the form:

$$G(p) = \frac{A_0 + A_1 p + A_2 p^2 + A_3 p^3}{B_0 + B_1 p + B_2 p^2 + B_3 p^3}$$

to represent the complete transfer function between demanded angle and achieved gyro angle. This makes no assumptions about the missile transfer function, but produces a polynomial having coefficients which are combinations of the various system parameters.

8.2 Optimization method

The optimization process was performed using an IBM 370 digital computer, the ground demand was used as the input to the polynomial and the cost function was the square of the difference between the polynomial output and the data from the air telemetry. The time sections used for this work were 2.5 seconds.

8.3 Results

The results obtained for the same time sections as used in 5 and 6 are shown in Table I. The variation of coefficient values for different time sections is greater than expected, (the only coefficients with small variations are B_2 and B_3).

The result of this work was the same as the conclusion to section 6, that the coefficients values were not identified accurately; the noise

Table I

Optimized coefficients of generalized model using data from flight trial.

Time	7.5-10.0	10.0-12.5	12.5-15.0	15.0-17.5
A_0	-0.893	-1.397	-0.651	-0.436
A_1	2.38×10^{-2}	0.84×10^{-2}	0.28×10^{-2}	0.02×10^{-2}
A_2	-0.245×10^{-4}	-4.766×10^{-4}	0.325×10^{-4}	0.517×10^{-4}
A_3	0.001×10^{-6}	-0.587×10^{-6}	-0.030×10^{-6}	-0.151×10^{-6}
B_0	1.0	1.0	1.0	1.0
B_1	0.794×10^{-1}	0.826×10^{-1}	0.324×10^{-1}	0.213×10^{-1}
B_2	0.317×10^{-2}	0.355×10^{-2}	0.309×10^{-2}	0.297×10^{-2}
B_3	0.350×10^{-5}	0.436×10^{-5}	0.306×10^{-5}	0.278×10^{-5}

introduced by the telemetry sets being a major contributor to the failure.

9. SOURCES OF NOISE

Three main sources of noise were present during various phases of this work:

(*a*) telemetry noise

(*b*) computing noise

(*c*) missile noise.

9.1 *Telemetry noise*

The main source of noise appears to have been from the telemetry system. Data from two different telemetry sets after being processed and merged are bound to suffer from the combined processes. It was observed from the hybrid computer work that one telemetry channel had been shifted relative to the other, this shift was removed before any of the work described here was performed.

9.2 *Computing noise*

Any analog computer has a basic noise level for its amplifiers, the noise level for the EAL 8812 amplifiers is 5mV. When the signal levels are in the range 10 - 100V this noise level is negligible and for the data preparation from the detailed model this condition existed. The optimization process was minimizing the error (as discussed in section 5) and during this process some of the signal levels become small so that the signal to noise ratio decreases and the noise becomes significant. The optimization process therefore tends to magnify the effect of amplifier noise.

9.3 Missile noise

A missile when in flight is subject to many effects which will result in noise on the gyro angle, but the major contribution comes from wind. The gyro data obtained from the flight trial will be contaminated by the missile response to whatever wind prevailed during the flight trial.

10. COMPARISON OF DIFFERENT RESULTS

The results obtained from the different phases of this work have been summarized within each section, comparisons between the various results show the effect of noise on the efficiency of identifying the polynomial coefficients.

10.1 Simplified model

The coefficients identified for the simplified model were not expected to be the same as those used in the detailed model since the approximations used would have an effect on the coefficient values. The results obtained using the simplified model with data from the detailed model showed that the approximations made in deriving the simplified model were good and that the variation of the coefficient values was small. The results obtained using the simplified model with the flight trials data show that the variation between coefficient values determined was significantly greater due to the presence of noise.

The different coefficients determined by the optimization process result in different transfer functions of the complete system. The variation of the system gain plotted against frequency, (Bode diagram) for the simplified model with the data from the detailed model is shown in Fig. 9. The plots show that the resonant peak of the system decreases as the time of flight increases. The low frequency gain is unity and the frequency cutoff is well defined. The Bode diagrams of the

593

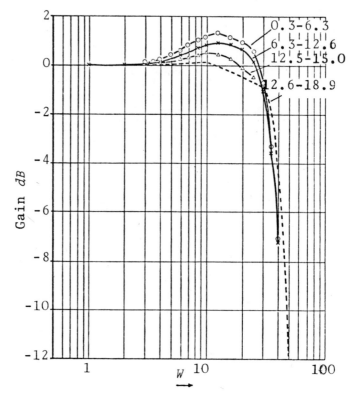

Fig. 9. Bode plots of simplified model with coefficients optimized using data from
detailed model

results obtained using the simplified model with
the flight trials data are shown in Fig. 11.
There is no pattern in either the resonant peak or
the frequency cut-off, although the low frequency
gain is always unity. The difference between the
results shown in Figs. 9 and 10 can be attributed
directly to the addition of noise on the data.

10.2 *General polynomial transfer function*

The results obtained with the general polynomial transfer function using the flight trial
data are shown plotted as a Bode diagram in Fig.11.

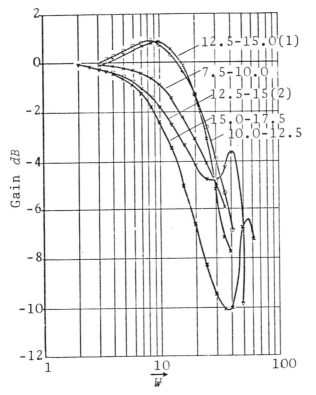

Fig. 10. Bode plots of simplified model with coefficients optimized using data from flight trial

The results show a wide variation in low frequency gain, a wide variation in resonant peak, but a common cut-off frequency.

11. CONCLUSIONS

11.1 *Forcing function*

The chosen forcing function appeared to satis-fy the requirement of exciting the loop sufficient-ly for the output to be measured accurately, yet keeping the system operating within the linear por-tion. The spectrum of the forcing function however

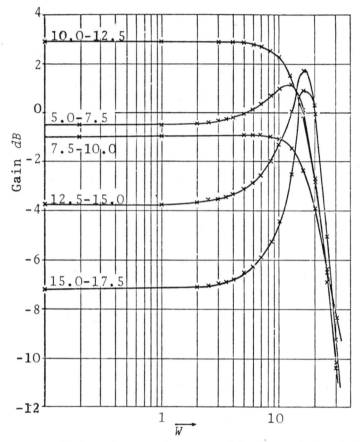

Fig. 11. Bode plots of generalized model with
coefficients optimized using data from
flight trial

had too small a low frequency content, due to the
separation of the spectral lines - see Fig. 12, to
enable the low frequency gain of the general polynomial transfer function to be determined.

11.2 *Simplified model*

The approximations used to derive the simplified model have been justified by the results
obtained using the data from the detailed model.

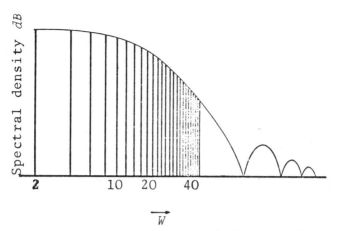

Fig. 12. Spectral content of forcing function

The use of the simplified model gave the correct low frequency gain due to the form of the model used, but the coefficients were not correctly identified when the flight trial data were used.

11.3 General polynomial transfer function

The general polynomial was not a good model to use since the forcing function contained too little information at the low frequency end to enable the coefficients to be determined accurately in the presence of the noise.

11.4 Flight trial data

The noise on the flight trial data must be reduced before any similar exercise may be accomplished.

REFERENCE

Nelder, J.A. and Mead, R. (1965) "A simplex method for function minimisation", *Computer J.*, **7**.

A Programming Approach to Urban Structure Planning

L. Pearl

(The Polytechnic, Wolverhampton)

SUMMARY

Urban land-use planning at the "structure plan" level may be thought of broadly as the spatial allocation of total population and total employment. Mathematical programming is little used in this situation on account of the difficulties of defining optimality.

In this paper an objective function is defined in terms of a "desirability index" which is a function of such zonal characteristics as population density, accessibility to work, maturity of landscape, topography etc. If D_i is the desirability index for zone i and p_i is the population of zone i, then the total desirability of any proposed plan is ΣDp. Since D is a function of population density and accessibility to work, ΣDp is a quadratic in the $2n$ variables $p_1, \ldots, p_n, e_1, \ldots, e_n$ where e_i is the number of jobs in zone i. In a structure planning situation, the constraints too, are difficult to formulate exactly. It is proposed that vaguely formulated constraints are appropriate in this situation, for example, "the population of zone i should be "near" a target value." The targets may be thought of as the planners' initial guess at an optimal plan.

By using a penalty function approach an iterative and interactive procedure is possible. At each iteration the planners having studied the proposed population and employment in each zone, are at liberty to change the targets or tightness of all or some of the constraints. Alternatively, they may use the output of the previous iteration as the targets for the next iteration. The planners' intuition therefore guides and is guided by the mathematics.

The basic algorithm is extremely simple, which means that sensitivity analyses are readily available. The problem of defining the parameters in the penalty function which govern the tightness of the constraints is also easily solved. The whole procedure is demonstrated using the data of Teesside Survey and Plan.

1. INTRODUCTION

Town and Country planning may be thought of as a set of instruments designed to attempt to determine and/or limit or encourage the economic and social life of a community. The two major fields are land-use planning and transportation planning. This paper is concerned basically with the former.

The nature of land-use planning is determined to a large extent by the various Planning Acts. For instance, the main planning tool introduced by the 1947 Act is the Development plan, which is based on a survey of the area under the control of the local authority. The data and analysis of the survey are used in the presentation of a broad picture of the expected development by private and public enterprises in the area over a period of about 20 years. The plan indicates the manner in which any land is to be used, and the stages in which any proposed development thereon is to be carried out. The plan may allocate land to such detailed uses as parks, air-fields, nature reserves, etc., or land may merely be designated as residential, agricultural or industrial. In general, the Development plan acts as a guide to public and private developers, pinpointing the market forces which are in existence or will come into being and in particular those forces which must be controlled. The controls over development act through the machinery of planning applications, inquiries, appeals etc.

In contrast to the Development plan of the 1947 Act, which was based on land-use maps, the 1968 Act introduced the Structure plan. This is a written policy on broad land-uses. Although the earlier type of Development plan was not intended as a rigid map indicating precise and irreversible planning decisions, there was a tendency for the plan to be treated as such. The lack of a map to accompany a structure plan almost guarantees a certain degree of flexibility. Under the new Act detailed planning is reserved for local "action areas". The 1947 Act called for 5-yearly reviews which in fact proved difficult in practice because of the requirement of prior surveys. The 1968 Act in contrast calls for a continuous monitoring.

The end product of the type of planning required by the acts may be deceptively simple in content. The land-use plan, which allocates total population and total employment among the constituent zones of the planning area, belies the extent of the analysis which produces it. In general, it is not obvious what forces are at work in an urban environment. The extent or nature of social forces is not fully understood, nor yet their interaction with the many economic forces. It is certainly not obvious which are the most efficient instruments for "control".

Two definite types of planning activity may be discerned in current practice, namely the evaluation of alternatives and the use of optimization models. The former approach is more widely used on account of the difficulties in defining optimality: see however Ben-Shahar *et al.* (1969), Laidlaw (1972), Lewis (1972), Schlager (1965) and Young (1972).

One of the first major planning studies undertaken in the spirit of the 1968 Act was Teesside Survey and Plan. This exercise was based on the "evaluation of alternatives" approach.

The initial step towards the production of the final land-use plan was the development of seven possible strategies, each corresponding to a particular spatial allocation of population and employment. Each strategy typified a particular type of structure. For example one strategy was a linear development in a westerly direction, another a satellite settlement south of Teesside. The major inputs to this stage of the analysis were the committed and autonomous growths in employment and population. The committed road programme also determined the choice of strategies to some extent. For instance the realignment of the A.19 meant that a strategy in a southerly direction had to be considered.

The strategies were compared on the basis of the cost and flexibility of the road system using standard transportation models. An innovation, however, was Teesplan's attempt to produce a single index to express the desirability of a zone from the point of view of its residents. The constituents of such an index are variables such as population density, accessibility to work, accessibility to open country, topography. A suitable linear function of these variables forms the index D_i which measures the desirability of zone i.

Although Teesplan used the index to evaluate each of their seven strategies it can be used as the basis of an optimization procedure as we show in the next section.

2. THE OPTIMIZATION PROCEDURE

An obvious objective function is the total cost to the community of the proposed allocation. If such quantities as demolition costs or redevelopment costs enter this function then the weight attached to the variable "population increase in zone i" is easily defined as the net cost per extra person. However, as soon as social costs are allowed to intrude, one is faced with the problem of defining trade-offs. If social costs are ignored, however, a programming approach might assign a flat uninteresting tract of land to housing, on account of the relatively small development costs. The desires of the potential residents for a "mature landscape", say, would not be taken into account. In many applications of linear programming, literal costs appear in the objective functions with social costs, for example environmental considerations, appearing as constraints. It would appear worthwhile to attempt to solve the allocation problem from the other viewpoint, namely to bring the residents' desires into the objective function and relegate costs to the constraints (Pearl (1974)). We will assume that our objective function is based on the desirability index, D.

As far as constraints are concerned, obviously total population and total employment do constrain the problem. Other constraints are less easily formulated in a precise manner. We will assume constraints may be formulated in the form that the optimal value of p_i (or e_i the number of jobs in zone i) should be *near* some particular target. This allows one to incorporate a genuine constraint such as zero development on land liable to flood, or more vaguely specified constraints. Such constraints may be related to an initial guess at the optimal plan. For instance, it may be felt the population of a zone with a current low density should be kept low, although it may be difficult to specify exactly what degree of increase, if any, over the existing and committed population, is to be allowed. In this circumstance one would choose a target equal to the existing and committed population.

The total problem may be formulated as follows.

If $\quad D_i = gp_i/L_i + hA_i + 2s_i, \quad i = 1,\ldots,n \quad$ (1)

where $\quad D_i$ = desirability of zone i

$\quad\quad p_i$ = population of zone i

$\quad\quad L_i$ = size of zone i

$\quad\quad A_i$ = an index of accessibility to work

$\quad\quad\quad = \Sigma_j e_j/c(d_{ij})$

$\quad\quad e_j$ = number of jobs in zone j

$\quad\quad c$ = a function representing a general cost of travel

$\quad\quad d_{ij}$ = distance between zone i and zone j

$\quad\quad s_i$ = a function of such variables as accessibility to shops, topography, age of housing, etc., for zone i

$\quad\quad n$ = number of zones

$\quad\quad g,h$ = parameters,

then the following optimization problem may be easily solved.

Choose $p_1,\ldots,p_n,\ e_1,\ldots,e_n$ to maximize ΣDp, the total desirability of the population of the region, subject to the constraints

p_i is to be near p_i^*, $i = 1,\ldots,n$ $\quad\quad$ (2)

e_i is to be near e_i^*, $i = 1,\ldots,n$ $\quad\quad$ (3)

$\Sigma e_i = E$ = total employment $\quad\quad\quad\quad$ (4)

$\Sigma p_i = P$ = total population $\quad\quad\quad\quad$ (5)

3. THE SOLUTION

A solution to the problem is to subtract from the maximand a penalty function.

Maximize $\quad w = \Sigma Dp - \Sigma(p_i-p_i^*)^2 f_i - \Sigma(e_i-e_i^*)^2 f_{i+n}$

$$- (\Sigma e_i - E)^2 F - (\Sigma p_i - P)^2 F \quad (6)$$

where f_i and F are parameters which govern the

"tightness" of the constraints.

The following procedure is now possible.

(i) Choose an "initial plan", *i.e.*, target values e_i^* and p_i^*. (In the absence of any definite desired values these may be taken as "existing and committed" values).

(ii) Maximize w to produce e_i^0 and p_i^0.

(iii) If any e_i^0 or p_i^0 is negative, increase the relevant f_i and proceed to step (ii), otherwise proceed to step (iv).

(iv) If the p_i^0 and e_i^0 are acceptable and no further increase in ΣDp is required, stop. If further increase is required, reset the f_i and proceed to step (ii). The new targets may be put equal to e_i^0 and p_i^0 if these are all acceptable, or with some changes if not.

The procedure is iterative and interactive - at each step (iv) there is the opportunity for planning considerations, other than total residential desirability to influence the procedure. For example a large p_i^0 for a particular zone may be unacceptable on account of the physical difficulties of possible building sites. In this case the new p_i^* would not be taken to be p_i^0, the original p_i^* perhaps being more appropriate.

The interactive phase may take a variety of forms. The recommended densities of certain key zones may be inspected one at a time, but in addition the whole of the recommended plan may be assessed with regard to transportation costs, development costs, the accessibility of firms to labour, etc. As a result of such analyses the targets of the next iteration will be a modification of the recommendations of the previous one.

The maximization of w is extremely simple since it is a quadratic in the unknown variables

$$i.e., \qquad\qquad w = 2\underset{\sim}{a}'\underset{\sim}{x} + \tfrac{1}{2}\underset{\sim}{x}'K\underset{\sim}{x} \qquad\qquad (7)$$

where

$$\underset{\sim}{x}' = (p_1,\ldots,p_n, e_1,\ldots,e_n)$$

$$\underset{\sim}{a}' = (f_1 x_1^* + FP + s_1,\ldots,f_n x_n^* + FP + s_n,$$

$$f_{n+1} x_{n+1}^* + FE,\ldots,f_{2n} x_{2n}^* + FE)$$

and

$$-K/2 = \begin{bmatrix} F+f_1' & F & \cdot & \cdot & F & H_{11} & H_{12} & \cdot & \cdot & H_{1n} \\ F & F+f_2' & \cdot & \cdot & F & H_{21} & H_{22} & & \cdot & H_{2n} \\ \cdot & \cdot & \cdot & \cdot & \cdot & \cdot & \cdot & & \cdot & \cdot \\ F & F & \cdot & \cdot & F+f_n' & H_{n1} & H_{n2} & \cdot & \cdot & H_{nn} \\ H_{11} & H_{12} & \cdot & \cdot & H_{1n} & F+f_{n+1} & F & \cdot & \cdot & F \\ H_{21} & H_{22} & \cdot & \cdot & H_{2n} & F & F+f_{n+2} & \cdot & \cdot & F \\ \cdot & \cdot & \cdot & \cdot & \cdot & \cdot & \cdot & & \cdot & \cdot \\ H_{n1} & H_{n2} & \cdot & \cdot & H_{nn} & F & F & \cdot & \cdot & F+f_{2n} \end{bmatrix} \quad (8)$$

$$(f_i' = f_i - g/L_i, \quad H_{ij} = -h/2c(d_{ij}))$$

The optimal value of $\underset{\sim}{x}$ is given by

$$-K\underset{\sim}{x}^0 = 2\underset{\sim}{a} \qquad\qquad (9)$$

Such a system, for $n = 50$, may be solved in about 90 seconds on an ICL 1903A computer.

4. CHOICE OF PENALTY PARAMETERS

An important part of the procedure is the choice of parameters f_i and F. They must be chosen such that the total population and total employment constraints are satisfied, but they must not be chosen so large that the penalty function dominates the maximand. In order to determine reasonable values analytically we must consider an artificial symmetrical problem.

Let $f_i = f$, $g = 0$, ($i.e.$, $f_i' = f_i = f$), $x_i^* = x^*$, $s_i = s$ ($i.e.$, $a_i = a$) for all i, and $d_{ij} = d$, ($i.e.$, $H_{ij} = H$) for all i,j.

By elementary operations on the system $-K\underset{\sim}{x} = 2\underset{\sim}{q}$ we can show that

$$p_1^0 = \ldots = p_n^0 = (FP + f x^*)/(nF + f + nH).$$

The discrepancy between Σp_i^0 and P is $HP/(F + f/n + H)$. Now if $P = 10^6$ and $H = 10^{-4}$ then if $F + f/n > 10^2$ the total population constraint will be satisfied.

In order to test whether the penalty dominates the maximand, let us introduce an exploitable situation thus. Let zone 1 be more attractive (as far as topography etc. is concerned) than the other zones which we assume are equally attractive, *i.e.*, $s_1 > s_2 = s_3 = \ldots = s_n = s$, say. By elementary operations we can show $p_2^0 = \ldots = p_n^0$ and that $p_1^0 - p_2^0 = (s_1 - s_2)/f$.

If f is much greater than $s_1 - s_2$ all the p_i^0's will be equal to P/n and the value of the maximand will be $((hA+s_1) + (hA+s)(n-1))P/n$. If f is chosen small enough p_1^0 will differ from p_2^0 and the maximand will be $(hA+s_1)p_1^0 + (hA+s)(n-1)p_2^0$. The difference between these two values is $(s_1-s_2)\times(P/n-p_1^0)$. In order to realize this potential gain, p_1^0 must be allowed to differ from p_2^0 by, say, up to 10^4, *i.e.*, f must be chosen to be 10^{-4}.

In order to satisfy the inequality $F+f/n>100$, obviously F must be chosen to be greater than 100.

5. STOCHASTIC CONSIDERATIONS

In equation (1) D_i is specified as a linear function of p_i and A_i (and hence of e_i) plus a function of exogenous variables. The form of this latter function is not prescribed but we will now assume that it is a linear function of k variables like age of housing, age of housing × topography, topography^2, etc. We could assume that $2\underset{\sim}{s} = S\beta$ where S is an $n \times k$ matrix of observed values of the k variables for the n zones, and β is a $k \times 1$ vector of coefficients. More realistically we will assume that desirability is a linear function of the exogenous and endogenous variables plus an error term, *i.e.*, $\underset{\sim}{D} = R\underset{\sim}{x} + S\beta + \underset{\sim}{\varepsilon}$, $E(\underset{\sim}{\varepsilon}) = 0$ where $\underset{\sim}{D} = (D_1,\ldots,D_n)'$ and R is a $n \times 2n$ matrix whose elements are functions of the coefficients g and h.

Ideally one would wish to calculate a "true" optimum based not only on the known values of R and β but also on the actual values of the ε's prevailing at the date of the plan. Now the solution to equation (9) may be written

$$\underset{\sim}{x}^0 = -K^{-1}2a$$
$$= -K^{-1}(\underset{\sim}{u} + (2s_1, \quad ,2s_n, 0, \quad ,0)')$$

where $\underset{\sim}{u}$ is a vector whose elements are functions of F, f_i, E, P and x_i^* only, *i.e.*, true $\underset{\sim}{x}^0 = -K^{-1}(\underset{\sim}{u} + (\beta'S' + \varepsilon',0))'$ but the $\underset{\sim}{x}^0$ that is calculated is based on the assumption that $\underset{\sim}{\varepsilon}$ coincides with its expectation, *i.e.*, calculated $\underset{\sim}{x}^0 = (-K)^{-1}\underset{\sim}{u} + (-K^{-1})(\beta'S',0)'$. The expected difference between these values is 0.

A further complication that may arise is that β is estimated by $\underset{\sim}{b}$, but if $\underset{\sim}{b}$ is unbiased for β then the same basic result remains true. However if all the coefficients g, h, $\beta_1,\ldots,\beta_k$ are estimated then because g and h feature in K^{-1} the result is not true. In these circumstances a knowledge of the possible bias would be desirable but a less sophisticated approach can provide answers to some of the questions that arise in this case. A simple sensitivity analysis based on $\partial x_i^0/\partial h$, $\partial x_i^0/\partial g$, $\partial x_i^0/\partial b_j$ would yield some information on the effect of "errors" in the coefficients. This approach is certainly appropriate when the estimation procedure is contentious, as for instance when a proxy for desirability is used.

If the system of equations (9) is solved by inverting $-K$ then the elements of the inverse can be used in the sensitivity analyses thus

$$\underset{\sim}{x}^0 = -K^{-1}\underset{\sim}{u} + -K^{-1}(\underset{\sim}{b}'S,0)$$

so $\partial x_i^0 / \partial b_j$ = jth element of $\underset{\sim}{k}_i S$ where $\underset{\sim}{k}_i$ = first n elements of the ith row of $-K^{-1}$.

6. THE PROCEDURE APPLIED TO TEESSIDE DATA

Teesside - 14 zone system

——— zone boundary; ····· boundary of Urban Teesside.

    Table I gives the results of three iterations of the procedure in the Teesside situation. The interactive facility is not used: the output of the first iteration provides the input to the next. Had the procedure been available at the time of Teesplan the planners would have been able to intervene after the first iteration.

Table I

	Iteration	Recommended population increases ('000's)				Recommended employment increases ('000's)			
		1	2	3	T^*	1	2	3	T^*
Zone									
	1	10	8	7	0	7	9	11	0
	2	15	20	25	0	11	19	26	0
Urban	3	6	1	-4	0	6	6	6	0
	4	9	7	5	32	3	2	0	8
	5	17	22	28	0	4	4	4	0
	6	13	14	15	0	5	6	7	0
	7	19	26	33	42	5	7	8	12
	8	14	16	19	81	0	-5	-10	16
Rural	9	7	3	-1	0	0	-1	-1	0
or	10	6	0	-5	0	-2	-3	-4	0
Semi-	11	24	36	49	5	5	6	7	3
Rural	12	6	1	-5	0	2	-1	-4	4
	13	6	1	-4	0	0	0	-4	0
	14	8	3	-1	3	-1	-1	-3	0
	$\Sigma Dp/\Sigma p$	20.6	20.9	21.1	20.6				

T^* increases proposed by Teesside Survey and Plan.

A possible intervention, for instance, may have resulted from a study of the proposed employment increase in zone 6. It may have been that increases of the order of 5000 jobs would of necessity have spoiled the visual features of the zone. The second iteration would therefore have commenced with the new employment target for zone 6 not equal to the output of the first iteration, but perhaps equal to the original target with an increase in the tightness of the relevant constraint.

Even without such interventions the results give some idea of the workings of the procedure.

The increasing sequence of total desirability represents a set of plans for which the average index ($\Sigma Dp/\Sigma p$) of the 700 000 people increases from 20.6 to 21.1. The magnitude of these increases may be appreciated by considering the nature of the calibration of the index. The ill-defined concept of desirability was made operational using a proxy variable "average zonal family income in 1966". (Numerical values for the coefficients in the index were obtained by regressing income on the dependent variables, population density, accessibility, maturity of landscape etc.)

The average value of the index in 1966 was 19.1. Of course the value for the proposed 1991 plan (21.1 for iteration 3) is not an income prediction, rather, one may loosely say that the "average" resident will live in a zone which has desirability equal to that enjoyed by families with an income of £21 per week in 1966.

The patterns of the proposed population increases can be explained by the nature of the objective function. The large population increases in zones 5, 6, 7 and 11 may be explained by the relatively high values of the variables topography, maturity of landscape, etc. In contrast, the large population increases in zones 1 and 2 are a consequence of high accessibility to work for these central zones.

The discrepancy between Teesplan and the procedure for the central zones may be explained by the artificial non-interactive nature of the example. The rigid trading off of low attractiveness and high population density against high accessibility to work would, in practice, be "overruled" by the interactive manipulation of targets and tightness parameters. Indeed the use of the procedure which exploits fully its interactive nature may provide a reasonable balance between two extremes.

On the one hand we avoid a thorough going

optimization procedure which might be seen to usurp a planner's function.

On the other hand we provide something more systematic than the evaluation of intuitively produced strategies. We have a procedure in which the mathematics guides and is guided by the planners' intuition and experience.

REFERENCES

Ben-Shahar, H., Mazor, M. and Pines, D. (1969) "Town Planning and Welfare Maximisation: A Methodological Approach", *Regional Studies,* **3**, 105-113.

Laidlaw, C.D. (1972) "Linear Programming for Urban Development Plan Evaluation", Praeger.

Lewis, J.P. (1972) "Linear Programming Models", *Built Environment,* **1**, 3.

Pearl, L. (1974) "A Land Use Design Model", *Urban Studies,* **11**, 3.

Schlager, K.J. (1965) "A Land Use Plan Design Model", *J. American Inst. Planners,* **3**, 2.

Young, W. (1972) "Planning: A Linear Programming Model", *G.L.C. Intelligence Unit Q. Bulletin,* 19.

SUBJECT INDEX